MW00651344

Programmable Microcontrollers with Applications

Programmable Microcontrollers with Applications

MSP430 LaunchPad with CCS and Grace

Cem Ünsalan

H. Deniz Gürhan

New York Chicago San Francisco
Athens London Madrid Mexico City Milan
New Delhi Singapore Sydney Toronto

McGraw-Hill Education books are available at special quantity discounts to use as premiums and sales promotions or for use in corporate training programs. To contact a representative, please visit the Contact Us page at www.mhprofessional.com.

Programmable Microcontrollers with Applications: MSP430 LaunchPad with CCS and Grace

1 2 3 4 5 6 7 8 9 0 DOC/ DOC 1 9 8 7 6 5 4 3

ISBN 978-0-07-183003-4
MHID 0-07-183003-0

This book is printed on acid-free paper.

Sponsoring Editor
Michael Penn

Editing Supervisor
Stephen M. Smith

Production Supervisor
Pamela A. Pelton

Acquisitions Coordinator
Amy Stonebraker

Project Manager
Sheena Uprety, Cenveo® Publisher Services

Copy Editor
Kevin Campbell

Proofreaders
Yamini Chadha and Anshu Sinha, Cenveo Publisher Services

Indexer
Robert Swanson

Art Director, Cover
Jeff Weeks

Composition
Cenveo Publisher Services

About the Authors

Cem Ünsalan, Ph.D., has worked in signal and image processing for 15 years. After getting a Ph.D. from The Ohio State University in 2003, he started working at Yeditepe University, Istanbul, Turkey, where he established the DSP Laboratory and has been teaching microprocessor and digital signal processing courses for 7 years. Dr. Ünsalan has published 16 journal articles and 4 international books. He holds one patent.

H. Deniz Gürhan received his B.Sc. from Yeditepe University. He is pursuing a Ph.D. in digital signal processing and embedded systems at the same university and works in the DSP Laboratory.

Contents

Preface

Smart systems have become inevitable parts of our lives. Every smart system needs an information processing unit. A microcontroller is a good candidate for such an operation. Therefore, a professional engineer or a fresh graduate should know how a microcontroller works. This book aims to introduce the working principles of a current-model microcontroller through applications. To do this, we need a specific microcontroller platform. The recently introduced Texas Instruments MSP430 LaunchPad is an excellent choice for this purpose. It is a compact platform with an MSP430 microcontroller on it.

The first step in understanding a microcontroller is to examine its construction. We devote three chapters to this issue. In Chap. 2, we start with a review of digital logic. Here we emphasize that the microcontroller is composed of logic gates in its basic form. In Chap. 3, we introduce data types used in a digital system. In this chapter, we provide the ways to represent positive and negative, fixed- and floating-point numbers in a microcontroller. We also explain what a word size and overflow mean. Then we consider the endian representations. In Chap. 4, we focus on the hardware of the MSP430 microcontroller. Therefore, we look at the central processing unit, memory, input and output ports, clocks and the timer modules, ADC and comparator modules, and the digital communication module. This chapter summarizes the properties of the MSP430 microcontroller to be considered throughout the book.

The second step in understanding a microcontroller is learning how to program it. To do so, we introduce Code Composer Studio (CCS) in Chap. 5. CCS is the unique environment in which to program TI microcontrollers and digital signal processors (DSP). We will use it to program the MSP430 in both C and assembly languages throughout the book. CCS is not only a programming environment. Using it, we can observe the status of the hardware components while the program is running. Therefore, it will be of great help in understanding the MSP430 in action. We also introduce the recent graphical peripheral configuration tool (Grace) under CCS. It will be of great help in the following chapters. In Chap. 6, we introduce the C programming techniques for our microcontroller. Here we first consider memory management and data types. Then we briefly overview basic C concepts. Although the C language may be sufficient for most applications, learning the assembly language is a must to understand the microcontroller. Therefore, we introduce the instruction set of the MSP430 in Chap. 7. We also look at the addressing modes and the usage of the stack in this chapter.

The third step in understanding a microcontroller is using it through different applications. To do so, we should know its properties in detail. In Chap. 8, we start with the digital input and output concepts. Therefore, we consider the configuration and usage of the input/output ports of the MSP430. At the end of the chapter, we pick the digital safe application and implement it step-by-step in both hardware and software. In Chap. 9, we focus on the interrupt concept, which is extremely important in event-driven programming. Therefore, we consider the occurrence of interrupts as well as the ways to handle them. To explain the interrupt concept further, we pick the washing machine application and implement it step-by-step at the end of the chapter. After interrupts, we consider time-based operations in Chap. 10. These concepts are also extremely important in applications. In this chapter, we start with the oscillators since they are the building blocks of the clocks. The MSP430 has more than one clock. We explore all these clocks and their usage areas. Another important topic in time-based operations is low-power modes. Effectively using them helps energy savings in applications. Therefore, we consider them next. In the same chapter, we also consider the usage of the watchdog timer and the timer modules. We pick the chronometer as the end-of-chapter application; we implement it step-by-step. In Chap. 11, we consider the processing of mixed signals. To do so, we start with the properties of analog and digital signals. Then we focus on the analog-to-digital conversion modules in the MSP430 microcontroller. Next, we focus on the digital-to-analog conversion. Since the MSP430 we are using does not have such a module, we use pulse width modulation instead. At the end of the chapter, we pick the non-touch paper towel dispenser as an application. As in all previous applications, we implement it step-by-step in both hardware and software. In Chap. 12, we focus on the digital communication module of the MSP430. Under this module, we consider the SPI, UART, and I^2C communication modes. As in the previous chapters, we pick a specific application and implement it step-by-step. In Chap. 13, we explore the flash memory of the MSP430. In Chap. 14, we provide sample applications on all topics considered. We picked these applications from real-life problems to show how a microcontroller can be used to solve them. In this final chapter, we provide the problem statement, equipment list, and circuit layout of each application. We expect the reader to implement these applications to master his or her knowledge in microcontroller-based system design.

In this book, we try to make all of these microcontroller concepts understandable to an undergraduate engineering student. Therefore, a professional engineer may also benefit from the book. Since we pick the MSP430 LaunchPad with the MSP430 microcontroller, the reader may find a wide variety of applications besides the ones considered in this book. However, the ones mentioned here will be the benchmark applications for the future. As a result, we expect the reader to become familiar with the microcontroller concepts in action.

Cem Ünsalan

H. Deniz Gürhan

Acknowledgments

The authors gratefully acknowledge the support of Texas Instruments in the framing and execution of this work through the European University Program. Most of the figures used in this book are the property of TI. They are used here with TI's permission.

Programmable Microcontrollers with Applications

1 Introduction

Chapter Outline

Microcontrollers are extensively used in our daily lives. Although they belong to the larger family of microprocessors, microcontrollers have certain distinctions. According to the consensus, a microprocessor does not contain a peripheral unit. On the other hand, a microcontroller should contain its peripherals to interact with the outside world. This property allows them to be used in most applications.

There are excellent books on microprocessors or microcontrollers. One type explains the theoretical and practical microcontroller concepts for a hypothetical system or for a microcontroller family (instead of a specific microcontroller). The idea here is to be less specific and more general. Therefore, the authors aim to explain the general concepts and ask the reader to apply them to the specific microcontroller he or she picks. Since a microcontroller family has a longer lifespan than a specific microcontroller, books in this group aim to be used over a longer time period.

The other type picks a specific microcontroller and explains its concepts. By default, the concepts explored in these books will be specific to the microcontroller at hand. As a result, they will not be general. In this book, we follow this approach and pick the TI MSP430 LaunchPad with an MSP430 microcontroller [12]. This may seem odd, but the focus of this book is explaining the microcontroller's concepts through applications. Therefore, it is must to pick a specific microcontroller and implement all the applications using it. We are aware that the digital electronics industry is dynamic and, most probably, a microcontroller will be obsolete within five or ten years. However, the newer members of the same microcontroller family will be based on the previous ones. Therefore, the current information about applications will be a valuable background for future microcontrollers. Also, other MSP430 microcontroller family members have similar properties. Therefore, the concepts considered in this book may be applied to them with minor modifications.

1.1 The TI MSP430 LaunchPad

There are several microcontroller platforms under different brands with various properties. In this book, we will focus on the TI MSP430 LaunchPad with an MSP430G2553 microcontroller. As we are writing this book, there are two versions of the TI MSP430 LaunchPad. These are revisions 1.4 (Rev.1.4) and 1.5 (Rev.1.5). We will cover both versions in the following chapters.

Throughout the book, we will refer to our microcontroller in two different ways. When we call it MSP430, this indicates that the explained concept is common to the other MSP430 family members also. We will call our microcontroller MSP430G2553 when we explain properties specific to it.

MSP430G2553 is a 16-bit microcontroller with 16 kB of memory. It has 16 general-purpose pins which can be used for digital input and output, timer applications, Analog-to-Digital Conversion (ADC), and digital communications. The MSP430 microcontroller family is designed to have ultralow power consumption. If these details do not mean much to you, do not worry. This book is written to explain these concepts.

We specifically picked the MSP430 LaunchPad platform shown in Fig. 1.1. TI introduced this platform as a unique coding and debugging environment for their

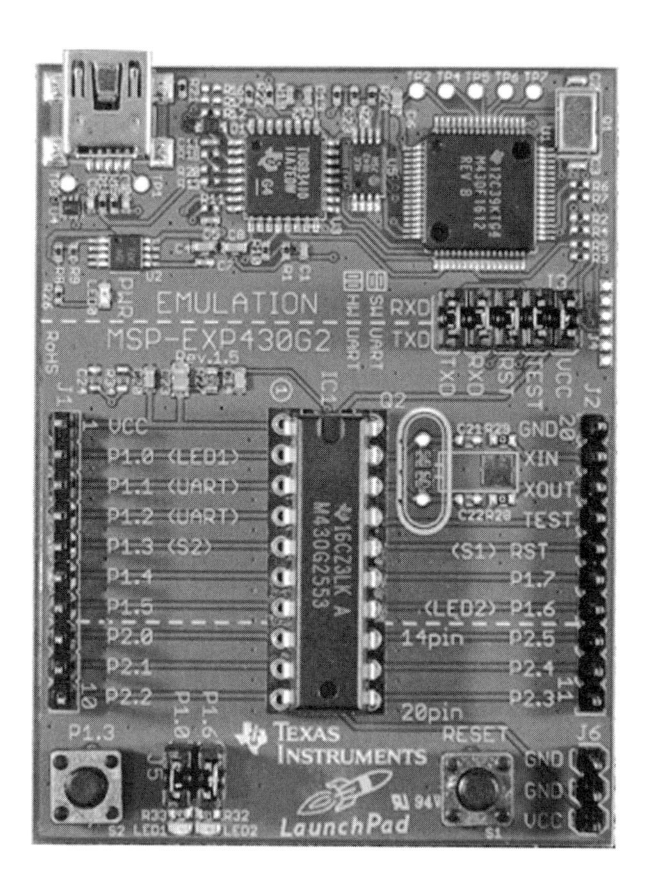

Figure 1.1

The TI MSP430 LaunchPad platform.

value-line microcontrollers. There is a USB connection on the MSP430 LaunchPad to communicate with the host PC. The coding environment is Code Composer Studio (CCS). CCS is the general environment for all TI devices. A code size–limited version of CCS is freely distributed by TI (at the time we are writing this book). The MSP430 LaunchPad platform is around $5, which is a reasonable price for such a microcontroller. C or assembly language may be preferred for coding. Throughout this book, we use both approaches to explain the concepts in detail.

1.2 Topics to Be Covered in This Book

The microcontroller is constructed from digital logic elements which are constructed from transistors. Therefore, Chap. 2 is a brief review of digital electronics. It emphasizes the physical properties of the microcontroller. Chapter 3 deals with the data types considered in this book. In a way, Chaps. 2 and 3 provide the background of the chapters to follow. Chapter 4 explores the hardware of the MSP430 microcontroller. Chapter 5 introduces the CCS environment. Also in this chapter, the graphical peripheral configuration tool (Grace) under CCS will be considered. Chapter 6 deals with the C programming concepts for the MSP430 microcontroller. This chapter serves two purposes. First, it is a review of the C programming language. Second, it gives insight on the C programming issues pertaining to hardware (since we can observe it through CCS). Chapter 7 deals with the MSP430 instruction set. With this chapter, we will start programming the MSP430 microcontroller via assembly language. Chapter 8 discusses digital input and output issues. We will introduce the concepts of how the microcontroller interacts with the outside world. Chapter 9 is on interrupts. This chapter will introduce the event-based programming concept. Chapter 10 is on timing-based operations. It will focus on the oscillators, clocks, low power modes, watchdog timer, and the timer module of the MSP430 microcontroller. Chapter 11 is on mixed-signal systems. Analog-to-digital and digital-to-analog conversion will be the main focus of this chapter. Therefore, analog signals can be processed on the digital MSP430 platform after mastering the concepts in this chapter. Chapter 12 introduces the basic digital communication methods through the MSP430 microcontroller. Chapter 13 is on flash memory programming. This book is on microcontrollers in action. Therefore, every concept explained in these chapters will have their related applications. Finally, Chap.14 is on applications in which more than one concept is used.

Sample codes in this book are available for readers on the companion website, www.mhprofessional.com/ProgrammableMicrocontrollers. Course slides for readers and instructors are available on the same website. The solution manual for instructors is also available on the companion website.

2 Review of Digital Circuits

Chapter Outline

Digital circuits are the essential parts of a microcontroller. Although the end user will never see them, all operations will be performed using these circuits. Hence, it is essential to know their physical properties. This chapter is a brief review of digital circuits and systems. A more detailed coverage of this topic can be found in [3, 5]. To note here, the circuits given in this chapter are not unique. There may be other circuits doing the same job. For consistency, we will take one subset and stick with it throughout the chapter.

2.1 Transistor as a Switch

A transistor is an active circuit element with three or four terminals. It can be used either as an amplifier or as a binary switch. For a digital circuit, the latter property is extremely important since all binary logic operations can be performed this way. Related to this, in a digital circuit the lowest level of information representation is done using a bit (binary digit). When we talk about the value of a bit (being either 0 or 1), we mean the voltage level on a transistor terminal is either high (V_{CC}) or low (ground).

There are two types of transistors: bipolar junction transistor (BJT) and metal oxide semiconductor field effect transistor (MOSFET). The BJT has three terminals: emitter (E), base (B), and collector (C). The current through the emitter and collector terminals is controlled by the current at the base terminal. On the other hand, the MOSFET has four terminals: gate (G), drain (D), source (S), and bulk. A voltage applied to the gate terminal forms a conducting channel between the drain and source terminals. A voltage applied between these terminals conducts the current on this channel. The bulk terminal is generally connected to the source in digital applications. Therefore, it is not shown in digital MOSFET representations. MOSFETs are preferred in digital systems due to their low power consumption and operation speed.

There are two MOSFET types based on their construction, N-channel MOSFET (NMOS) and P-channel MOSFET (PMOS). When the voltage between the gate and the source (V_{GS}) is 0, NMOS acts like an open switch and cannot conduct current between the source and drain terminals. Under the same setup, PMOS acts like a closed switch and conducts the current between the source and drain terminals. When V_{GS} equals V_{CC}, PMOS acts like an open switch. Here, NMOS acts like a closed switch. Therefore, by applying a suitable voltage level to the gate, the current flow between the drain and gate can be controlled. These scenarios are shown in Fig. 2.1. In this figure, PMOS is distinguished from the NMOS by a bubble in its gate terminal.

Complementary metal oxide semiconductor (CMOS) is a special technology that uses NMOS and PMOS transistors on the same substrate. Digital circuits are generally built using CMOS technology due to their minimal power consumption. For consistency, we will only consider CMOS-type logic gates in this chapter.

2.2 Logic Gates from Transistors

As mentioned in Sec. 2.1, by applying a suitable voltage to the gate of a transistor, the current flow (hence the voltage) between its drain and source can be controlled. This will lead to the development of digital logic gates: NOT, NAND,

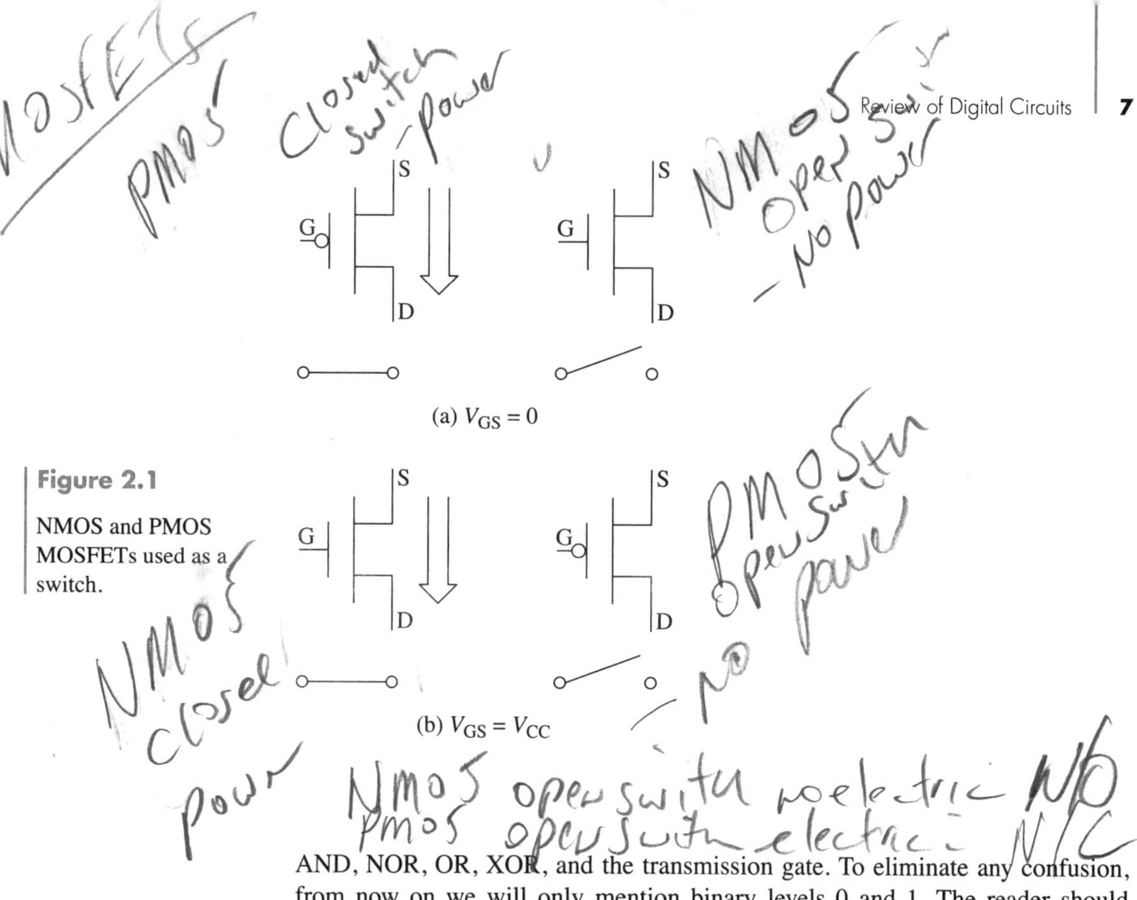

Figure 2.1
NMOS and PMOS MOSFETs used as a switch.

AND, NOR, OR, XOR, and the transmission gate. To eliminate any confusion, from now on we will only mention binary levels 0 and 1. The reader should remember that they correspond to voltage levels ground (low) and V_{CC} (high), respectively.

2.2.1 *NOT Gate*

The NOT gate is actually an inverter. It has a single input and output. When the input of the NOT gate is 0, its output is 1. When its input is 1, the output is 0. The symbol for the NOT gate is given in Fig. 2.2.

The NOT gate consists of one NMOS and one PMOS transistor as shown in Fig. 2.3*a*. In the same figure, the working principle of the NOT gate at the transistor level is given. In the first scenario (Fig. 2.3*b*), the logic level 1 (V_{CC}) is applied to the input. In the NOT circuit, the NMOS transistor turns on and the PMOS transistor turns off. As a result, the NMOS transistor sinks current from the output node. Therefore, it goes to the logic level 0 (ground). In the second scenario (Fig. 2.3*c*), the logic level 0 (ground) is applied to the input. Now, the NMOS transistor turns off and the PMOS transistor turns on. As a result, the PMOS sources current to

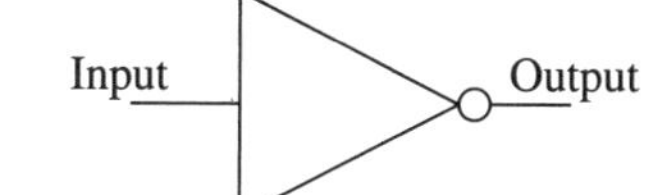

Figure 2.2
The NOT gate symbol.

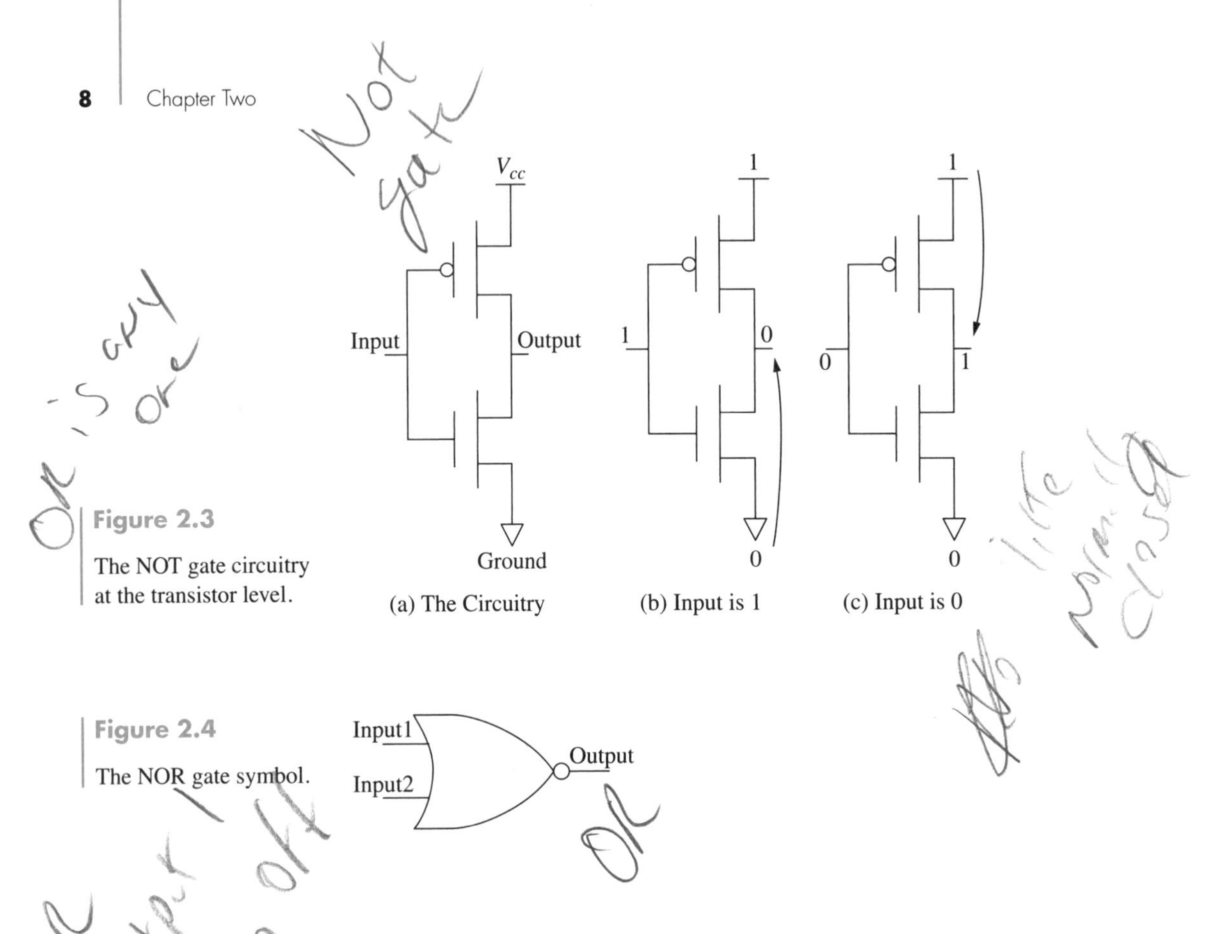

Figure 2.3 The NOT gate circuitry at the transistor level.

Figure 2.4 The NOR gate symbol.

the output node. Therefore, it goes to the logic level 1 (V_{CC}). These two scenarios clearly show the NOT operation at the transistor level.

2.2.2 *NOR, OR Gates*

The NOR (NOT-OR) is the next logic gate to be considered. Its symbol is given in Fig. 2.4. The truth table for the NOR gate is given in Table 2.1. As can be seen in this table, the NOR gate gives logic level 1 when all its inputs are logic level 0. For all other input combinations, the output is 0.

The transistor-level circuitry of the NOR gate is given in Fig. 2.5. As can be seen in this figure, the CMOS NOR gate is constructed by serially connected PMOS and parallel connected NMOS transistors. When the inputs of the NOR gate are at logic level 0, the PMOS transistors are open. They source current to the output node. Therefore, the output goes to logic level 1. But when one of the inputs

Table 2.1 The truth table for the NOR gate.

Input1	Input2	Output
0	0	1
0	1	0
1	0	0
1	1	0

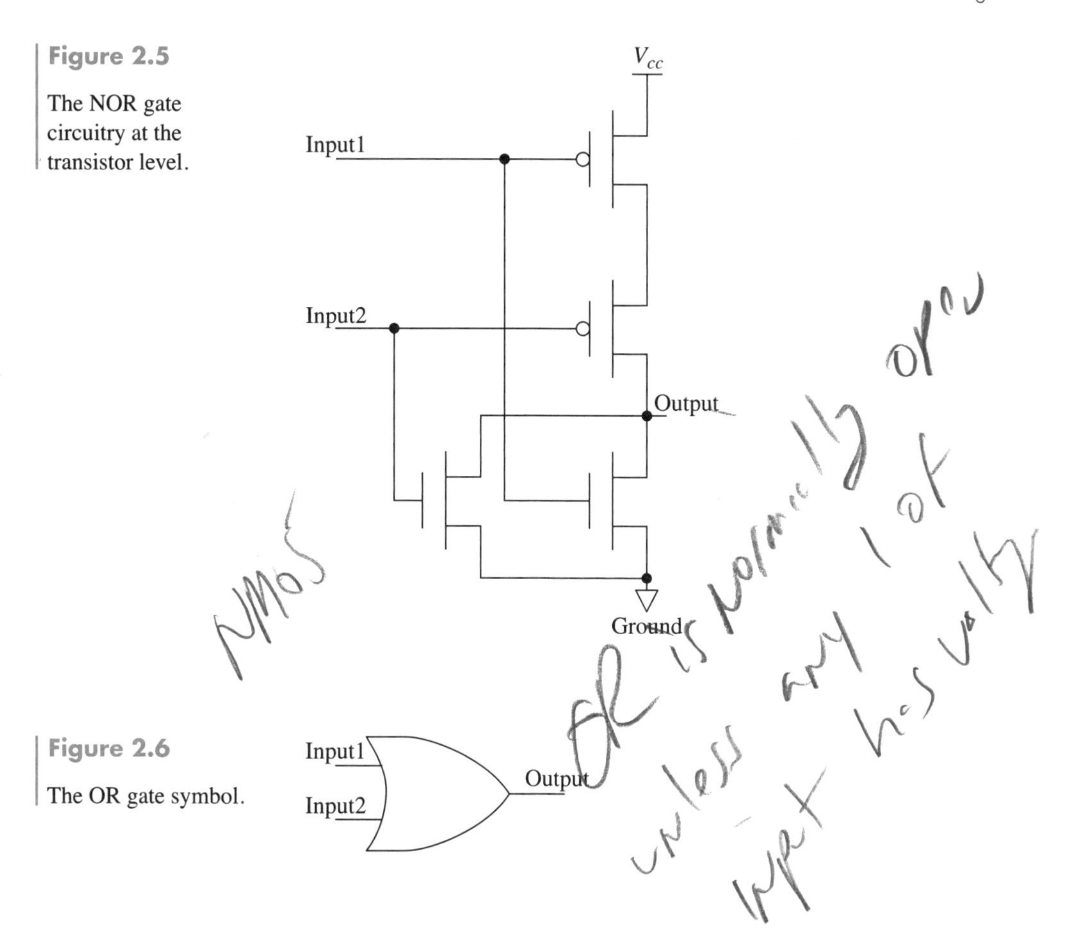

Figure 2.5

The NOR gate circuitry at the transistor level.

Figure 2.6

The OR gate symbol.

is at logic level 1, one of the PMOS transistors is closed and the other is open. Therefore, the current is sinked from the output node and it goes to logic level 0.

The OR gate is just the NOR with a NOT connected to its output. Its symbol is given in Fig. 2.6. The truth table for the OR gate is given in Table 2.2. As can be seen in this table, the OR gate gives logic level 0 when all its inputs are at logic level 0. For all other input combinations, the output is 1.

Based on the preceding definition, the transistor level circuitry of the OR gate is given in Fig. 2.7. As can be seen in this figure, to obtain the OR gate only one inverter is added to the output of the NOR gate.

Table 2.2

The truth table for the OR gate.

Input1	Input2	Output
0	0	0
0	1	1
1	0	1
1	1	1

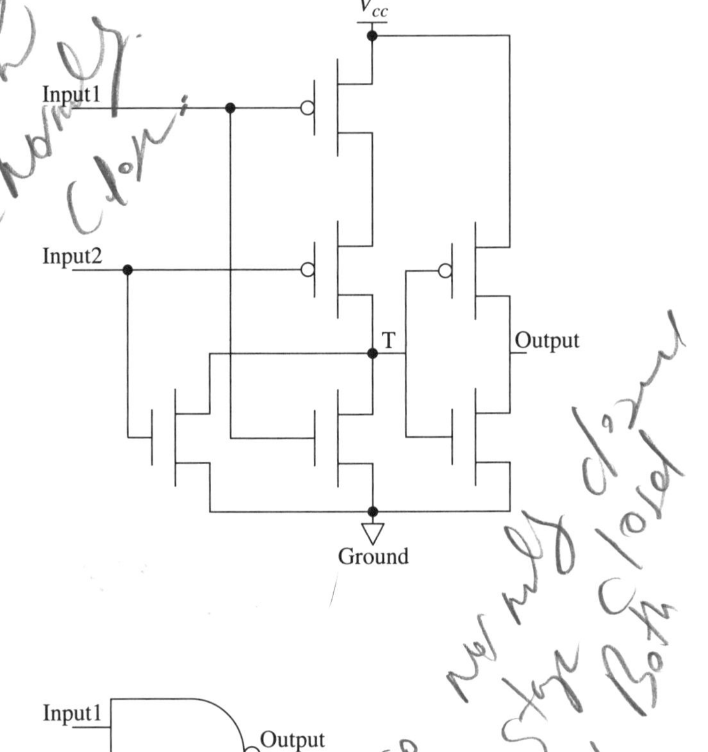

Figure 2.7

The OR gate circuitry at the transistor level.

Figure 2.8

The NAND gate symbol.

2.2.3 *NAND, AND Gates*

The NAND (NOT-AND) is the next logic gate to be considered. Its symbol is given in Fig. 2.8. The truth table for the NAND gate is given in Table 2.3. As can be seen in this table, the NAND gate gives logic level 0 when all its inputs are at logic level 1. For all other input combinations, the output is 1.

The transistor-level circuitry of the NAND gate is given in Fig. 2.9. As can be seen in this figure, the CMOS NAND gate is constructed by parallel connected PMOS and serially connected NMOS transistors. When the inputs of the NMOS transistors are at logic level 1, they are open and sink current from the output node.

Table 2.3

The truth table for the NAND gate.

Input1	Input2	Output
0	0	1
0	1	1
1	0	1
1	1	0

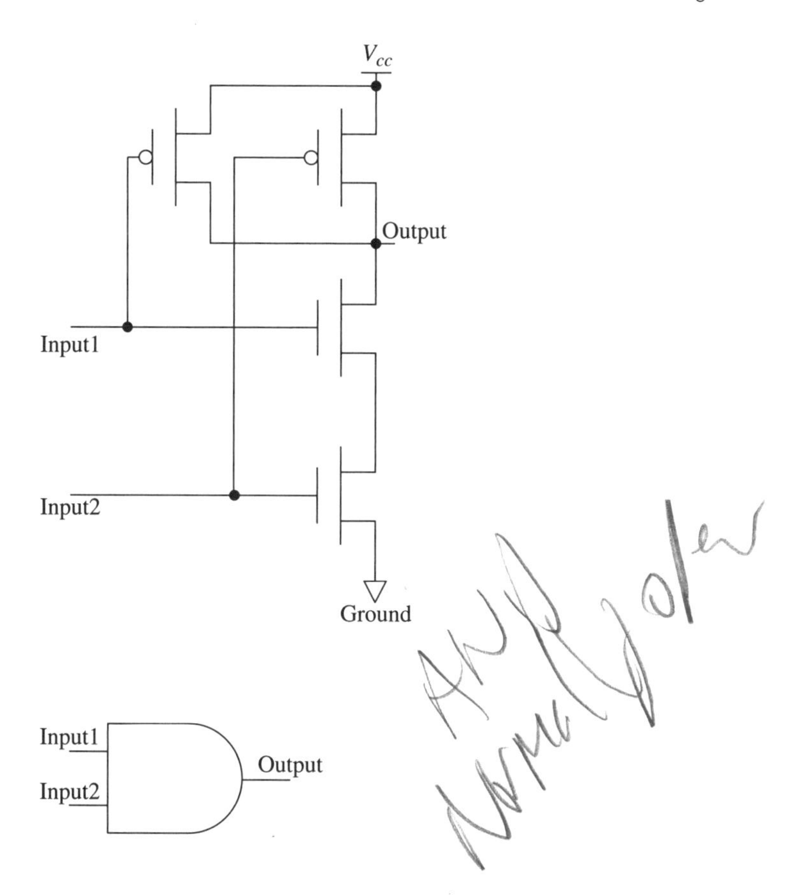

Figure 2.9

The NAND gate circuitry at the transistor level.

Figure 2.10

The AND gate symbol.

Therefore, the output goes to logic level 0. But when one of these inputs is at logic level 0, one of the NMOS transistors is closed and one of the PMOS transistors is open. Hence, the current is sourced to the output node. Therefore, the output goes to logic level 1.

The AND gate is just the NAND with a NOT connected to its output. Its symbol is given in Fig. 2.10. The truth table for the AND gate is given in Table 2.4. As can be seen in this table, the AND gate gives logic level 1 when all its inputs are at logic level 1. For all other input combinations, the output is 0.

Based on the preceding definition, the transistor-level circuitry of the AND gate is given in Fig. 2.11. As can be seen in this figure, only one inverter is added to the output of the NAND gate to obtain the AND gate.

Table 2.4

The truth table for the AND gate.

Input1	Input2	Output
0	0	0
0	1	0
1	0	0
1	1	1

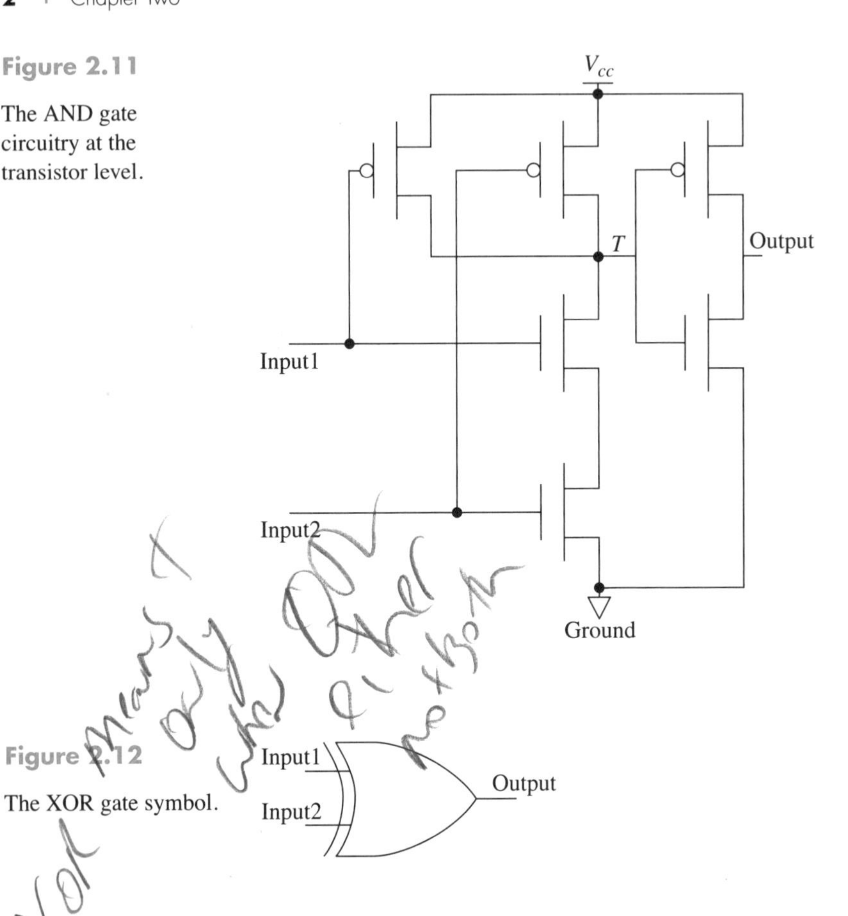

Figure 2.11

The AND gate circuitry at the transistor level.

Figure 2.12

The XOR gate symbol.

2.2.4 *XOR Gate*

The logic gate to be considered in this section is XOR (Exclusive-OR). It gives logic level 1 when its inputs have different logic levels. When the inputs of the XOR gate are the same, it gives logic level 0. The symbol for the XOR gate is given in Fig. 2.12. The truth table for this gate is given in Table 2.5.

The transistor-level circuitry of the XOR gate is given in Fig. 2.13. As can be seen in this figure, if both inputs are the same, the output is at logic level 0. Let's consider two examples. In the first example, both inputs are at logic level 0. Only the rightmost NMOS pair will be open. They will sink current from the output node. Hence, the output will go to logic level 0. In the second example, the

Table 2.5

The truth table for the XOR gate.

Input1	Input2	Output
0	0	0
0	1	1
1	0	1
1	1	0

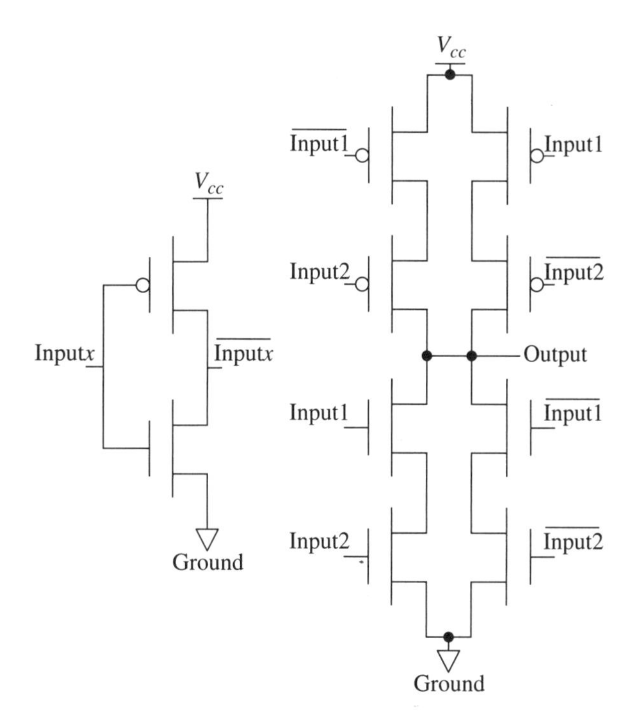

Figure 2.13

The XOR gate circuitry at the transistor level.

Input1 is at logic level 1 and the Input2 is at logic level 0. In this situation, only the leftmost PMOS pair will be open. They will source current to the output node; hence, the output will go to logic level 1.

2.2.5 *The Transmission Gate*

The transmission gate is a complementary CMOS switch constructed by parallel connected NMOS and PMOS transistors as shown in Fig. 2.14. This gate either passes or stops the current between its input and output terminals (source and drain), depending on the control terminal (the gate). The symbol for the transmission gate is given in Fig. 2.15.

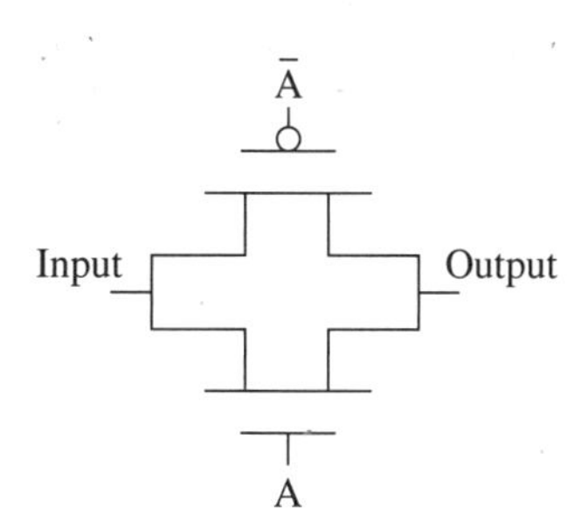

Figure 2.14

The circuitry of the transmission gate.

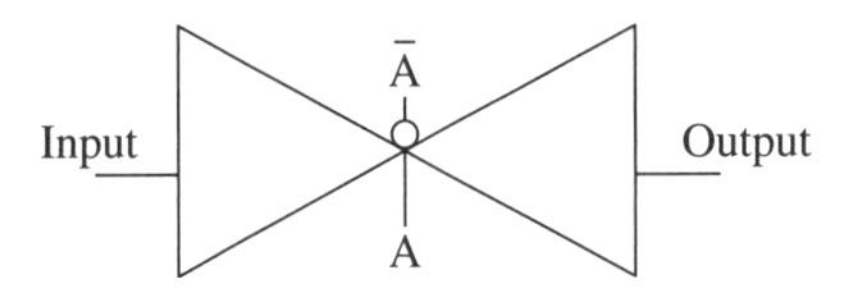

Figure 2.15

The symbol for the transmission gate.

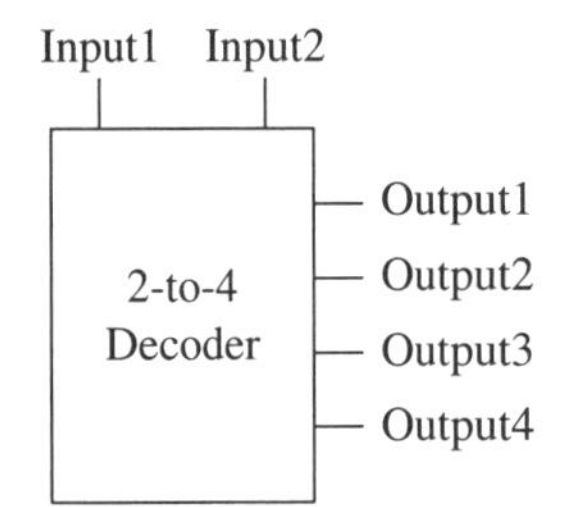

Figure 2.16

The two-to-four decoder symbol.

2.3 Combinational Circuits from Gates

Logic gates introduced in Sec. 2.2 can be used to construct combinational circuits. In these, the output is not affected by previous input values. In other words, there is no memory in combinational circuits. In this section, we will consider the decoder, multiplexer, and adder as combinational circuits.

2.3.1 *The Decoder*

The basic function of a decoder is to decode its input and give a specific output corresponding to its input. In general, a decoder has N inputs and 2^N outputs to cover all input combinations. The symbolic representation of a two-to-four decoder is given in Fig. 2.16. Its truth table is given in Table 2.6.

The decoder can be constructed by AND and NOT gates. For the two-to-four decoder, the circuit diagram at the logic gate level is given in Fig. 2.17. As can be seen in this figure, the two-to-four decoder is constructed by using two NOT and four AND gates. If we consider Output1, it gives logic level 1 only when Input1 and Input2 are 0. This input combination sets all other output pins to logic level 0.

2.3.2 *The Multiplexer*

The multiplexer (MUX) is a combinational logic circuit that transfers data coming from several inputs to a single output. Therefore, it can be used to select a specific

Table 2.6

The truth table for the two-to-four decoder.

Input1	Input2	Output1	Output2	Output3	Output4
0	0	1	0	0	0
0	1	0	1	0	0
1	0	0	0	1	0
1	1	0	0	0	1

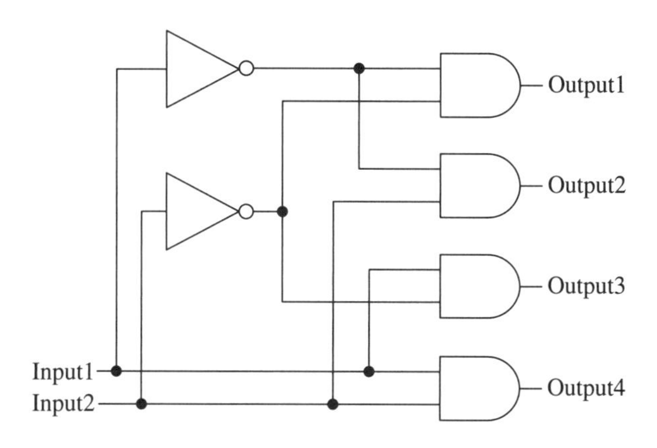

Figure 2.17

Circuit diagram of the two-to-four decoder at the gate level.

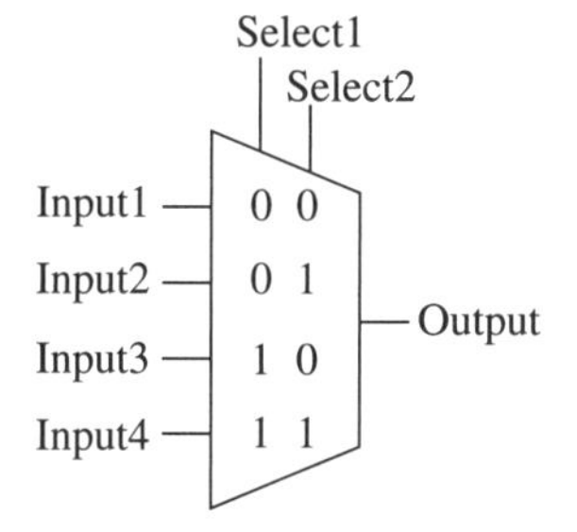

Figure 2.18

The MUX symbol with two select pins.

input from a group of inputs and feed it to output. To perform this task, the multiplexer has N select pins, 2^N input pins, and one output pin. The symbol for a multiplexer with two select pins is given in Fig. 2.18. The truth table for this MUX is given in Table 2.7.

A four-to-one multiplexer (with two select pins) at the gate level is given in Fig. 2.19. It can be easily seen that only one AND gate is enabled for each select input sequence. For instance, the first AND gate is enabled when Select1 and Select2 are 0. All other AND gates are disabled for this sequence. Hence, only Input1 appears at Output.

Table 2.7

The truth table for the MUX with two select pins.

Select1	Select2	Output
0	0	Input1
0	1	Input2
1	0	Input3
1	1	Input4

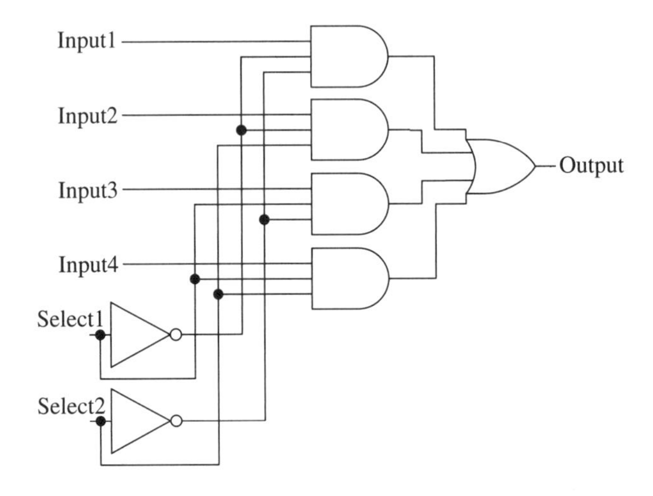

Figure 2.19

Circuit diagram of the four-to-one multiplexer built from basic logic gates.

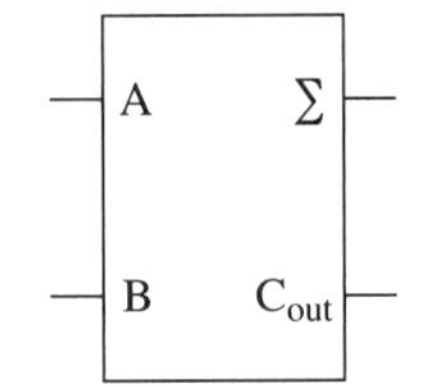

Figure 2.20

Symbol for the half adder (for single-bit addition).

2.3.3 *Adders*

Addition is the most important arithmetic operation in a digital system since all other arithmetic operations can be performed using it. There are two basic adder types, half and full. The half adder (for single-bit addition) has two inputs and two outputs. It adds the input bits and gives the sum and carry bits as output. The symbol for the half adder is given in Fig. 2.20. The truth table for this half adder is given in Table 2.8.

As can be seen in Table 2.8, the carry bit is logic level 1 when both input bits are at logic level 1. This corresponds to the AND operation. The sum bit ($\sum$) has logic level 1 when input bits have different logic levels. This corresponds to the XOR operation. Based on these observations, the gate-level representation of the half adder can be constructed as given in Fig. 2.21.

Table 2.8

The truth table for the half adder (for single-bit addition).

A	B	$\sum$	C_{out}
0	0	0	0
0	1	1	0
1	0	1	0
1	1	0	1

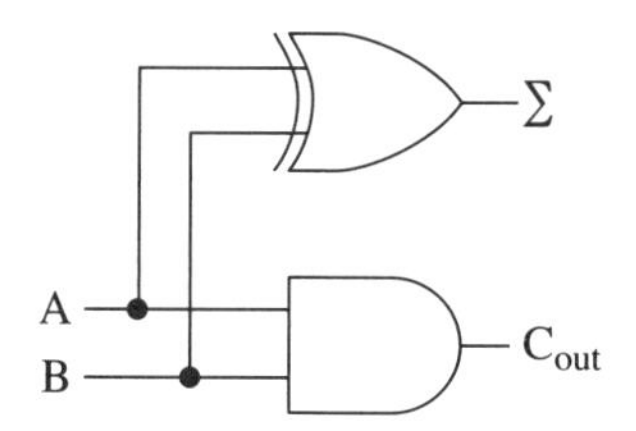

Figure 2.21

Circuit diagram of the half adder at the gate level.

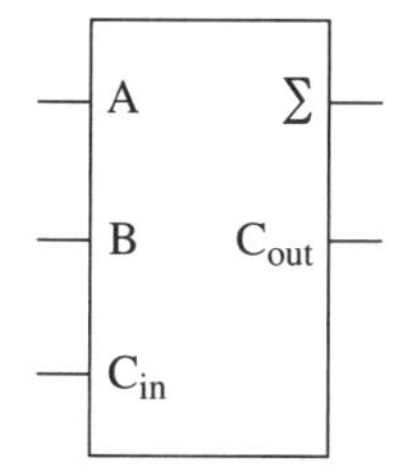

Figure 2.22

Symbol for the full adder (for single-bit addition).

The half adder does not take the carry bit into account in addition. This causes problems when adding binary numbers with more than one digit. The full adder is introduced to overcome this problem. Besides having two input pins, the full adder also has a carry-in pin. The symbol for the full adder (for adding one digit only) is given in Fig. 2.22. The truth table for this full adder is given in Table 2.9.

As in the half adder, the gate-level circuit diagram for the full adder can be constructed by analyzing Table 2.9. The final constructed circuit diagram for the full adder is given in Fig. 2.23.

To add binary numbers with more than one digit, full adders can be used in parallel. The most popular setup for this operation is the 4-bit parallel adder. Its block diagram is given in Fig. 2.24. As can be seen in this figure, every digit of the two binary numbers is added separately from the least significant digit (A1, B1) to the most significant digit (A4, B4).

Table 2.9

The truth table for the full adder (for single-bit addition).

A	B	C_{in}	Σ	C_{out}
0	0	0	0	0
0	1	0	1	0
1	0	0	1	0
1	1	0	0	1
0	0	1	1	0
0	1	1	0	1
1	0	1	0	1
1	1	1	1	1

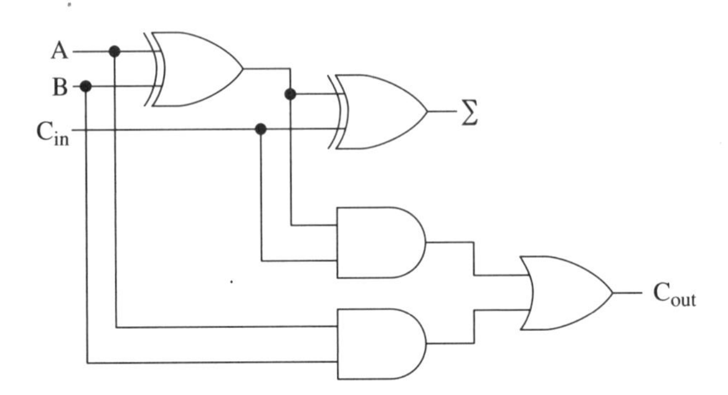

Figure 2.23

Circuit diagram for the full adder at the gate level.

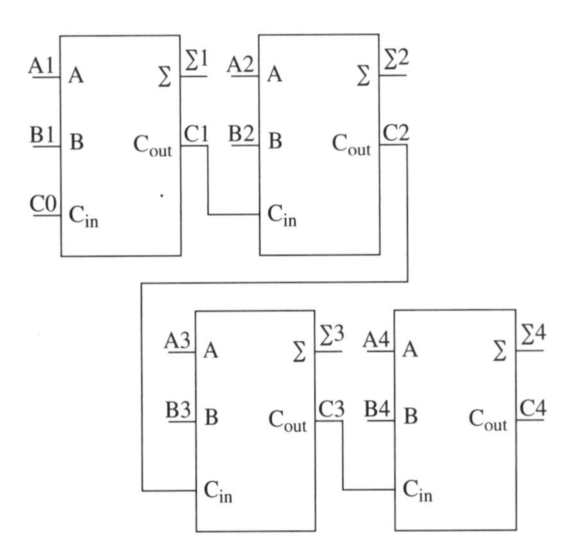

Figure 2.24

Block diagram of the 4-bit parallel adder.

2.4 Sequential Circuits from Gates

In contrast to combinational circuits, the output of a sequential circuit depends not only on the current inputs, but also on the previous inputs (or outputs). This dependence on the past requires data storage. As we will see in the following sections, data storage can be achieved by feedback loops added to combinational circuits.

2.4.1 *Latches from Gates*

A latch is a basic storage element which can store 1 bit of data. An SR latch formed with two cross-coupled NAND gates is shown in Fig. 2.25. This setup is called an active-low input SR latch. As can be seen in this figure, the SR latch has two inputs, Set (S) and Reset (R). It has two outputs, Q and $\overline{Q}$. In fact, Q and $\overline{Q}$ are complements of each other.

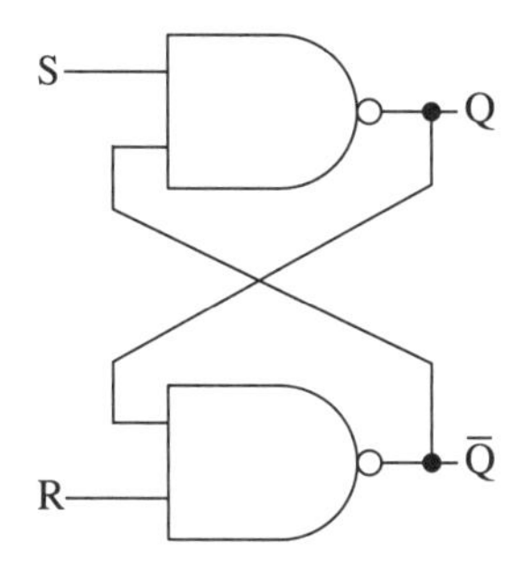

Figure 2.25

The SR latch at the gate level.

Table 2.10

The truth table for the SR latch.

S	R	Q	$\overline{Q}$
0	0	U	U
0	1	1	0
1	0	0	1
1	1	**Q**	$\overline{\mathbf{Q}}$

The truth table for the SR latch is given in Table 2.10. In this table, U stands for undefined. As can be seen in this table, when the input S has logic level 0 and R has logic level 1, the output Q will be at logic level 1. When S has logic level 1 and R has level 0, Q will be at logic level 0. When both S and R have logic level 1, the SR latch stays in its previous state. In other words, it stores the previous bit level. When S and R are at logic level 0, a contradiction occurs. In this case, both Q and $\overline{Q}$ should be at logic level 1. However, as mentioned before, Q should be the complement of $\overline{Q}$. Therefore, in this input combination the output will be undefined due to the race conditions in the circuit [5]. In order to prevent this undesired condition, the D latch is introduced.

The D latch has one input and two outputs. Its circuit diagram at the gate level is given in Fig. 2.26. As can be seen in this figure, a NOT gate is added between the S and R inputs. Therefore, they can never be at logic level 0 at the same time. Hence, the contradiction can be avoided.

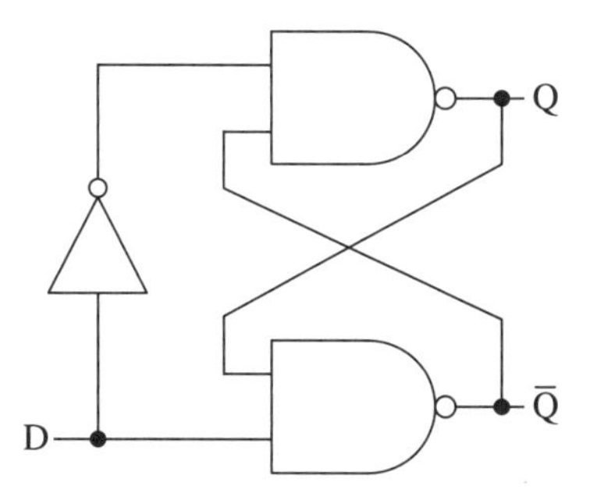

Figure 2.26

The D latch at the gate level.

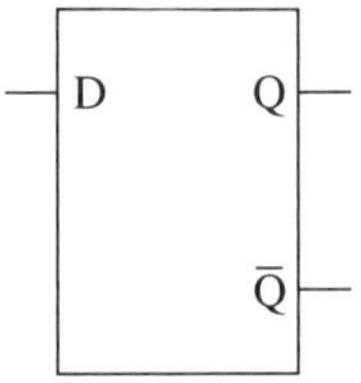

Figure 2.27

Symbol for the D latch.

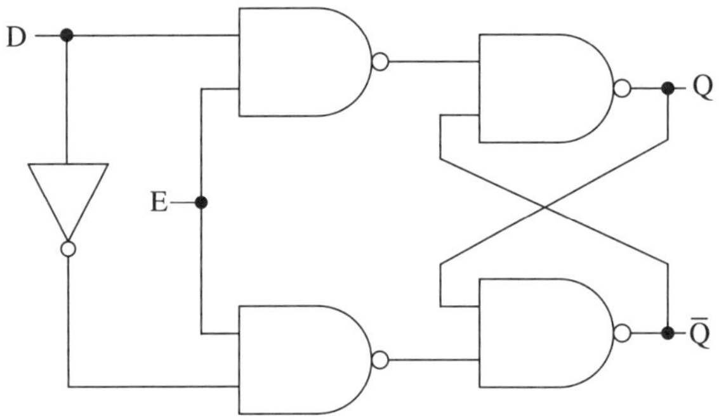

Figure 2.28

Gated D latch at the gate level.

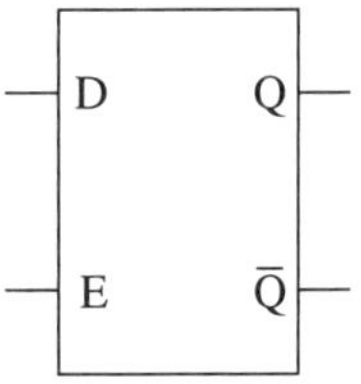

Figure 2.29

Symbol for the gated D latch.

The truth table for the D latch is the same as that for the SR latch without the undesired contradiction condition. In summary, when D has logic level 0, Q will be at logic level 0. When D has logic level 1, Q will be at logic level 1. Therefore, the D latch simply stores 1 bit of information. The symbol for the D latch is given in Fig. 2.27.

Latches are sensitive to their inputs all the time. Hence, sometimes disabling inputs is a desired property. A gated latch (latch with an enable input) is used for this purpose. This latch cannot be used until the enable input activates the latch. The circuit diagram of a gated D latch is given in Fig. 2.28. The symbol for the gated D latch is given in Fig. 2.29.

2.4.2 *Flip-Flops from Gates*

The flip-flop is also a 1-bit storage element. The difference between the flip-flop and the latch is the method for changing states. Flip-flop changes its state only at the rising or falling edge of the clock signal. Even if the input changes after the clock edge, its state still remains unchanged. There are basically four main types of flip-flops: SR, D, JK, and T.

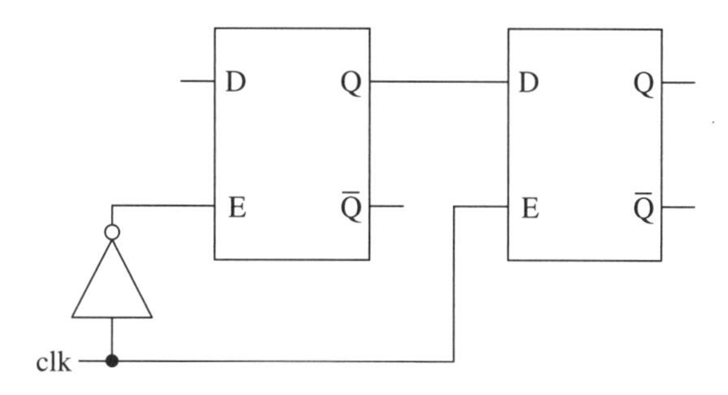

Figure 2.30

Block diagram of the master-slave D flip-flop.

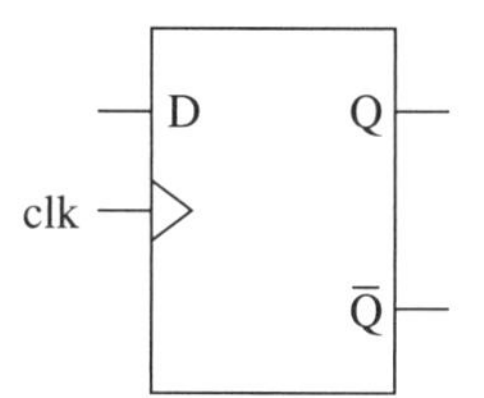

Figure 2.31

Symbol for the master-slave D flip-flop.

The D-type flip-flop can be constructed by connecting two gated D latches as shown in Fig. 2.30. The first and second latches are called master and slave in this setup. Therefore, this configuration is called a master-slave D flip-flop. The symbol for the D-type flip-flop is given in Fig. 2.31.

2.4.3 Counters from Flip-Flops

The counter is the basic building block of timer modules in a microcontroller. As the clock signal is fed to the counter, it changes its state. The number of flip-flops used in a counter indicates its capacity. For example, a 3-bit counter is built by three flip-flops. Hence it has eight states. In other words, it can count from zero to seven. The symbol for a counter is given in Fig. 2.32.

There are two counter types, asynchronous (ripple) and synchronous. In the ripple counter, the clock signal is fed only to the first flip-flop. The remaining flip-flops are clocked in a chain. A block diagram of a 3-bit ripple counter is given in Fig. 2.33. On the other hand, in the synchronous counter all flip-flops are clocked with the same clock signal. A block diagram of a 3-bit synchronous counter (with T flip-flops) is given in Fig. 2.34.

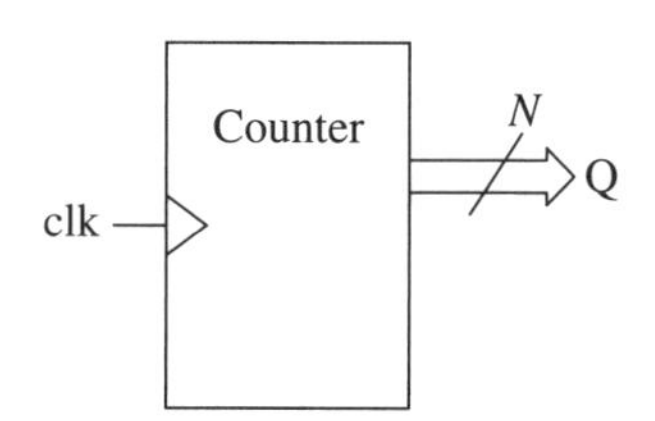

Figure 2.32

Symbol for a counter.

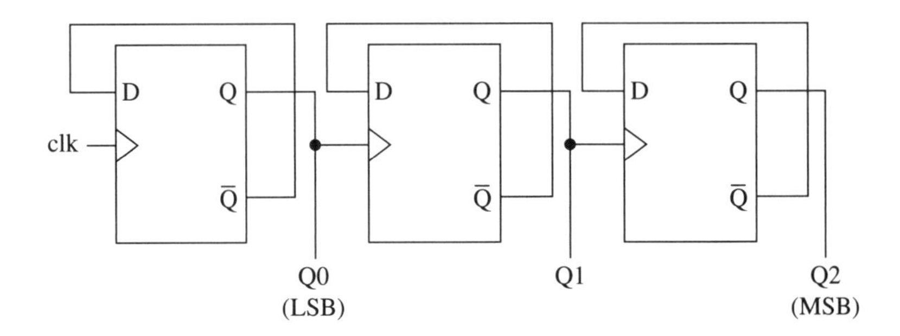

Figure 2.33

Block diagram of a 3-bit ripple counter.

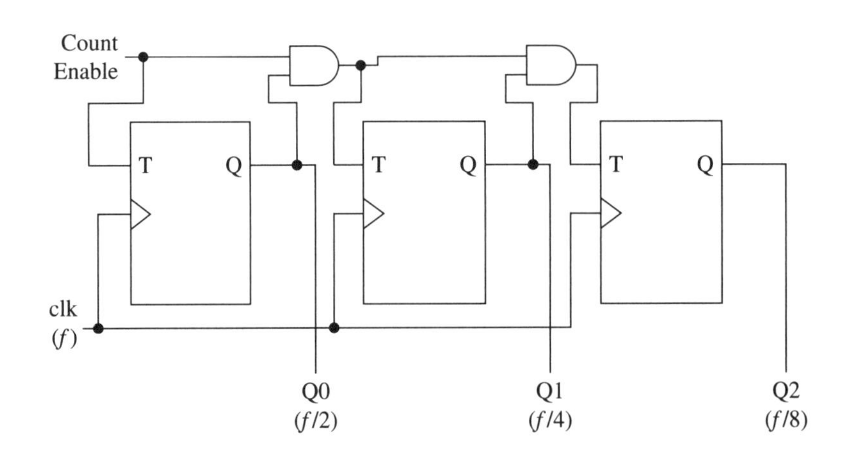

Figure 2.34

Block diagram of a 3-bit synchronous counter.

The synchronous counter can also be used as a frequency divider. This operation can be seen in Fig. 2.34. The input clock frequency, f, is divided by a power of two in each flip-flop. Hence, the output of each flip-flop becomes Q0 = $f/2$, Q1 = $f/4$, and Q2 = $f/8$. This property will be extensively used in the timer module of the microcontroller.

2.4.4 *Register from Flip-Flops*

Register is an N-bit storage element constructed by N flip-flops. A block diagram of a 4-bit register is given in Fig. 2.35. As can be seen in Fig. 2.35, in the register every flip-flop changes its content by a trigger from the clock. Therefore, the 4-bit data is stored to this register sequentially. The symbol for an N-bit register is given in Fig. 2.36.

2.4.5 *Shift Register from Flip-Flops*

There are four shift register types: serial in/serial out, parallel in/serial out, parallel in/parallel out, and serial in/parallel out. A block diagram of the serial in/serial out shift register is given in Fig. 2.37. As can be seen in this figure, the serial in/serial out shift register is constructed by a group of flip-flops connected as a chain. Hence,

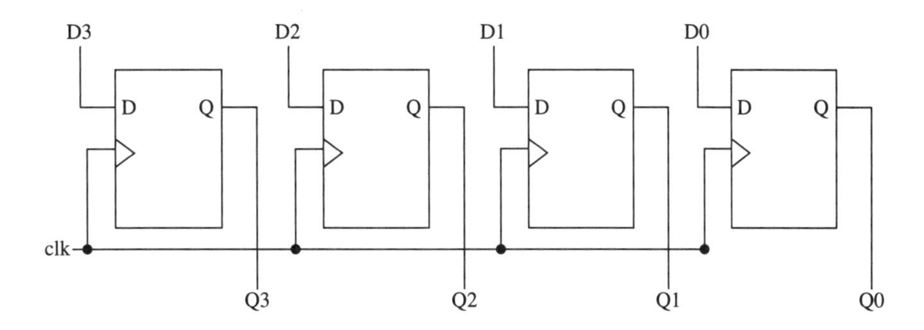

Figure 2.35

Block diagram of a 4-bit register.

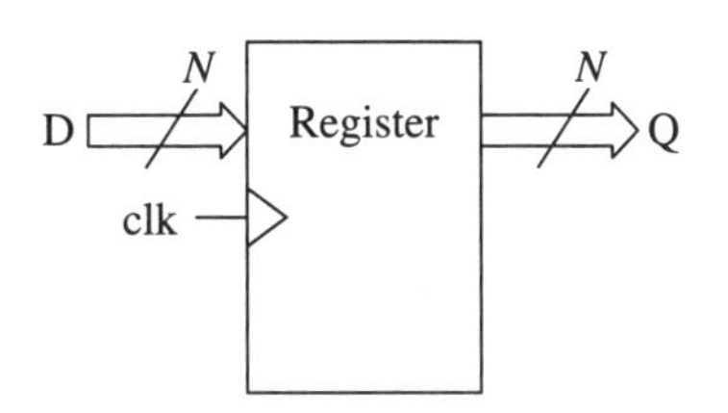

Figure 2.36

Symbol for an N-bit register.

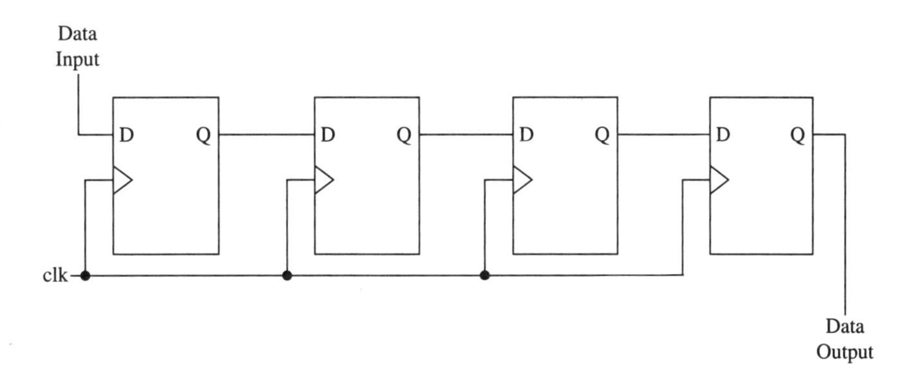

Figure 2.37

Block diagram of the serial in/serial out shift register.

the output of one flip-flop is connected to the input of the next flip-flop. In this setup, all flip-flops are driven by the same clock source.

As can be seen in Fig. 2.37, the new data bit is received from the Data Input pin. The last data bit is shifted out from the Data Output pin with every clock signal. Therefore, this generates a delay of N clock cycles for the data. Here, N is the number of flip-flops in the shift register. Similarly, parallel in/serial out and serial in/parallel out shift registers are also commonly used for the communication between serial and parallel interfaces.

Figure 2.38

Block diagram of a simple memory with four registers.

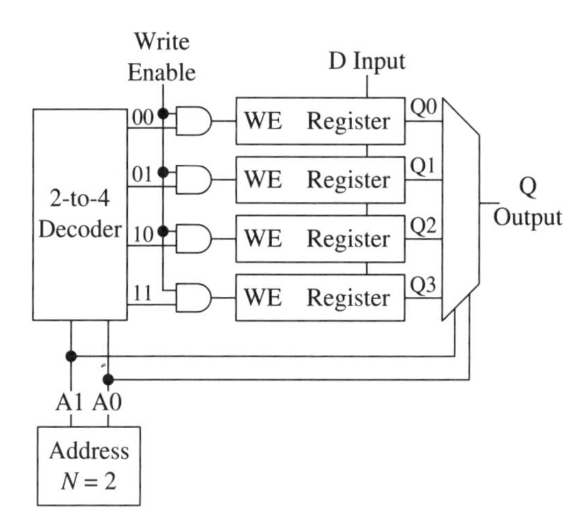

2.4.6 *Memory from Registers*

Registers can be combined to form memory blocks. Therefore, the memory can be called as a collection of addressable register locations. In fact, recent memory systems are constructed by a different technology (as in flash, to be considered in Chap. 13). However, to briefly explain the working principles of a general memory block, the discussion in this section is necessary.

A memory block consists of three key parts: address decoder, memory cells (registers), and output selector (MUX). A block diagram of a simple memory (with four registers) can be seen in Fig. 2.38. In this setup, each register can store a block of bits (such as 8, 16, 32 bits). This also shows the number of parallel flip-flops used in the register. When the write enable (WE) signal is at logic level 1, only one of the AND gates is enabled. The data is written to the specific register selected by the address decoder. When the WE is at logic level 0, the MUX chooses one address line and the data is read from that register.

The symbol of a generic memory block is given in Fig. 2.39. As can be seen in this figure, the memory block has N address wires, which can be used to reach 2^N separate address locations inside the memory block. Here, N represents the address to be reached within the memory. D represents the data (with length M) to be written to the memory. Q is the data to be read from the memory.

2.5 Summary

Digital circuits are the basic building blocks of a microcontroller. Although the user will not deal with them in a practical application, he or she should know them to understand the working principles (as well as limitations) of the microcontroller. Therefore, we reviewed digital circuits in this chapter. We started with transistors and formed binary logic gates using them. We then used these to form combinational and sequential digital circuits. These will be the basic building blocks of

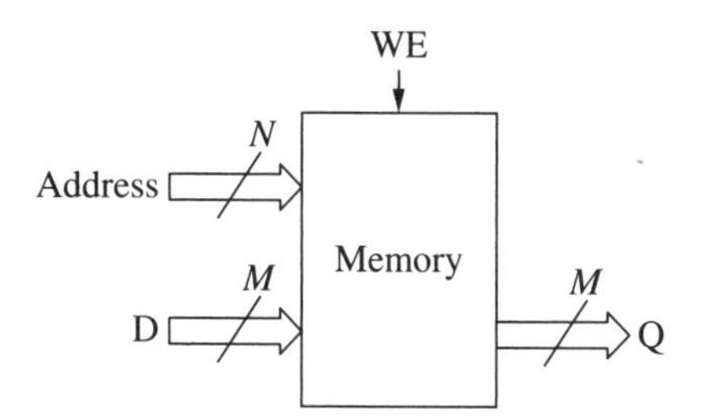

Figure 2.39

Symbol for a generic memory block.

the microcontroller modules such as the arithmetic logic unit, timer, and analog-to-digital converter.

2.6 Problems

2.1 What is the minimum switching time for a recent CMOS transistor?

2.2 What is the value of V_{CC} for the recent CMOS transistor-based gates?

2.3 Construct an SR latch with NOR gates.

2.4 How can we make

a. a multiplication operation, if we only have a full adder at hand?

b. a division operation, if we only have a full adder at hand?

2.5 How can we construct a counter with range 0–FFFFh?

2.6 Give an example of binary multiplication by 2 using a shift register.

2.7 Give an example of binary division by 4 using a shift register.

2.8 How can we form a frequency divider (with division of 2^0 to 2^7) using flip-flops?

2.9 What does address space mean in a memory block?

2.10 Construct a 16-bit memory space using gated D latches as the basic building blocks. Do not forget to add control circuitry for WE.

3 Data Types

Chapter Outline

In this chapter, we review the basic data types and representations in MSP430. We first consider the binary representation. Then, we explore the fixed- and floating-point representations of binary numbers. We next focus on the word size and overflow issues. The next topic we consider is the endian representation. We also consider the American Standard Code for Information Interchange (ASCII) characters and the MSP430 data types. The representations considered in this chapter will be used extensively in the following chapters.

3.1 Number Representations

In our daily lives, we use the decimal number system. This representation associates the weight (powers of 10 here) of the digit with its location. Here, the least significant digit gets the weight 10^0, the next one gets 10^1, and so on. Using this form, we can represent an entity in a systematic way. Therefore, a decimal number 255 means we have $2 \times 10^2 + 5 \times 10^1 + 5 \times 10^0$. If we want to represent a decimal number with fractional parts, we follow the same strategy. Now, the weights of the digits in the fractional part become 10^{-1}, 10^{-2}, and so on starting from the dot (separating the integer and fractional parts) from left to right. As an example, the decimal number 1.25 corresponds to $1 \times 10^0 + 2 \times 10^{-1} + 5 \times 10^{-2}$.

The binary number representation is more suitable for digital systems, since they only use two levels, 0 and 1 (represented by two voltage values in the transistor level as explained in Chap. 2). Here, each binary level is called a bit (binary digit). Eight bits correspond to 1 byte, 1024 bytes to 1 kilobyte (kB), 1024 kilobytes to 1 megabyte (MB), and 1024 megabytes to 1 gigabyte (GB).

The binary number representation has weights in powers of two: $2^0, 2^1, 2^2, \ldots, 2^N$. For the fractional parts, the weights become $2^{-1}, 2^{-2}, 2^{-3}$ and so on, starting from the dot separating the integer and fractional parts. In a binary number, the bit with the highest weight is called the most significant bit (MSB). The bit with least weight is called the least significant bit (LSB).

In this book, if we want to represent a number different from the decimal representation, we will add an appropriate suffix to it. For binary numbers, this suffix will be "b." For hexadecimal numbers, this suffix will be "h." Finally, for the octal numbers, this suffix will be "q."

Conversion between decimal and binary numbers can be done using successive division and multiplication operations. For detailed information, please see [5]. Here, we provide two examples. We can represent the decimal number 255 in binary form as $1 \times 2^7 + 1 \times 2^6 + 1 \times 2^5 + 1 \times 2^4 + 1 \times 2^3 + 1 \times 2^2 + 1 \times 2^1 + 1 \times 2^0$. Or in short form 11111111b. Similarly, for the decimal number 1.25, we have the binary representation $1 \times 2^0 + 0 \times 2^{-1} + 1 \times 2^{-2}$. In short, the binary representation becomes 1.01b.

Although binary numbers are natural for digital systems, their representation may not be practical. Hexadecimal numbers can be used instead for a more compact representation. Here, there are 16 digits as (0, 1, 2, 3, 4, 5, 6, 7, 8, 9, A, B, C, D, E, F). The binary number 11111111b can be represented in hexadecimal form as FFh. The decimal number 1.25 can be represented in hexadecimal form as 1.4h. For conversions between binary and hexadecimal representations, please see [5].

3.2 Negative Numbers

There may be negative numbers in operations. Although in ordinary arithmetic we put a negative sign in front of the number, we do not do so in a digital system. Three methods are available for representing both positive and negative numbers in a digital system. These are: signed bit, one's complement, and two's complement representations.

The first representation mimics the ordinary practice (negative sign in front of the number) by a sign bit in the MSB of the number. In this representation, a positive number will have a sign bit of 0. A negative number will have a sign bit of 1. Hence the name signed bit representation. Although this method seems straightforward, it is not very effective since addition and subtraction operations may need extra circuitry.

The second representation is based on the bit complement operation. Here, the negative number is represented by the bit complement of the corresponding positive number. Therefore, this representation is called the one's complement. In this representation, no extra bit is assigned to the sign bit. However, the arithmetic operations are not straightforward in this representation. For a more detailed explanation, please see [5].

The third form of negative number representation is based on two's complement. Here, the negative number is first represented in one's complement form. Then the result is incremented by 1. Two's complement representation is used for representing negative numbers in the MSP430. Let's say we have the binary number 01001100b. We can obtain its negated version in two's complement form in two steps. First, we obtain its one's complement representation, 10110011b. Adding 1 to the result gives us the two's complement of this number, 10110100b.

Two's complement representation has a major advantage. Subtracting two binary numbers can be rephrased as adding the first number to the two's complement of the second. Therefore, only one adder circuit (introduced in Chap. 2) is needed for both addition and subtraction operations. The resulting representation also keeps the sign information. Therefore, the need for an extra sign bit is also eliminated.

Let's consider two subtraction examples. The first one is subtracting the binary number 00111111b from 01000000b. First, we obtain the two's complement of 00111111b as 11000001b. Adding 11000001b to 01000000b gives 100000001b. As can be seen, the result can be represented by 9 bits. In other words, an overflow occurred. We will explore the overflow issue in Sec. 3.4. If the overflow occurs, we should discard it and the result is final. That is, the subtraction operation results in 00000001b. If we subtract the number 01000000b from 00111111b, we follow the same steps and obtain 11111111b. There is no overflow. Therefore, the result is negative and represented in two's complement form. We can check it by obtaining the two's complement of the first subtraction result, 00000001b. As can be seen, the two's complement representation simplifies life for us.

3.3 Fixed- and Floating-Point Representations

The binary numbers to be processed may also have fractional parts. In Sec. 3.1, we distinguished the integer and fractional parts of such numbers by a dot. In a digital system, this is not possible. Instead, there are two different methods to represent binary numbers with integer and fractional parts. These are fixed- and floating-point representations.

3.3.1 *Fixed-Point Representation*

The number of bits assigned to the integer and fractional parts is fixed in this representation. Hence the name fixed-point representation. This has various advantages. This method is easy to implement since the number of bits assigned to the integer and fractional parts is fixed. Also, the numbers in this form can be processed faster.

Following TI's representation, we can show an unsigned (no sign bit) fixed-point number as UQp.q. Here, U represents the unsigned bit notation, p+q = n shows the number (p and q being the integer and fractional parts) [2]. We provide some fixed-point representation formats in Table 3.1.

As an example, let's consider the decimal number 255.25. The first step to represent this number in fixed-point representation is finding the binary (or hexadecimal) representation of the integer and fractional parts separately. The integer part can be represented as FFh. The fractional part can be represented as 4h. Assume that we would like to represent this number in UQ16 form. Therefore, there will be no fractional part. The number of bits to be assigned to the integer part will be 16. The resulting number will be 00FFh. Zeros added to the left of the number will not affect its value. They will satisfy the fixed-point representation format. If the UQ16.16 fixed-point representation is used for the same number, then the integer part of 255.25 will be the same as 00FFh. The fractional part will be 4000h. Here, zeros are added to the right of the number so that the value of the fractional part will not be affected. The fixed-point representation of the number will be 00FF4000h. As can be seen, there is no separator between the integer and fractional part of the number. Knowing that the number is in UQ16.16 format, we can easily extract the integer and fractional parts (since we know the number of bits assigned to each).

In a similar manner, we can also represent signed numbers. In this form, the MSB is reserved for the sign bit. We provide three signed bit formats for the fixed-point representation in Table 3.2. Similar to the unsigned bit representation, the fixed-point number will be in the form Qp.q.

Table 3.1 Fixed-point, unsigned number representation formats.

Format	Minimum	Maximum	Resolution	No. bits for p	No. bits for q	No. total bits
UQ16.	0	$2^{16} - 1$	1	16	0	16
UQ.16	0	$1\text{-}2^{-16}$	2^{-16}	0	16	16
UQ16.16	0	$2^{16} - 1$	2^{-16}	16	16	32

Table 3.2

Fixed-point signed number representation formats.

Format	Minimum	Maximum	Resolution	# bits for p	# bits for q	# total bits
Q15.	-2^{15}	$2^{15}-1$	1	15	0	16
Q.15	-1	$1-2^{-15}$	2^{-15}	0	15	16
Q15.16	-2^{15}	$2^{15}-2^{-16}$	2^{-16}	15	16	32

3.3.2 *Floating-Point Representation*

The fixed-point representation is easy to implement and process. However, it has a major drawback. The number of bits assigned to the integer and fractional parts is always fixed. This causes limitations both in the range of numbers to be represented and their resolution. The floating-point representation can be used to overcome these problems. As the name implies, the number of bits assigned to the integer and fractional parts is not fixed in this representation. Instead, the assigned number of bits differs for each number, depending on its significant digits. Therefore, a much wider range of values can be represented by this form.

In floating-point representation, a binary number with fractional parts will be shown as $N = (-1)^S \times 2^E \times F$. Here, S stands for the sign bit, E represents the exponent value, and F stands for the fractional part. The floating-point number N is saved in the memory as $X = SEF$.

To represent the floating-point number $N = (-1)^S \times 2^E \times F$, the number should be normalized such that the integer part will have one digit. For ease of binary representation, the exponent will be biased by $2^{(e-1)} - 1$, where e is the number of bits to be used for E in the given format. Finally, a certain number of bits will be assigned to the S, E, and F, depending on the standard format used for representation. The IEEE 754 standard is used by most digital systems in floating-point representation. This standard is summarized in Table 3.3.

Let's take three examples to explain the floating-point representation. In the first example, we will have 255.25. We will follow the following itemized procedure to obtain its floating-point representation.

- Decide on the format: Let's pick the Half format for this example.
- Represent the integer and fractional parts of the decimal number in binary form: Our number becomes 11111111.01b.
- Decide on the sign bit S: Since the number is positive, $(-1)^0 = 1$, S = 0b.

Table 3.3

The IEEE 754 standard for floating-point representation.

Format	Exponent bias	No. bits for *S*	No. bits for *E*	No. bits for *F*	No. total bits
Half	15	1	5	10	16
Single	127	1	8	23	32
Double	1023	1	11	52	64
Quad	16383	1	15	112	128

- Normalize the number such that the integer part will have one digit: Our number becomes 1.111111101×2^7b.
- Find the exponent value: For the half format, the exponent bias is 15. Therefore, the exponent will become $E = 15 + 7 = 22$ with bias. Or in binary form, $E =$ 10110b.
- Find the fractional part: Our fractional part (after normalization) was 111111101b. Since 10 bits should be used to represent the fractional part of the number in half format, $F =$ 1111111010b. Remember, since this is the fractional part, we add the extra zero to its right so that the value of the number will not be affected.
- Construct $X = SEF$: Finally, $X =$ 0101101111111010b. Or in hexadecimal form, $X =$ 5BFAh.

The next example is representing −255.25 in single floating-point format. Again, the itemized conversion is as follows:

- Decide on the format: Let's pick the Single format for this example.
- Represent the integer and fractional parts of the decimal number in binary form: Our number becomes 11111111.01b.
- Decide on the sign bit S: Since the number is negative, $(-1)^1 = -1$, $S =$ 1b.
- Normalize the number such that the integer part will have one digit: Our number becomes 1.111111101×2^7b.
- Find the exponent value: For the single format, the exponent bias is 127. Therefore, the exponent will become $E = 127 + 7 = 134$ with bias. Or in binary form, $E =$ 10000110b.
- Find the fractional part: Our initial fractional part (after normalization) was 111111101b. Since 23 bits should be used to represent the fractional part of the number in single format, $F =$ 11111110100000000000000b. Remember, since this is the fractional part, we add extra zeros to its right so that the value of the number will not be affected.
- Construct $X = SEF$: Finally, $X =$ C37F4000h in hexadecimal form.

The last example is representing 8751.135 in half format. The itemized conversion is as follows.

- Decide on the format: Let's pick the Half format for this example.
- Represent the integer and fractional parts of the decimal number in binary form: Our number becomes 10001000101111.001000101000b.
- Decide on the sign bit S: Since the number is positive, $(-1)^0 = 0$, $S =$ 0b.
- Normalize the number such that the integer part will have one digit: Our number becomes $1.0001000101111001000101000 \times 2^{13}$b.
- Find the exponent value: For the half format, the exponent bias is 15. Therefore, the exponent will become $E = 15 + 13 = 28$ with bias. Or in binary form, $E =$ 11100b.
- Find the fractional part: Our initial fractional part (after normalization) was 0001000101111001000101000b. Since 10 bits should be used to represent the

fractional part of the number in half format, it becomes $F = 0001000101$b. Unlike previous examples, we had to discard some bits in the fractional part at this step. This is due to the number of bits that can be used.

- Construct $X = SEF$: Finally, $X = 0111000001000101$b. Or in hexadecimal form, $X = 7045$h.

3.4 The Word Size and Overflow

The number of bits that can be processed by a microcontroller at once is called its word size. The word size is 16 bits or 2 bytes for MSP430. If an arithmetic operation results in more than 16 bits, an overflow occurs. The overflowed bit should be saved somewhere in the microcontroller. Under the MSP430 it will be saved in the carry bit of the status register (to be explained in Chap. 4). Therefore, the exact result of the operation will not be lost (until the next operation). The overflow should be taken into account especially in assembly programming to be explored in Chap. 7.

3.5 Little and Big Endian Representations

Sometimes, the data to be saved in the microcontroller's memory may be larger than its word size. Hence, the large data must be partitioned and saved in successive memory locations. For such cases, there are two representations: *little endian* and *big endian*. The least significant bits of the data are saved first in the little endian representation. In the big endian representation, the most significant bits are saved first.

As an example, let's consider the floating-point representation of −255.25 in the previous section. In single format, the representation was $X =$ C37F4000h, which needs 4 bytes (or two words). Let's label two successive memory locations I and II. In the little endian representation, I will hold 4000h and II will hold C37Fh. In the big endian representation, I will hold C37Fh and II will hold 4000h. As a reminder, the endian representation for each microcontroller is fixed. The little endian representation is preferred for the MSP430. We will observe its effect in the following chapters.

3.6 ASCII Characters

We do not only process numbers in microcontrollers. For some applications, we may need to handle characters and symbols as well. As we know, everything in the microcontroller is represented in binary form. The ASCII code is introduced to represent characters and symbols in binary form. ASCII stands for the American Standard Code for Information Interchange. The ASCII code for characters and symbols is given in Table 3.4. In this table, LSB stands for least significant byte and MSB stands for most significant byte. To represent a specific character (or a symbol), its corresponding code should be given. Let's assume that we would like to represent the @ symbol. The ASCII code for the symbol @ can be obtained as 40h from Table 3.4.

Table 3.4 ASCII code table.

	LSB															
	0	1	2	3	4	5	6	7	8	9	A	B	C	D	E	F
0	NUL	SOH	STX	ETX	EOT	ENQ	ACK	BEL	BS	HT	LF	VT	FF	CR	SO	SI
1	DLE	DC1	DC2	DC3	DC4	NAK	SYN	ETB	CAN	EM	SUB	ESC	FS	GS	RS	US
2		!	”	#	$	%	&	’	(	)	*	+	‘	-	.	/
M 3	0	1	2	3	4	5	6	7	8	9	:	;	<	=	>	?
S 4	@	A	B	C	D	E	F	G	H	I	J	K	L	M	N	O
B 5	P	Q	R	S	T	U	V	W	X	Y	Z	[	\	]	^	–
6	`	a	b	c	d	e	f	g	h	i	j	k	l	m	n	o
7	p	q	r	s	t	u	v	w	x	y	z	{	|	}	~	DEL

3.7 Summary

The data to be stored in the microcontroller will be in binary form. In this chapter, we focused on the representations of binary numbers. First, we focused on the methods to represent negative binary numbers. We explored the sign bit, one's complement, and two's complement representations for negative numbers. Then, we considered the problem of representing binary numbers with fractional parts. We explored the fixed- and floating-point representations. We also provided examples on converting decimal numbers (with a fractional part) to these forms. Then we explored the word size and overflow issues. The word size for the MSP430 is 16 bits. Related to this, we focused on the little and big endian representations. They help us to save data larger than the word size. Finally, we looked at the ASCII table to represent characters in binary form.

3.8 Problems

3.1 The MSP430 microcontroller uses two's complement representation in subtraction operations. Calculate the following (using pencil and pen) in binary arithmetic:

a. FFFFh+0005h
b. FFFFh-0005h
c. 0005h-FFFFh

3.2 Is MSP430 a fixed- or a floating-point microcontroller?

3.3 Find the fixed-point representation of the number 315.2342 in formats

a. UQ16.
b. UQ.16.
c. UQ16.16.

3.4 Find the fixed-point representation of the numbers −315.2342 and 315.2342 in formats

a. Q15.
b. Q.15.
c. Q15.16.

3.5 You are given four numbers: 13.25, 15.50, 17.50, and 19.25. Find the hexadecimal representation of these numbers in fixed-point UQ16.16 format.

3.6 Find the floating-point representation of the numbers −315.2342 and 315.2342 in formats
a. half.
b. single.
c. double.

3.7 We will only have an approximation in representing the number 8751.135 in half floating form. What is the difference between the actual number and this approximation?

3.8 Find the floating-point representation of the number 8751.135 in single form. Will there be an approximation here?

3.9 Find the floating-point representation of the number π in half form.

3.10 Pick two numbers and calculate their sum in binary arithmetic. Make sure that there is an overflow.

3.11 Pick two numbers and calculate their difference in binary arithmetic. Make sure that there is an overflow.

3.12 We want to save the hexadecimal number CBBCh in a microcontroller with the word size of 2 bytes. How do we write this number
a. if the memory organization is in little endian form?
b. if the memory organization is in big endian form?

3.13 Which endian representation does MSP430 use?

3.14 We want to store the numbers considered in Prob. 3.5 (in UQ16.16 format) in the memory of the MSP430G2553. Let's assume that the lowest possible memory location to be used is 0200h. Start filling these numbers (in hexadecimal form) from the lowest possible memory address allowed. Take into account the microcontroller's endian representation.

3.15 The MSP430 microcontroller keeps a floating-point number (in single format) in two successive memory locations (let's say 0200h and 0202h for this problem) as 522Bh and 449Ah. What is this number in decimal form?

3.16 The ASCII codes given in Table 3.4 are called regular. What happens if we want to represent regional characters like ü, ı, and ç?

4 MSP430 Architecture

Chapter Outline

The aim in this chapter is to familiarize you with the hardware architecture of the MSP430 microcontroller. Modules in this architecture will be explored in the following chapters in detail. We will start with the general layout of the MSP430G2553 architecture. Then we will focus on the central processing unit (CPU), memory, input and output ports, clocks and the timer module, analog-to-digital conversion (ADC) and comparator modules, digital communication module, and other modules.

4.1 General Layout

The functional block diagram of the MSP430G2553 microcontroller is given in Fig. 4.1. We will use this representation in grouping blocks. As can be seen in this figure, the MSP430G2553 has a 16-MHz CPU. It has 16 kB flash and 512 bytes of RAM. It has two input and output ports named P1 and P2. It has a clock system, two timer modules, and a watchdog timer. It has ADC and comparator modules. It has a digital communication module with universal serial communication interface (USCI). It has a brownout protection module, the memory address bus (MAB) and the memory data bus (MDB), and interface modules (JTAG, spy-bi-wire, emulation).

4.2 Central Processing Unit

The MSP430 CPU has a 16-bit reduced instruction set computing (RISC) architecture, with 27 physical and 24 emulated instructions. We will explore this instruction set in Chap. 7 in detail. The CPU is based on Von-Neumann architecture such that the data and instructions are treated the same through the MAB and MDB. More detail on this architecture can be found in [6]. The block diagram of the MSP430G2553 CPU is given in Fig. 4.2. The CPU can be further divided into the following submodules based on this diagram.

4.2.1 *Arithmetic Logic Unit*

The arithmetic logic unit (ALU) performs the arithmetic and logical operations. The two arithmetic operations in this module are addition and subtraction. Subtraction is done using two's complement form as explained in Sec. 3.2. Comparison of two numbers can be performed by the ALU also. Logical operations AND, OR, and XOR can be done bitwise within the ALU. There is neither multiplication nor division operation defined in the MSP430 instruction set. Therefore, the programmer should form suitable algorithms for this purpose in assembly language. Besides providing the result of an operation, the ALU also sets the status bits (flags) based on the operation. This will be explained next.

4.2.2 *CPU Registers*

The MSP430 has 16 registers, each having 16-bit storage capacity. Four of these registers (R0, R1, R2, R3) have dedicated usage. The remaining 12 registers are general purpose. All of these 16 registers can be directly accessed through software. Next, we explain the registers with dedicated usage in detail.

The R0 register is called the **program counter (PC)**. It points to the next instruction to be read from memory and executed by the CPU. In storing the

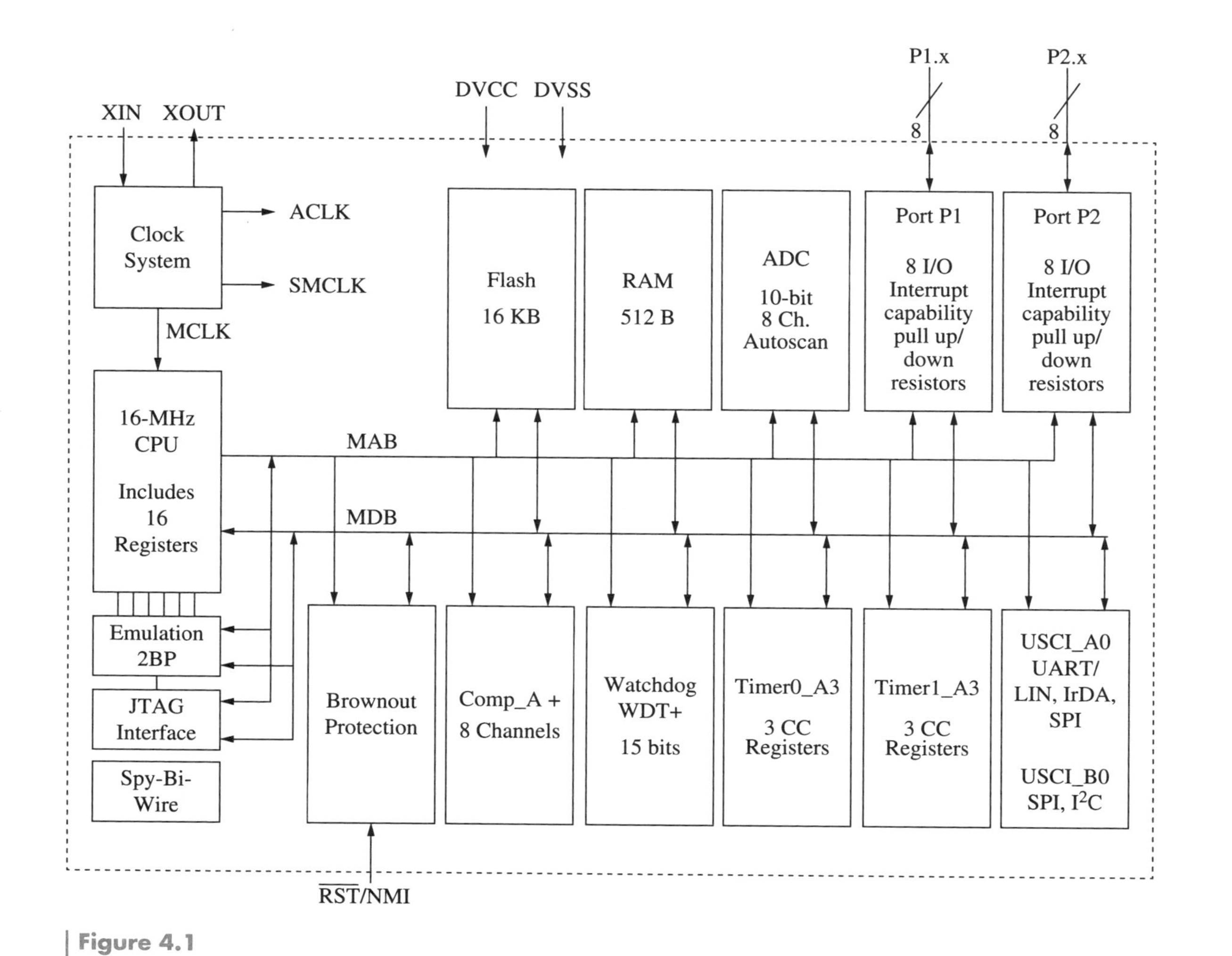

Figure 4.1
Functional block diagram of the MSP430G2553 microcontroller.

Figure 4.2

Block diagram of the MSP430G2553 CPU.

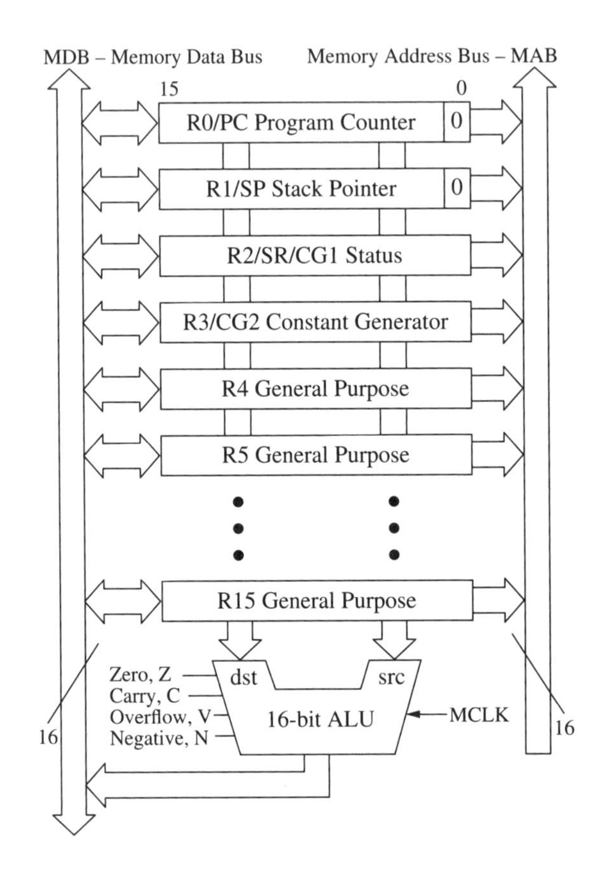

instructions to the memory, even-numbered locations are always used. Therefore, the PC is always incremented by multiples of two. The first instruction to be executed is special. As a reset occurs (either at the startup or during operation), the CPU goes to the reset vector address (to be explained in Chap. 9). This address keeps the address of the first line of the code.

The R1 register is called the **stack pointer (SP)**. It is mainly used to handle stack operations (to be explained in Sec. 7.4). It is also used in the interrupt and function calls. As in the PC, SP also keeps a memory address (of the stack). The SP should be defined at the beginning of the assembly programs.

The R2 register is called the **status register (SR)**. It stores the status and control bits as given in Table 4.1.

In Table 4.1, CG1 stands for the constant generator (to be explained next). The V bit (flag) represents the overflow. This bit is set when an overflow occurs in the

Table 4.1

The status register bits.

Bits	15 - 9	8	7	6	5	4	3	2	1	0
	Reserved for CG1	V	SCG1	SCG0	OSCOFF	CPUOFF	GIE	N	Z	C

ALU. The SCG1, SCG0, and OSCOFF bits are used for clock operations. These will be explored in detail in Chap. 10. Similarly, setting the CPUOFF bit disables the CPU. The GIE bit should be set to enable general interrupts (to be explored in Chap. 9). The N bit (negative flag) is set when the ALU operation gives a negative result. The Z bit (zero flag) is set when the ALU operation gives zero. Finally, the C bit (carry flag) is set when a carry occurs in the ALU operation.

Register R3 and the most significant seven bits of register R2 (SR) are reserved for constant generators (CG1/CG2). These are responsible for generating six constants (0004h, 0008h, 0000h, 0001h, 0002h, FFFFh) used in the microcontroller. These constants are used in emulated instructions (to be explored in Sec. 7.1).

The remaining registers R4–R15 are general purpose. They can be used to store data, address pointers, or index values. They can be accessed with byte or word instructions. It is advantageous to use them in assembly programming since they are on the CPU. We will explore these issues in Chap. 7 in detail.

4.3 Memory

The MSP430G2553 has a 16-bit address bus (MAB). Therefore, it can map 64 kB of memory space. A 16-bit address bus allows direct access and branching throughout the entire memory range. The MSP430G2553 also has a 16-bit data bus (MDB). This allows direct manipulation of word-based arguments. To note here, each memory location is formed by 1 byte of data, and the CPU is capable of addressing the data value either at the byte or word level. Words are always stored and retrieved from even addresses. This even address keeps the least significant byte. The following odd address keeps the most significant byte. This is the little endian representation explained in Sec. 3.5. The data can be accessed from either an even or an odd address in byte operations.

The MSP430G2553 has two types of memory. These are RAM and flash. The RAM is used for temporary storage. Hence it is suitable for keeping variables. The flash is a nonvolatile memory. It can still keep the data when power goes off. Hence, it is primarily used to store the code to be executed. We will further explore the flash memory in Chap. 13.

4.3.1 *The Memory Map*

The memory map does not just represent the RAM and flash in the MSP430 architecture. It also represents the interrupt and reset vector table, special function registers (SFRs), and peripheral modules. Therefore, the input and output ports (to be considered in Chap. 8) are treated as memory addresses. Based on these definitions, the memory map of the MSP430G2553 is given in Table 4.2.

As can be seen in Table 4.2, the highest 64 bytes of address space (between FFFFh and FFC0h) are used for interrupt and reset vector tables. These will be explored in detail in Chap. 9. The next 16320 bytes of address space (between FFBFh and C000h) are used for code memory. Here, constants are also saved. We will discuss this issue in Chap. 6. The 256 bytes (between 10FFh and 1000h) are used for information memory. The calibration data for peripherals are stored in this memory area. These were all in the flash area of the memory. The 512 bytes between 03FFh and 0200h of the memory are reserved for the data. Hence, the local

Table 4.2
MSP430G2553 memory map.

Address	Size (Bytes)	Type	Usage
FFFFh-FFC0h	64	Flash	Interrupt and reset vector table
FFBFh-C000h	16320	Flash	Code memory
10FFh-1000h	256	Flash	Information memory
03FFh-0200h	512	RAM	Data memory
01FFh-0100h	256	Peripheral	16-bit registers
00FFh-0010h	240	Peripheral	8-bit registers
000Fh-0000h	16	Peripheral	8-bit SFR

and global variables are saved here. This area is also used for the stack operations to be explained in Sec. 7.4. This part of the memory is from the RAM. The next 256 bytes of address space (between 01FFh and 0100h) are used for 16-bit peripheral registers. The next 240 bytes of memory space (between 00FFh and 0010h) are used for 8-bit peripheral registers. The lowest 16 bytes of memory space (between 000Fh and 0000h) are used for 8-bit SFRs. These will be explored next.

4.3.2 *Peripheral and Special Function Registers*

In the following chapters, we will study the digital input and output, interrupts, timers, analog-to-digital conversion, and digital communication modules of the MSP430 microcontroller. For all these, some parameters should be adjusted through their control registers. In fact, all these control registers are kept in the lowest 512 bytes of memory as peripheral registers (16- and 8-bit) and SFRs. Interrupt enable 1 (IE1), interrupt enable 2 (IE2), interrupt flag 1 (IFG1), and interrupt flag 2 (IFG2) are defined under SFRs. We will explore all these peripheral registers and SFRs in later chapters. Here, we would like to emphasize that these control registers are also kept in the memory.

4.4 Input and Output Ports

The microcontroller can interact with the outside world through its input and output ports. Here, the processed data can be analog or digital. MSP430G2553 has 16 pins (arranged in two ports as port P1 and P2) to be used for input and output. All the pins can be used as input or output. They can also be used for both analog and digital signals. Therefore, they are called general purpose input and output (GPIO). We will consider these in detail in the following chapters. Here, we would like to mention one important issue. The input and output ports will be simply taken as memory addresses through peripheral registers mentioned in the previous section. This is generally called memory mapped input-output (memory mapped I/O). Therefore, reading or writing data to input-output ports is simply like reading or writing data to a specific memory address.

4.5 Clocks, the Timer, and Watchdog Timer Modules

The MSP430G2553 has one clock, two timers, and a watchdog timer module. We will explore their properties in detail in Chap. 10. Here, we briefly overview them.

4.5.1 *Clocks*

As mentioned in Chap. 3, the ones and zeros in the code level correspond to the two different states of the transistor at the physical level. There is a certain time needed to switch from one state to the other. Therefore, the operations within the microcontroller are done in clock cycles to prevent transition problems. With each clock cycle, the processor performs an action that corresponds to an instruction phase. Besides, timers and peripheral modules may also need other clock signals to operate. Therefore, the MSP430G2553 has more than one clock source. These are called the master clock (MCLK), sub-main clock (SMCLK), and auxiliary clock (ACLK). These clocks are also based on different oscillators. We will consider all these in Chap. 10 in detail.

4.5.2 *The Timer and Watchdog Timer Modules*

MSP430G2553 has two timers on it. These can be programmed for timing and capture/compare operations. In fact, the timer is a counter. There is also a watchdog timer module which needs specific consideration. It resets the CPU in periodic time intervals to eliminate any unexpected program failures (causing infinite loops). This module can also be used as a timer. We will explore all timer modules in detail in Chap. 10. Till then, we will have a code line to disable the watchdog timer in all our programs.

4.6 ADC and Comparator Modules

MSP430 can process analog signals as well. This is done in two ways. First, the comparator can be used such that the analog input voltage is compared with a reference voltage. Depending on the comparison, the output of the comparator will be either zero or one. This can be represented by 1 bit. Second, the ADC module can provide the digital form of the analog signal fed to the microcontroller. MSP430G2553 has a 10-bit ADC module. We will explore both the comparator and the ADC modules in detail in Chap. 11.

4.7 The Digital Communication Module

The MSP430 has a digital communication module called the USCI. This module supports universal asynchronous receiver/transmitter (UART), serial peripheral interface (SPI), and inter integrated circuit (I^2C) communication modes. We will explore these issues in detail in Chap. 12.

4.8 Other Modules

In this section, we summarize the MSP430 modules that we will not explore in detail in the following chapters. These are the brownout protection module, emulation logic with spy-bi-wire interface module, and the JTAG interface module.

The brownout protection module provides the proper internal reset signal to the device during power on and off. The interface modules provide the communication link between the microcontroller and the host computer. More detail on these modules can be found in [17].

4.9 The Pin Layout of the MSP430G2553

The pin layout of the MSP430G2553 microcontroller is given in Fig. 4.3. The general usage area of the GPIO pins in this figure is explained in Table 4.3. As can be seen in this table, each pin can be used for various purposes. We will

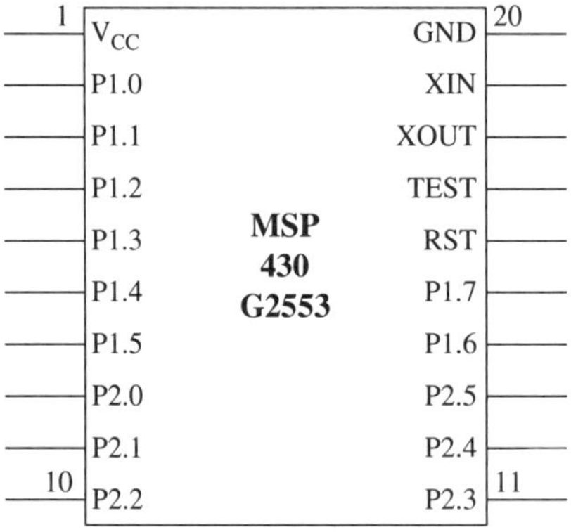

Figure 4.3

Pin layout of the MSP430G2553.

Pin	Left	MSP 430 G2553	Right	Pin
1	V_{CC}		GND	20
	P1.0		XIN	
	P1.1		XOUT	
	P1.2		TEST	
	P1.3		RST	
	P1.4		P1.7	
	P1.5		P1.6	
	P2.0		P2.5	
	P2.1		P2.4	
10	P2.2		P2.3	11

Table 4.3

Pin usage table for MSP430G2553.

Pin	Port Name	Usage Area
1	V_{CC}	Source voltage
2	P1.0	Digital I/O, Interrupt, Timer, ADC
3	P1.1	Digital I/O, Interrupt, Timer, ADC, Digital Commmunication
4	P1.2	Digital I/O, Interrupt, Timer, ADC, Digital Commmunication
5	P1.3	Digital I/O, Interrupt, ADC
6	P1.4	Digital I/O, Interrupt, Timer, ADC, Digital Commmunication
7	P1.5	Digital I/O, Interrupt, Timer, ADC, Digital Commmunication
8	P2.0	Digital I/O, Interrupt, Timer
9	P2.1	Digital I/O, Interrupt, Timer
10	P2.2	Digital I/O, Interrupt, Timer
11	P2.3	Digital I/O, Interrupt, Timer
12	P2.4	Digital I/O, Interrupt, Timer
13	P2.5	Digital I/O, Interrupt, Timer
14	P1.6	Digital I/O, Interrupt, Timer, ADC, Digital Commmunication
15	P1.7	Digital I/O, Interrupt, ADC, Digital Commmunication
16	Reset	Non-maskable interrupt
17		
18	P2.7	Digital I/O, Interrupt, External Crystal
19	P2.6	Digital I/O, Interrupt, Timer, External Crystal
20	V_{SS}	Ground voltage

explore each property separately in the following chapters. In this table, we only summarize the usage areas of the pins to be considered in this book. Other usage areas of the mentioned pins can be found in [17].

4.10 Summary

We explored the architecture of the MSP430 microcontroller in this chapter. Here our focus was on hardware modules. Although we will explore each module in detail, seeing all together with their interactions provides for better insight. Therefore, we considered the CPU first. We explored the arithmetic logic unit and registers in it. Then, we considered the memory. The most important point here is the memory map of the microcontroller. Therefore, we explored it in detail. Next, we considered the ports of the MSP430. Then, we briefly reviewed the clock, timer, and watchdog timer modules. These will be the main focus of time-based operations. We next considered the ADC and comparator modules. We then briefly reviewed the digital communication module. Finally, we provided the pin layout of the MSP430. Our aim was to summarize the usage of each pin by different modules.

4.11 Problems

4.1 Pick another microcontroller and compare it with the MSP430 in terms of architecture.

4.2 What do program counter, status register, and stack pointer mean? Where are they kept in the MSP430 microcontroller?

4.3 What do flash and RAM mean? In the MSP430G2553, what is the size of the flash and RAM?

4.4 How many clocks does the MSP430G2553 microcontroller have?

4.5 What is the size of the MAB for the MSP430? What is the the maximum addressable memory location with this MAB?

4.6 What should be the size of the MAB to address 4 gigabytes of memory (with word size of 64 bits)?

4.7 What does memory mapped input and output mean?

4.8 According to the memory map of the MSP430G2553 microcontroller,

a. What can be the maximum code size to be processed? Here, assume that 100 bytes are kept for storing the data.

b. What can be the maximum stack size?

c. What can be the maximum data size to be stored? Here, assume that the code to process this data needs 512 bytes.

5 Code Composer Studio

Chapter Outline

Code Composer Studio (CCS) is the unique environment for TI's embedded processors. Although a new version of CCS is introduced every year, we believe that the reader should become familiar with at least one CCS version. Therefore, we pick the most recent version of CCS (version 5.3) in this book [11]. We believe that, even if a new version of CCS is introduced in the future, it will not be totally different from this version. We will start with the setup of CCS. Then, we will deal with creating C and assembly projects under CCS. On these, we will explore CCS properties during code execution. We will also explore the new Graphical Peripheral Configuration Tool (Grace) in this chapter. Grace will allow us to configure the hardware of the MSP430 graphically.

5.1 Setup

The official version of CCS for the MSP430 is freely available on the TI website (as we are writing this book). Although this is a code size limited version, it is sufficient for our purposes. Next, we will explain how to download and install it. To note here, the following steps are for a Windows 7–based PC. For Linux installation, please see the TI website.

5.1.1 *Downloading and Installing CCS*

Before starting to download CCS, you need to have a TI account. You can get it from http://www.ti.com. After you register, you will be eligible to download the latest version of CCS through the website http://processors.wiki.ti.com/index.php/Download_CCS. There are two options in the download website. You can select either the website or the off-line installer. In the first option, you can install CCS with required configurations from the Internet directly. If you select the second option, all the installation documents will be downloaded to your computer. Then, you can install CCS from these. The following steps are the same for both installing options.

1. Click on the executable file and start the installation.
2. Accept the license agrement then click Next in the first window.
3. A window asking for the installation directory will appear. Use the default location or create a new folder.
4. In the Setup Type window, select Custom to arrange the setup configuration (instead of installing the complete set).
5. In the Processor Support window, select the only MSP430 Low Power MCUs.
6. In the Select Components window, click Next without changes.
7. In the Select Emulators window, un-select MSP430 Parallel Port FET.
8. The CCS Installation Options window will appear. It will provide a list of documents that will be installed. Click Next and start the installation.

5.1.2 *Hardware Setup*

The MSP430 LaunchPad comes with the MSP430G2553 microcontroller on it. This microcontroller is programmed with a demo software which toggles the onboard red and green LEDs in a sequence. When the MSP430 LaunchPad is connected to the PC through the USB, the driver installation starts first. After installation, the demo starts automatically. This indicates that the hardware is working properly.

5.1.3 Starting CCS, Opening a Workspace, and Choosing the License

When CCS starts for the first time, a window appears so that the location of a workspace folder can be configured. Either use the default workspace folder or change the location by clicking the Browse button. This workspace folder keeps the project and settings files after CCS is closed. Therefore, the same projects and settings will be available when CCS is opened again. The workspace is saved automatically when CCS is closed. The Licence Setup Wizard window appears as the workspace settings are done. Select CODE SIZE LIMITED (MSP430). Then click Finish. Now, CCS is ready to run. A window should appear as given in Fig. 5.1.

5.1.4 CCS Perspectives

There are two perspectives for CCS. These are the CCS Edit and CCS Debug. CCS opens in the Edit perspective every time it starts. This perspective is used for creating projects, building them, and observing errors in them. The CCS Edit perspective is switched to the CCS Debug perspective automatically when the created project is debugged. This perspective is used for debugging projects. It

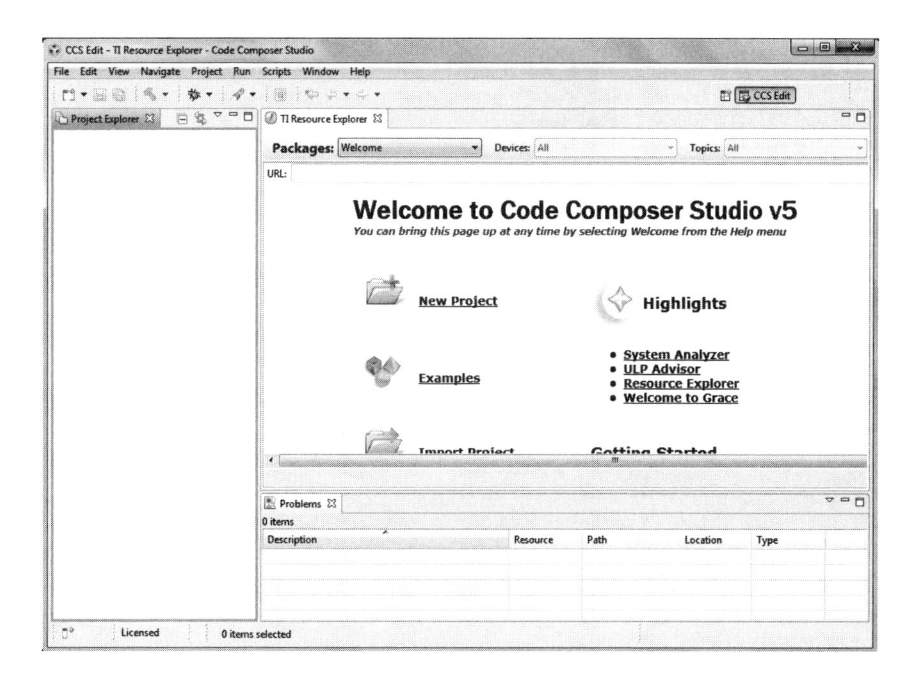

Figure 5.1

The CCS window.

can also be used for observing the hardware (such as registers and memory) or software (such as variables and disassembly) during code execution. The user can switch between these two perspectives using the small icons in the upper right corner of the main CCS window.

5.2 Creating a C Project

A C project contains source, header, and include files. CCS generates an executable output file (with extension .out) from these. This file is used by the MSP430. This section is about creating a C project starting from the beginning.

5.2.1 *A New C Project*

To create a new project in CCS, click File → New → CCS Project. A new window will pop up as in Fig. 5.2. In this window, write a project name and select Variant as shown in Fig. 5.2 for the MSP430G2553 microcontroller. Finally, select Empty Project (with main.c) and click Finish. After these steps, the project will be created

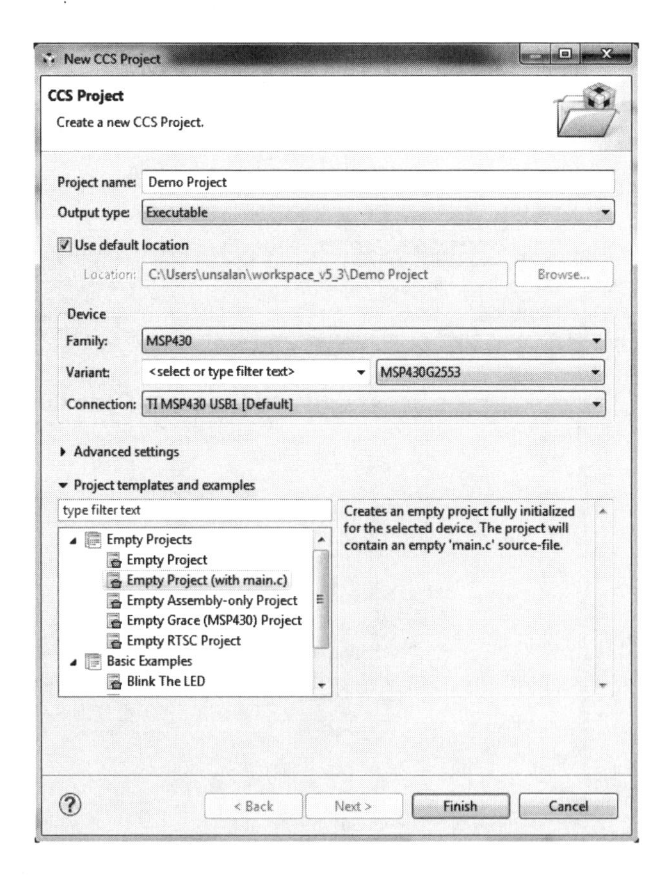

Figure 5.2 Creating a C project.

with the source file named main.c. The generated project should be seen in the Project Explorer window.

When the project is created, an information window about The Ultra-Low-Power Advisor (ULP Advisor) appears. Low power consumption is crucial for MSP430 devices. The ULP Advisor gives valuable information on how to use this property most effectively. The information can be seen from Infos in the Problems window.

When the project is created, the compiler optimization runs automatically. This can remove unused variables and statements. Therefore, it can affect the debugging process. There are two options to prevent this. If compiler optimization is necessary, variables must be declared as volatile. If compiler optimization is unnecessary, the optimization can be disabled by the following steps. In the Project Explorer window, right-click on the Project and select Properties. In the pop-up window, select Optimization and set Optimization Level to off as shown in Fig. 5.3.

After writing the C code in main.c, save it by clicking the Save button in the upper left corner of the CCS menu. In this section, we will use the code given in Listing 5.1. Here, the included header file is msp430.h. Through it, the compiler automatically selects the header file for the MSP430 version. We provide the header file for the MSP430G2553 in the Appendix.

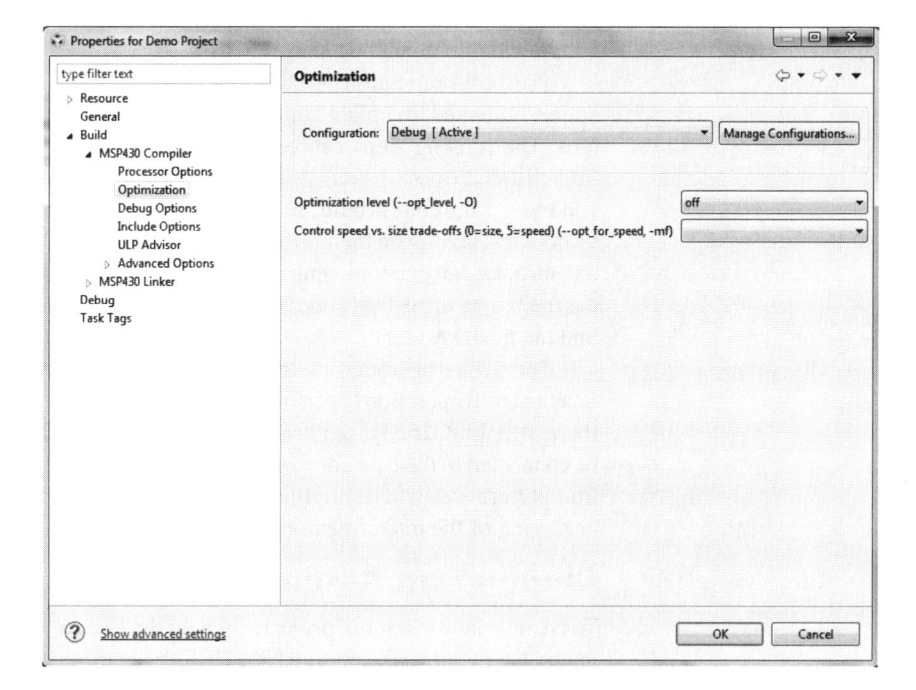

Figure 5.3 Disabling optimization.

Listing 5.1 The first C program for MSP430.

```
#include <msp430.h>

int d = 0;

void main(void)
{
 WDTCTL = WDTPW|WDTHOLD;

 int a = 1;
 float b = -255.25;
 char c = 'c';
 d = d+1;

 while(1);
}
```

5.2.2 Creating a Header File

For some projects, a header file may be needed. To add it to the project, right-click on the project in the Project Explorer window and select New → Header File; or click File → New → Header File. Give the generated header file a name like header.h and click Finish. An empty window will open for the header file. As this header file is added to the project, do not forget to add a line `#include "header.h"` in the main C code.

5.2.3 Building and Loading the Project

There are two buttons on the horizontal toolbar of CCS for code generation. The first one is the Build button with a hammer shape. The second one is the Debug button with the green bug shape.

The build operation is basically used for error detection. When the Build button is clicked, the main source code is linked with all other source and header files. The running steps can be observed in the Console window. The warnings (in yellow), errors (in red), and infos (in blue) can be observed in the Problems window. As the code is built, code sections with warnings and errors can be reached by double-clicking on them in the Problems window. Do not forget that sometimes one mistake can generate multiple errors. Double-clicking on the error may direct the user to an error-free code line. For such cases, examine the code carefully to find the mistake.

The debug operation includes the build. When the Debug button is clicked, first the build operation is performed. Then, CCS loads the code to the target device (here MSP430G2553). To perform this operation, the MSP430 LaunchPad should be connected to the host computer. After the debug operation is complete, the CCS Edit preference switches to the CCS Debug preference. The code is run until the beginning of the main function.

5.3 Creating an Assembly Project

To create a new assembly project, click on the File → New → CCS Project on the main CCS menu. In the pop-up New CCS Project window, select Empty Assembly-only Project and click Finish. After these steps, the project will be created with the source file named main.asm. The generated project should be seen in the Project Explorer window. We will use the sample code given in Listing 5.2. We will see the instructions and directives used here in Chap. 7.

Listing 5.2 The first assembly program for MSP430.

```
 .cdecls C,LIST,"msp430.h"

 .text
 .retain
 .retainrefs

RESET
 mov.w #WDTPW|WDTHOLD,WDTCTL
 mov.w #__STACK_END,SP

 mov.b #11h,R4
 mov.w #00AAh,R5
 and.w R4,R5

 jmp $

;------------------------
;Stack Pointer definition
;------------------------
 .global __STACK_END
 .sect .stack

;-------------------
;Interrupt Vectors
;-------------------
 .sect RESET_VECTOR
 .short RESET
 .end
```

5.4 Program Execution

As the main code (either written in C or assembly) is debugged, the next step is its execution. Thc buttons for the program execution are placed in the Debug window as shown in Fig. 5.4. The name of each button can be observed by moving the cursor over it. These buttons and their functions are explained briefly in the list below.

- **Resume**: Resumes the execution of code from last location of the program counter. When it is pressed, execution continues until a breakpoint or a suspend button press.
- **Suspend**: Halts the execution of the code. All windows used to observe software and hardware parts are updated with recent data.
- **Step Into**: Executes the next line of the code. If this line calls a subroutine, the compiler just executes the next line in the subroutine then stops.

Figure 5.4 Program execution menu.

- **Step Over**: Executes the next line of the code. If this line calls a subroutine, the compiler executes the whole subroutine then stops.
- **Assembly Step Into**: Executes the next assembly instruction. If this instruction calls a subroutine, the compiler just executes the next instruction in the subroutine then stops.
- **Assembly Step Over**: Executes the next assembly instruction. If this instruction calls a subroutine, the compiler executes the whole subroutine then stops.
- **Step Return**: Completes the execution of the subroutine.
- **Reset CPU**: Resets the target microcontroller. It works similar to the reset pin. When it is clicked, registers of the device return to their default states.
- **Restart**: Returns the program counter to the beginning of the loaded program.

5.4.1 *Inserting a Breakpoint*

To stop the execution of the program in a specific code line, a breakpoint should be added. Right-click on the code line to place the breakpoint and select Breakpoint (Code Composer Studio). There are three types of breakpoint options under this item: Breakpoint (Software Breakpoint), Hardware Breakpoint, and Watchpoint as shown in Fig. 5.5.

A software breakpoint is an instruction which is placed at the breakpoint address to halt the code execution. A hardware breakpoint is an address value which halts the code execution when the PC matches this value. A watchpoint is actually a special kind of hardware breakpoint. It is based on a specified data value. The program halts when the code generates it during execution.

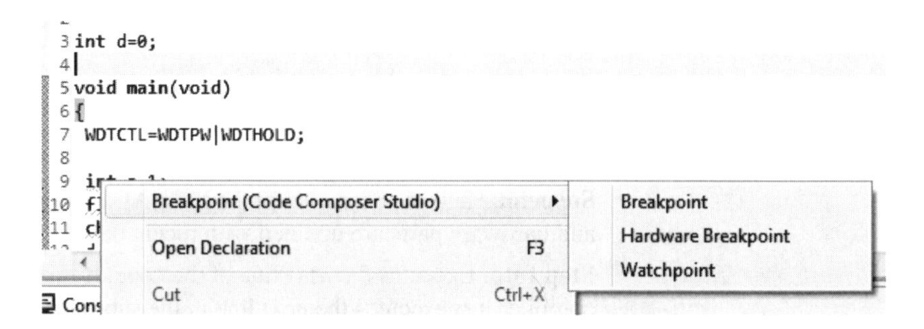

Figure 5.5 Adding a breakpoint.

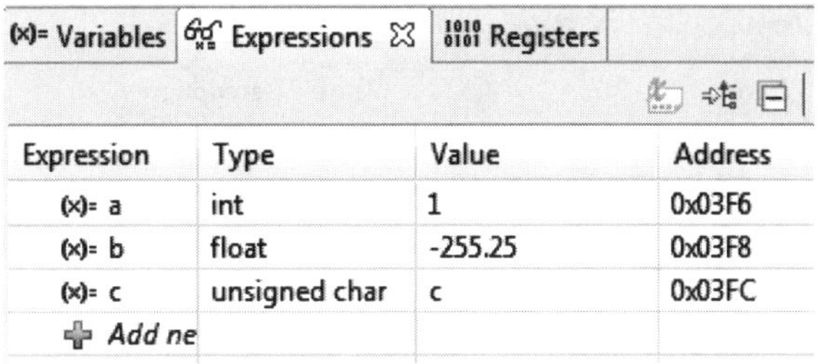

Figure 5.6

Adding a Watch Expression.

(x)= Variables	Expressions	Registers	
Expression	**Type**	**Value**	**Address**
(x)= a	int	1	0x03F6
(x)= b	float	-255.25	0x03F8
(x)= c	unsigned char	c	0x03FC
Add ne			

There are two ways to alter the inserted breakpoint. First, the breakpoint can be deleted by toggling it. To do so, right-click on the breakpoint. Then, select the Toggle Breakpoint from the pop-up window. Second, the breakpoint can be disabled. To do so, select the Disable Breakpoint from the same pop-up window.

5.4.2 *Adding a Watch Expression*

In CCS, the Expressions window (as shown in Fig. 5.6) can be used to observe selected variables. In order to add a variable to it, select the variable to be observed and right-click on it. Then, click Add Watch Expression. Also the Add new expression can be clicked in the Expressions window and the name of the variable can be entered in the box. Do not forget to halt the execution process to observe the changes in the selected variables.

Global variables can also be observed in the Expressions window either by using the Add new expression button or by right-clicking on the Expressions window then selecting Add Global Variables. However, defining a global variable alone is not enough to arrange a memory location for the compiler. The global variable must be used in the main code.

Local variables can also be observed in the Variables window as in Fig. 5.7. They are already listed. Both Expressions and Variables windows are opened automatically after the debugging process. In case of their absence, they can also be opened from the View menu.

5.5 Observing Hardware under CCS

CCS is not a simple C or assembly compiler. Through it, we can also observe the hardware status of the microcontroller. In this section, we explore how to observe the key hardware elements of the MSP430.

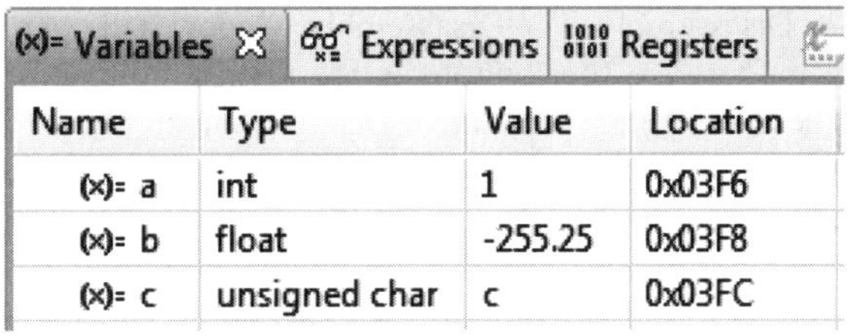

Figure 5.7

Observing local variables in the Variables window.

(x)= Variables	Expressions	Registers	
Name	**Type**	**Value**	**Location**
(x)= a	int	1	0x03F6
(x)= b	float	-255.25	0x03F8
(x)= c	unsigned char	c	0x03FC

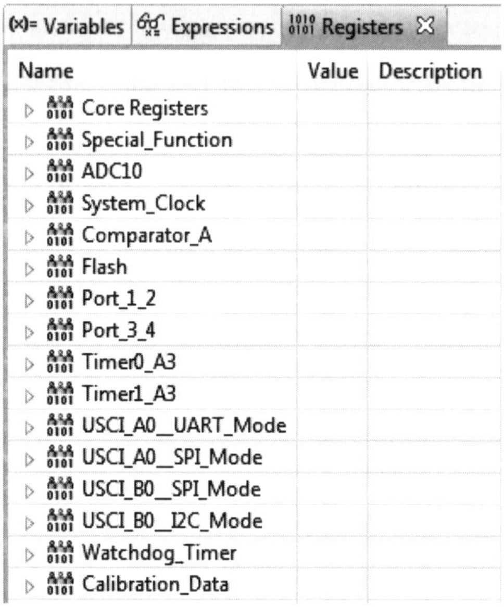

Figure 5.8

Observing registers.

(x)= Variables | Expressions | Registers

Name	Value	Description
Core Registers		
Special_Function		
ADC10		
System_Clock		
Comparator_A		
Flash		
Port_1_2		
Port_3_4		
Timer0_A3		
Timer1_A3		
USCI_A0__UART_Mode		
USCI_A0__SPI_Mode		
USCI_B0__SPI_Mode		
USCI_B0__I2C_Mode		
Watchdog_Timer		
Calibration_Data		

5.5.1 *Registers*

As we mentioned in Chap. 4, MSP430 registers include the program counter, stack pointer, status register, and general-purpose registers. All other registers to control peripherals and special functions are listed separately. To observe the status of these registers, the Registers window (under the View menu) can be used. This is shown in Fig. 5.8.

5.5.2 *Memory*

To observe the memory contents, click View → Memory Browser. Write the starting address of the memory location to be observed in the empty box as shown in Fig. 5.9. The machine language equivalent of the code can also be observed by using the memory browser window.

5.5.3 *Disassembly*

CCS also allows observing the assembly code corresponding to the compiled C code. To do so, the disassembly window should be opened by clicking the View → Disassembly. The Disassembly window will open as shown in Fig. 5.10.

There are four buttons in the vertical column of the Disassembly window. These are: Link with Active Debug Context, Show Source, Assembly Step Into, and Assembly Step Over. Assembly Step Into and Assembly Step Over buttons are already explained in Sec. 5.4. When the Link with Active Debug Context button is pressed, a blue arrow will appear at the left horizontal column of the Disassembly window to follow assembly code execution. The Show Source button may be used to link every C code line with the corresponding assembly line.

Figure 5.9

Observing memory.

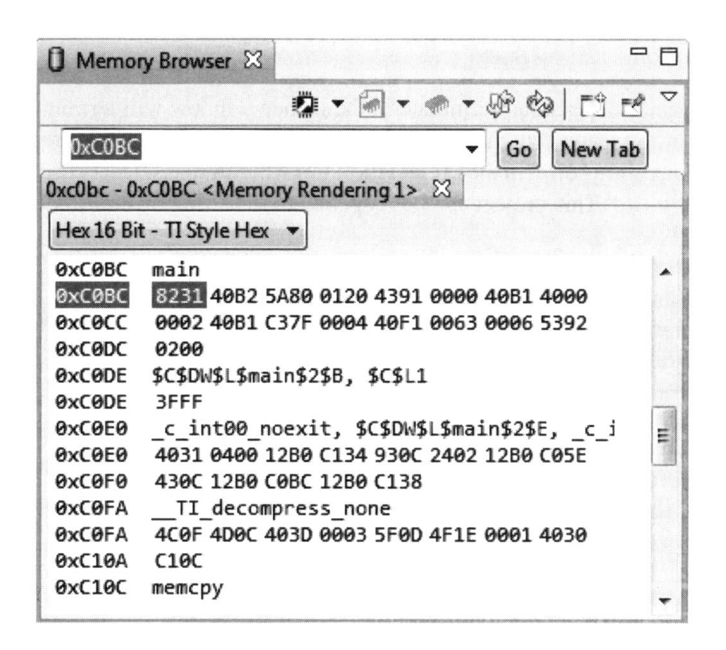

Figure 5.10

Disassembly window.

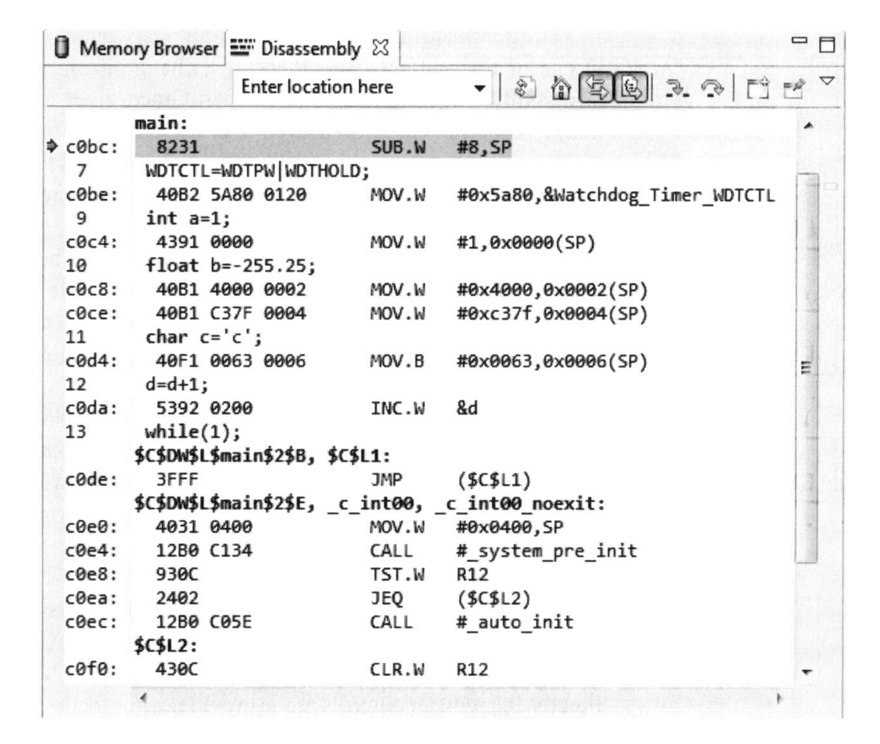

5.6 Terminating the Debug Session and Closing the Project

Clicking Terminate in the Debug window will terminate the active debug session and switch the CCS Debug preference to the Edit preference. Right-click on the project in the Project Explorer window and select Close Project to close the project. This project can be reopened by selecting Open Project.

5.7 Graphical Peripheral Configuration Tool (Grace)

Grace is a user-friendly graphical user interface (GUI) tool to enable and configure peripherals visually. In this book, we use Grace version v2.0. In order to create a Grace-based project, click File → New → CCS Project. In the pop-up New CCS Project window, select Empty Grace (MSP430) Project. All other steps are the same as those for creating a C project.

When the project is created, it is opened in the main.cfg configuration window with a Welcome preference. When the Device Overview button at the top of this window is clicked, the peripheral interface opens as in Fig. 5.11. In this figure, there are white and blue boxes. White boxes represent inaccessible blocks. Blue boxes represent accessible peripheral blocks. Some blue boxes have green check marks on them. This indicates that these blocks are enabled and configured without error.

All peripheral blocks work synchronously under Grace. Therefore, a change in one of them affects the others. If a change in one block is not feasible (due to a conflict with another block), then Grace gives a warning with a red cross mark. When the cursor is moved over the red cross mark, a yellow line pops up and explains the reason for the conflict. An example of such a case is given in Fig. 5.12. In this figure, a conflict exists on P1.0. Therefore, Grace put a red cross on it.

The supply voltage for the microcontroller can be changed under Grace. To do so, the user should select the desired voltage value from the drop-down menu on the left of the DVCC. In this menu, voltage values ranging from 1.8 V to 3.6 V can be selected based on the application. Generally, 3.6 V is picked to prevent any undesired problems.

In the Device Overview View, clicking a peripheral directs the user to the peripheral's window. In this window, the "enable *the name of the peripheral* in my configuration" check box should be checked. Otherwise, Grace assumes that the peripheral will not be used. After this operation, four modes will show up on the top of the window: overview, basic user, power user, and register controls. The overview mode gives brief information and basic examples about the usage of Grace for this peripheral. The basic user mode is very useful for beginners. Most of the configurations can be done in a simple way with this mode. The power user mode contains detailed configuration settings. These are for advanced users. Finally, the register controls mode provides direct access to the peripheral registers. In the following chapters, we will deal with each peripheral (and its Grace modes) in detail.

Figure 5.11

Grace, the Device Overview window.

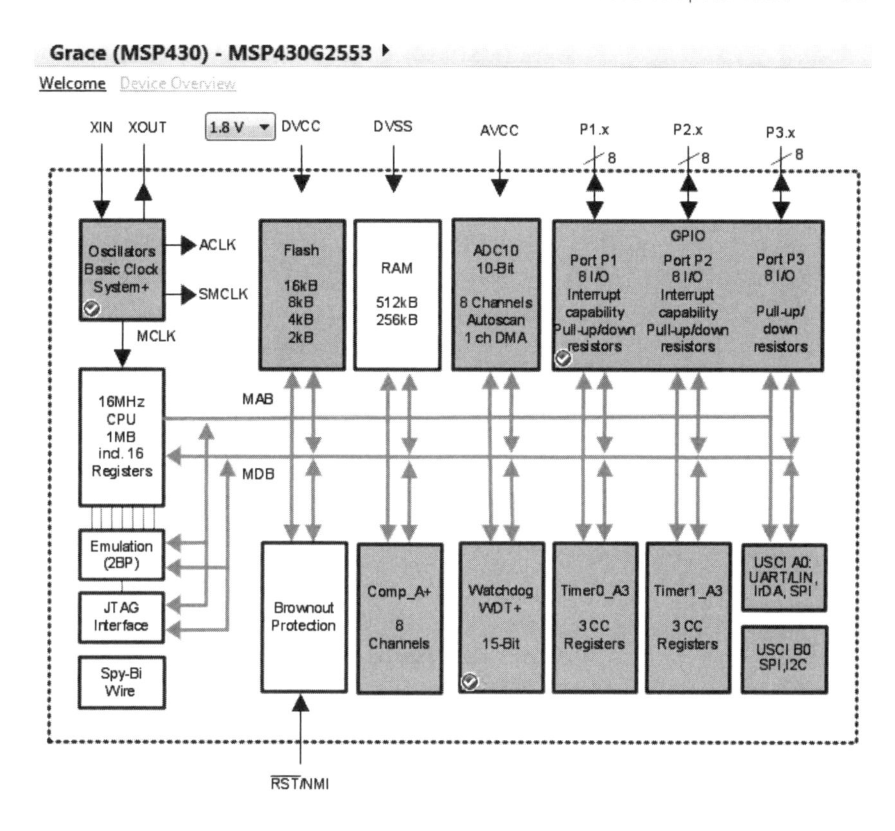

As we mentioned previously, Grace is just a GUI. When the Grace project is created, a main.c file is also formed under the project. When all the peripherals are configured under Grace, the user can run debug. CCS wraps up all the configuration settings under a function `Grace_init()` and adds it to main.c. The header file `Grace.h` containing the Grace-related definitions is also included in main.c. CCS also indicates where the code should be added under the main.c. The user can add his or her codes in this area. An example of such a main file generated by Grace is given in Listing 5.3.

Listing 5.3 The main.c file generated by Grace.

```
/*
 * ======== Standard MSP430 includes ========
 */
#include <msp430.h>
/*
 * ======== Grace related includes ========
 */
#include <ti/mcu/msp430/Grace.h>
/*
 *  ======== main ========
```

Figure 5.12 An example of a conflict on P1.0.

Grace (MSP430) ▸ GPIO - Pinout 20-TSSOP/20-PDIP

Overview Pinout 32-QFN Pinout 20-TSSOP/20-PDIP Pinout 28-TSSOP Power User P1/P2 P3

TEXAS INSTRUMENTS MSP430G2553

Function	Pin name	Pin	Pin	Pin name	Function
	DVCC	1	20	DVSS	
GPIO Input	P1.0	2	19	P2.6	XIN
GPIO Input	P1.1	3	18	P2.7	XOUT
GPIO Input	P1.2	4	17	TEST/SBWTCK	
GPIO Input	P1.3	5	16	RST/NMI/SBWTDIO	
GPIO Input	P1.4	6	15	P1.7	GPIO Input
GPIO Input	P1.5	7	14	P1.6	GPIO Input
GPIO Input	P2.0	8	13	P2.5	GPIO Input
GPIO Input	P2.1	9	12	P2.4	GPIO Input
GPIO Input	P2.2	10	11	P2.3	GPIO Input

```
 */
int main(void)
{
 Grace_init();
// Activate Grace-generated configuration

// >>>>> Fill-in user code here <<<<<

 return (0);
}
```

5.8 The Terminal Window

CCS has an internal terminal program to communicate with the MSP430 LaunchPad through the universal asynchronous receiver/transmitter (UART) communication mode. We will use this property in Chap. 12. The terminal window can be accessed by clicking on the View → Other→Terminal. The terminal window will open as shown in Fig. 5.13.

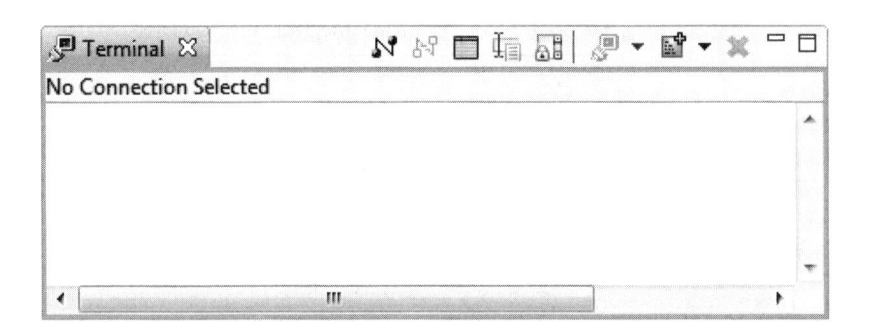

Figure 5.13 The terminal window.

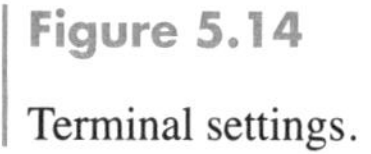

Figure 5.14

Terminal settings.

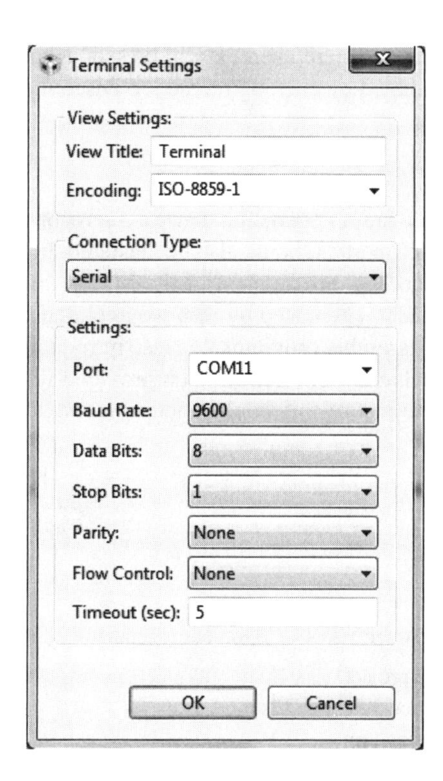

To establish a serial communication link, the configuration settings should be done first. To do so, the Terminal Settings window will be used as shown in Fig. 5.14. The communication type should be selected as Serial first. Then the desired baud rate, data bits, stop bits, and parity values can be set. The flow control and timeout settings can be left unchanged.

The port number used by the MSP430 LaunchPad can be found under Windows 7 using the following steps. First, right-click on the computer icon. Select the properties option from the list. This will open up a new window. In this window, open the Device Manager. The Ports(COM&LPT) in the list gives the port number MSP430 LaunchPad is using.

After checking these settings, the link between the terminal and the MSP430 LaunchPad can be established by clicking the green Connect mark in the terminal window. The code on the MSP430 must be debugged and run to use the terminal window. Let's assume that the code can send and receive data. Data sent from the MSP430 LaunchPad can be observed in the terminal window. It can be cleared anytime by right-clicking and selecting the Clear Terminal option. Data can also be sent from the terminal in two different ways. In the first option, the user can click anywhere in the terminal window. Then, any character pressed on the keyboard can be sent as soon as the key is pressed. In the second option, the toggle command input field can be used. To do so, the user should right-click on the terminal window and select the Toggle Command Input Field. A subwindow appears below. The text to be sent can be entered here. It will be sent as the Enter key is pressed. The link can be terminated by clicking on the red Disconnect mark.

5.9 Summary

Knowing its hardware is not enough to use a microcontroller. The coding environment with all its properties should also be mastered. In this chapter, we introduced the CCS as the coding environment of the MSP430 microcontroller. We started with installing CCS. Then, step by step we explored its usage in debugging and executing C and assembly programs. We also introduced Grace (GUI environment of CCS) in this chapter. The information provided in this chapter will be extensively used in the rest of this book. Therefore, we strongly suggest you master it.

5.10 Problems

5.1 Download the latest version of CCS to your computer and install it.

5.2 Create an empty C project.

a. Add the code block given in Listing 5.1.
b. Debug the code and run it.
c. Observe the local and global variables.
d. Add breakpoints and observe their effects.
e. Obtain the assembly code corresponding to the C code.

5.3 Create an empty assembly project.

a. Add the code block given in Listing 5.2.
b. Debug the code and run it.
c. Observe the register values before and after the program is run.
d. Add breakpoints and observe their effects.

MSP430 Programming with C

Chapter Outline

In this chapter, we consider the C programming issues for MSP430. Therefore, we strongly suggest you refresh your knowledge of C concepts [1]. Here, we will briefly review the basic C concepts from the microcontroller perspective. In other words, we will see how Code Composer Studio (CCS) acts while compiling C code. Therefore, we will first deal with memory management issues. Then we will consider C data types. Next we will briefly cover the arithmetic and logic operations. Then we will consider the control structures. We will also focus on arrays and pointers. Finally, we will consider miscellaneous issues that we will see in later chapters.

6.1 Memory Management

As we saw in Table 4.2, the memory locations for both code and the data are well defined for the MSP430 microcontroller. In this chapter, we will consider these issues on actual examples. Since we are using CCS as the compiler, we will see how it manages memory for both code and data.

6.1.1 *The Code*

After compiling the C code, CCS places the main code block in the flash memory starting from the memory address C000h up to FFBFh. Therefore, the C code written for the MSP430G2553 cannot be larger than 16,320 bytes. Let's consider the sample C code given in Listing 6.1. After following the steps given in Chap. 5, we can generate a C project under CCS from this code block.

Listing 6.1 The sample C program.

```
#include <msp430.h>
 int a = 1;
void main(void)
{
 WDTCTL = WDTPW|WDTHOLD;
 int b = 2;
 int c;
 c = a+b;
 while(1);
}
```

The user can observe how the C code given in Listing 6.1 is placed in memory by using the Disassembly window. We provide the Disassembly window from C000h to C02Ch and C0E8h to C106h in Fig. 6.1 for this sample code. As can be seen in this figure, CCS places the initialization data for the code starting from memory location C000h. The actual code is placed starting from memory location C0E8h.

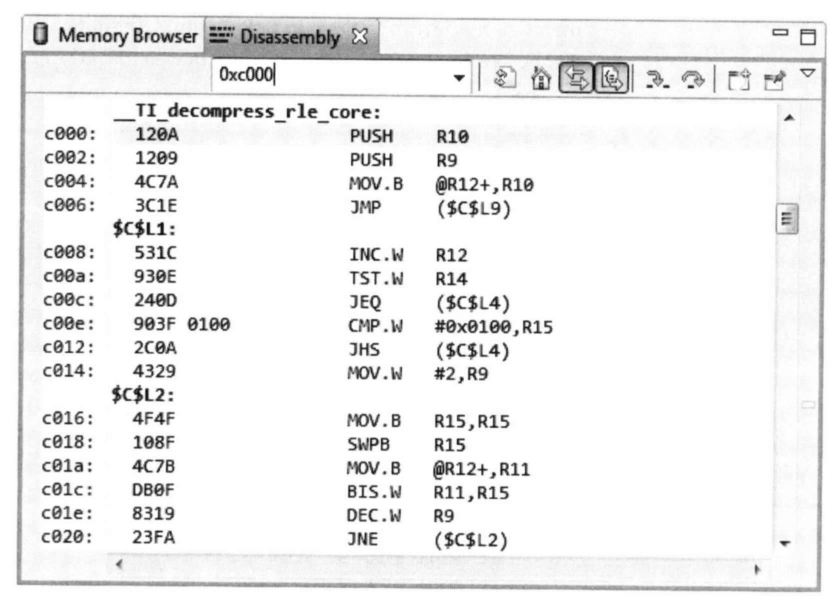

(a) Memory map from C000h to C02Ch

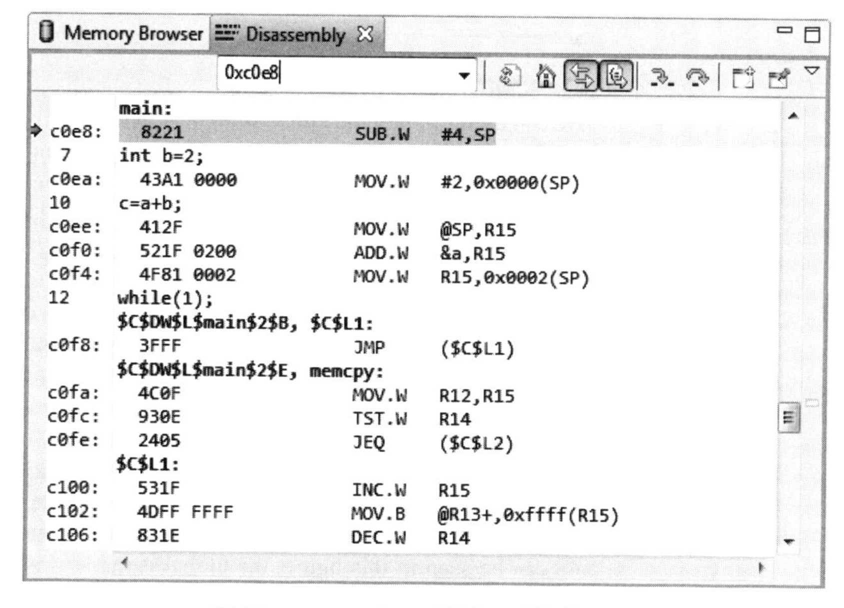

(b) Memory map from C0E8h to C106h

Figure 6.1

Memory contents (from C000h to C106h) observed in the Disassembly window.

We can arrange the sample code in Listing 6.1 such that the addition operation is done in a function. We provide the modified C code in Listing 6.2. We provide the memory map from C05Eh to C084h and C094h to C0B2h for this code block. As can be seen in this example, the function is placed in memory after the main code.

Listing 6.2 The sample C program, with a function.

```
#include <msp430.h>

 int sum(int d, int e);

void main(void)
{
 WDTCTL = WDTPW|WDTHOLD;

 int a = 1;
 int b = 2;
 int c;

 c = sum(a,b);
 while(1);
}

 int sum(int s1,int s2){
 int s = s1+s2;
 return s;
}
```

6.1.2 Local and Global Variables

We can define a variable either as *local* or *global* in the C language. As the name implies, the global variable is available to all program blocks. However, the local variable is only available to the function it is defined in. CCS keeps local and global variables in two different memory locations. The global variables are kept starting from the lowest possible memory address (0200h) in the RAM. As a new global variable is added, the memory address is incremented and the new variable is saved. On the other hand, local variables are kept in the stack (0400h). Based on the definition of the stack, local variables are saved from top to bottom. Here, it is important to note that the C language takes the `main()` as a function. Hence, the variables defined within the main function are also treated as local. Therefore, a global variable should be defined before the `main()`.

We reconsider the sample code given in Listing 6.1 to show the difference between local and global variables. Here, the variable `a` is defined as global. The variables `b` and `c` are defined as local. We show the Expressions window in Fig. 6.3. As can be seen in this figure, the global variable `a` is kept in 0200h (the lowest data address). The local variables `b` and `c` are kept in 03FAh and 03FCh (in the stack) as expected.

6.2 C Data Types

A variable declaration in the C language means a memory location. Therefore, the first issue in a declaration is deciding the size of the memory to be used. The second issue is the format to be used in this assigned memory location. These two

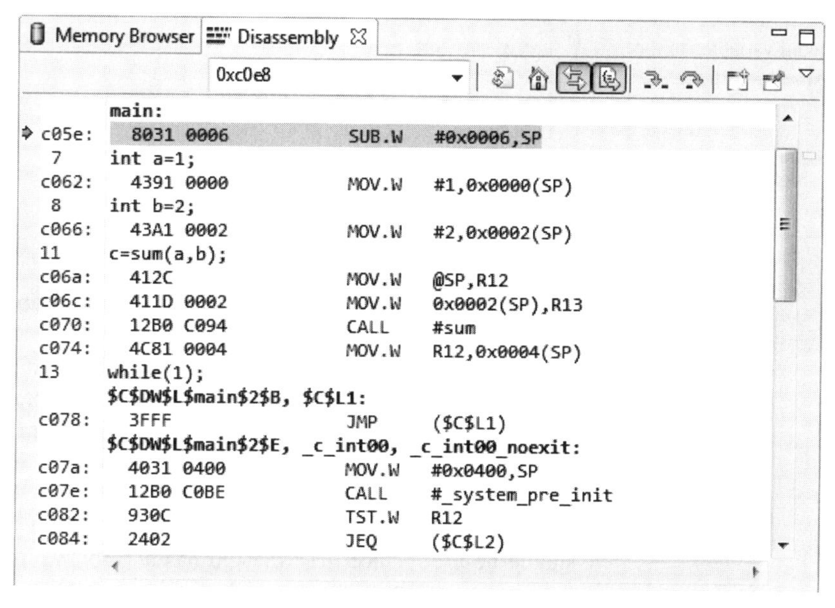

(a) Memory map from C05Eh to C084h

Figure 6.2

Memory contents (from C05Eh to C0B2h) observed by the Disassembly window.

```
Memory Browser   Disassembly
0xc0e8
16      int sum(int s1, int s2){
        sum:
c094:    8031 0006        SUB.W   #0x0006,SP
c098:    4D81 0002        MOV.W   R13,0x0002(SP)
c09c:    4C81 0000        MOV.W   R12,0x0000(SP)
17      int s=s1+s2;
c0a0:    4D0C             MOV.W   R13,R12
c0a2:    512C             ADD.W   @SP,R12
c0a4:    4C81 0004        MOV.W   R12,0x0004(SP)
19      }
c0a8:    5031 0006        ADD.W   #0x0006,SP
c0ac:    4130             RET
        __mspabi_func_epilog, __mspabi_func_epilog_7:
c0ae:    4134             POP.W   R4
        __mspabi_func_epilog_6:
c0b0:    4135             POP.W   R5
        __mspabi_func_epilog_5:
c0b2:    4136             POP.W   R6
        __mspabi_func_epilog_4:
```

(b) Memory map from C094h to C0B2h

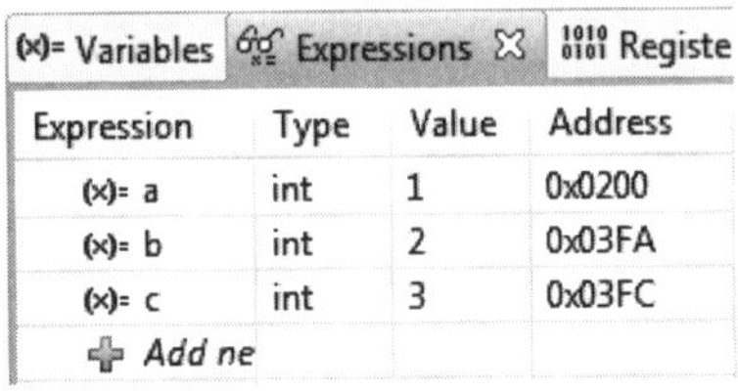

Figure 6.3

Observing local and global variables in the Expressions window.

(x)= Variables | Expressions | Registe

Expression	Type	Value	Address
(x)= a	int	1	0x0200
(x)= b	int	2	0x03FA
(x)= c	int	3	0x03FC
Add ne			

issues are handled by predefined data types for variables. We provide the C data types under CCS in Table 6.1.

As can be seen in Table 6.1, there are three main data types under CCS. The first group consists of *characters*. They can be represented in either signed or unsigned form. The number of bits assigned to them is eight. The ASCII characters corresponding to these numbers are given in Table 3.4. The second group includes *short, int,* and *long* (signed and unsigned). These data types need 16 or 32 bits. While saving the 32-bit data type *long*, the little endian representation should be used, since the MSP430 has a 16-bit word length. The data types in this group cannot represent numbers with fractional parts. The third group consists of *float* and *double*. These are the only possible representations for numbers with fractional parts. For these, the floating-point representation with the single format (given in Table 3.3) is used. Here also the little endian representation is used.

We provide examples for the mentioned C data types in Listing 6.3. Here, we define five global variables with different types having positive or negative values. To note here, we redefine the character a within the program for the compiler to run properly. Besides this there is no other purpose.

Table 6.1

C data types under CCS.

Data Type	No. of bits	Representation	Min. Value	Max. Value
signed char	8	ASCII	−128	127
char, unsigned char	8	ASCII	0	255
short, signed short	16	Two's complement	−32768	32767
unsigned short	16	Binary	0	65535
int, signed int	16	Two's complement	−32768	32767
unsigned int	16	Binary	0	65535
long, signed long	32	Two's complement	−2147483648	2147483647
unsigned long	32	Binary	0	4294967295
float, double, long double	32	IEEE 32-bit	1.175495e-38	3.40282346e+38

Listing 6.3 The C program for data types.

```
#include <msp430.h>

 char a = '@';
 short b = -1;
 int c = 2;
 long d = 3;
 float e = 12.3;
 float f = -255.25;
void main(void)
{
 WDTCTL = WDTPW|WDTHOLD;

 a = '@';

 while(1);
}
```

As we run the C code in Listing 6.3, we get the memory map given in Fig. 6.4. This memory map provides us with valuable information. First, the character @ (in variable a) is saved in memory by its ASCII code 0040h (as given in Table 3.4). The negative short variable b (with value −1) is kept as FFFFh, which is the two's complement representation of −1. The integer variable c with value 2 is saved as is. All the previous variables were occupying 2 bytes (one word). Hence, the endian representation is not used for them. However, the variable d is defined in long type. This requires 4 bytes (two words). As can be seen in the memory map, the variable d with value 3 is kept in two parts. The first part 0003h is kept in memory location 0206h and the second part 0000h in memory location 0208h. The little

Figure 6.4 The memory map for the program showing C data types.

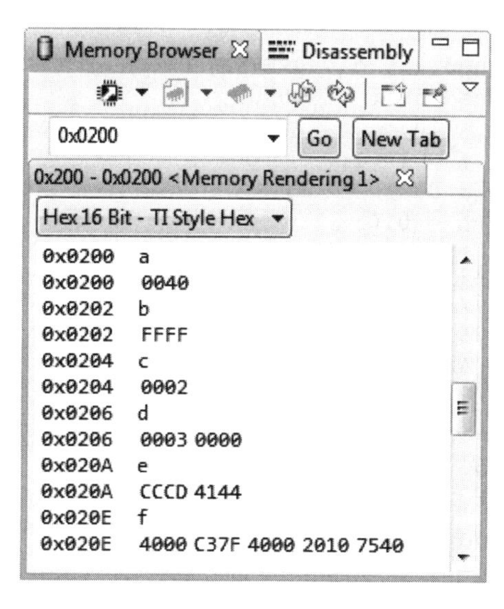

endian representation can be clearly seen here. This is also the case for the float variable e (with value −12.3). The hexadecimal representation (in terms of the single floating format) for this number is 4144CCCDh. This is kept in 4 bytes in the little endian format such that CCCDh is kept in memory location 020Ah and 4144h is kept in memory location 020Ch. Finally, the float variable f is specifically set as −255.25. The hexadecimal representation obtained here justifies the result in Sec. 3.3.2.

6.3 Arithmetic and Logic Operations

We have addition, multiplication, subtraction, division, mode, and remainder arithmetic operations in C language. These may seem straightforward. However, in Chap. 7, we will observe that besides addition and subtraction, none of the other operations can be performed in the MSP430 assembly language. Therefore, the C language simplifies life for us.

It is important to mention that in the C language, the overflow cannot be observed directly. Therefore, the programmer should take the overflow into account. In a similar manner, the result of an arithmetic operation is converted to the assigned variable type. The other important point is that we can define hexadecimal numbers by a prefix 0x in the C language. In Listing 6.4, we provide such examples.

Listing 6.4 Arithmetic operations.

```
#include <msp430.h>

void main(void)
{
 WDTCTL = WDTPW|WDTHOLD;

 int a = 32767;
 int b;
 unsigned int c = 0xFFFF;
 unsigned char d = 0x00;
 int e = 10;
 float f = 10.1;
 int g = 0;
 float h = 0.0;

 a += 1;
 b = 17/2;
 c + = 0x0001;
 d - = 0x01;
 e /= 0;
 f /= 0;
 g /= g;
 h /= h;

 while(1);
}
```

Figure 6.5

The expressions window after arithmetic operations.

(x)= Variables | Expressions | Registers

Expression	Type	Value	Address
(x)= a	int	-32768	0x03EA
(x)= b	int	8	0x03EC
(x)= c	unsigned int	0	0x03EE
(x)= d	unsigned char	0xFF (Hex)	0x03F0
(x)= e	int	0	0x03F2
(x)= f	float	1.#QNAN	0x03F4
(x)= g	int	-1	0x03F8
(x)= h	float	1.#QNAN	0x03FA
Add ne			

After the C code in Listing 6.4 is run, the variables can be observed in the Expressions window as given in Fig. 6.5. As can be seen in this figure, adding 1 to the integer variable a (which was initially 32767) caused an overflow. Therefore, we see −32768 for a instead of 32768 in the Expressions window. The overflow bit in the status register has changed. However, we should use implicit functions for the MSP430 to observe it. We will see this in Sec. 6.6. For the integer variable b, we assigned a float by a division operation. Here, only the integer part is saved as can be seen in the Expressions window. The variables c and d keep hexadecimal values. The first one is defined as an unsigned integer. Hence it can keep 2 bytes (one word) of data. The second one is defined as an unsigned char. It can keep 1 byte of data. To observe these values in the Expressions window, we should adjust the format of the number by right-clicking on the variable and selecting the "number format" option. The overflow can also be observed for the variables c and d. The variables e and f indicate what happens when a division by zero occurs. The result is zero for the integer variable e. Therefore, the programmer cannot detect division by zero. However, the result becomes 1.#QNAN for the float variable. This indicates division by zero under CCS. After a 0/0 division, the integer variable g becomes −1. Again the result becomes 1.#QNAN for the float variable h. As can be seen, detecting division by zero or 0/0 is easier for the float variables.

We can perform bitwise logic operations (*and*, *or*, *xor*, *not*) in C. In performing these, only byte (or word) level operations can be done. We provide examples on the usage of logic operations in Listing 6.5.

Listing 6.5 Logic operations.

```
#include <msp430.h>

void main(void)
{
 WDTCTL = WDTPW|WDTHOLD;

 unsigned char a = 0x02;
 unsigned char b = 0xFF;
```

```
unsigned char c,d,e,f;

c = a|b;
d = a&b;
e = a^b;
f = ~a;

while(1);
}
```

After the C code in Listing 6.5 is run, the variables can be observed in the Expressions window as given in Fig. 6.6. As can be seen in this figure, all the logic operations are done on a bit basis. It is also possible to observe a variable in binary form in the Expressions window by adjusting its number format. This representation may help the programmer to observe the result of the logic operation in a more descriptive manner.

6.4 Control Structures

We have the condition check and loop operations under control structures. However, we will not explore these in detail. As we mentioned previously, the reader should consult C books for them.

6.4.1 *Condition Check*

There are two options if a condition is to be checked within a program. The first one is the `if` or `if else` statement. The usage of these is straightforward. One of the two options is selected based on a binary decision. The second one is the `switch` statement. It should be used if more than two options are available. The condition checks will be inevitable in our applications. Either we will check the status of a switch, or we will observe the analog voltage level in a pin. We will perform appropriate actions based on the obtained values. We will need condition check statements for these and similar cases.

Figure 6.6

The Expressions window after logic operations.

(x)= Variables | Expressions | Registers

Expression	Type	Value	Address
(x)= a	unsigned char	0x02 (Hex)	0x03F8
(x)= b	unsigned char	0xFF (Hex)	0x03F9
(x)= c	unsigned char	0xFF (Hex)	0x03FA
(x)= d	unsigned char	0x02 (Hex)	0x03FB
(x)= e	unsigned char	0xFD (Hex)	0x03FC
(x)= f	unsigned char	11111101 (Binary)	0x03FD

6.4.2 Loops

If we want to execute a code block more than once, we can use loop operations. We have three options for loop operations in the C language: `for`, `while`, and `do while`. They differ in terms of their starting and stopping conditions. We suggest the reader check them in a C book.

We will use loop statements in various applications. However, we have one standard usage which may seem odd. We will have an infinite loop line at the end of most our codes, or the main program block will be kept in an infinite loop. This can be performed by a code line `while(1)` or `for(;;)`. The main reason for using such an infinite loop is as follows: For almost all microcontroller applications, the code should run indefinitely. In other words, the program should not end after the first run. To perform this, we will let the microcontroller stay in an infinite loop without exiting the program.

6.5 Arrays and Pointers

Arrays and pointers deserve specific consideration in the C language. When an array is defined in C, it is treated like a pointer. Therefore, in this section we consider arrays and pointers together.

Pointers and pointer arithmetic are one of the most confusing topics in C. Fortunately, in our case we can observe the memory map of the MSP430 directly under CCS. This will help us to understand the usage of the pointers and their arithmetic.

Let's start with the pointer definition. In Listing 6.6, we define a global integer variable `a` and initially assign 3 to it. Next, we define a global pointer named `a_pointer` with integer type. We provide the Expressions window (after the code is run) in Fig. 6.7. As can be seen, the variable `a` with value 3 is stored in the memory address 0200h. The address of the variable `a` is stored in the pointer `a_pointer` with the code line `a_pointer = &a;`. Therefore, the pointer keeps the memory address. In the following line, we change the entry of this memory address to `*a_pointer = 5;`. As can be seen in the Expressions window, this changes the value of the variable `a`.

Listing 6.6 Pointer usage example.

```
#include <msp430.h>

 int a = 3;
 int *a_pointer;

void main(void)
{
 WDTCTL = WDTPW|WDTHOLD;

 a_pointer = &a;
 *a_pointer = 5;

 while(1);
}
```

Figure 6.7

The Expressions window for the pointer example.

(x)= Variables | Expressions | Registers

Expression	Type	Value	Address
(x)= a	int	5	0x0202
▷ ➡ a_pointer	int *	0x0202	0x0200

Let's consider an array with five integers and observe the operations on it. We provide such a code in Listing 6.7. We first define a global integer array `a` with entries {1, 2, 3, 4, 5}. We provide the Expressions window (after the code is run) in Fig. 6.8. As can be seen, the `a` array is in fact saved by its starting address 0200h. We assign this address to the pointer `a_pointer` in the code line `a_pointer = a;`. Then, we reach the fourth element of the array and change it to zero by incrementing the pointer value by 3. There are two important issues here. First, the array can be processed as if it is a pointer as mentioned above. Second, increments and decrements are done relative to the pointer type in pointer arithmetic. Therefore, the code line `a_pointer + =3;` incremented the pointer value by 6 (3×2 bytes), since each integer occupies 2 bytes.

Listing 6.7 Array usage example.

```
#include <msp430.h>

 int a[5] = 1,2,3,4,5;
 int *a_pointer;

void main(void)
{
 WDTCTL = WDTPW|WDTHOLD;

 a_pointer = a;
 a_pointer + = 3;
 *a_pointer = 0;

 while(1);
}
```

Figure 6.8

The Expressions window for the array and pointer example.

(x)= Variables | Expressions | Registers

Expression	Type	Value	Address
◢ a	int[5]	0x0200	0x0200
(x)= [0]	int	1	0x0200
(x)= [1]	int	2	0x0202
(x)= [2]	int	3	0x0204
(x)= [3]	int	0	0x0206
(x)= [4]	int	5	0x0208
▷ ➡ a_pointer	int *	0x0206	0x020A

Finally, we consider dynamic arrays formed by pointers. In Listing 6.8, we form a dynamic array using a pointer. Here, we first define the global integer variable a and the integer pointer a_pointer. We fill the successive memory locations starting from the address of the variable a by using a for loop. We provide the memory map (after the code is run) in Fig. 6.9. As can be seen in this example, initially we did not define the size of the array (formed by the pointer). We can adjust the array size on the fly since we are using pointer arithmetic. The only disadvantage here is that the programmer is responsible for memory management (allowable memory size and other memory entries).

Listing 6.8 The dynamic array using a pointer.

```
#include <msp430.h>

 int a = 0;
 int *a_pointer;

void main(void)
{
 WDTCTL = WDTPW|WDTHOLD;

 int count;

 a_pointer = &a;

 for(count = 1; count<10; count++)
 {
 a_pointer++;
 *a_pointer = count;
 }

 while(1);
}
```

6.6 Miscellaneous Issues

In this section, we briefly summarize some miscellaneous issues in the C language for the MSP430. The first issue is the define statement and the const declaration. When a constant is defined by the # define keyword, CCS converts all the affected code lines to the final value in the compiling process. Therefore, no more calculations are done during execution. In a similar manner, if a global variable is defined as constant by the const keyword, the data it contains is saved in the code memory block.

In this and the following chapters, we may need to reach the lowest hardware components such as registers. In the C language this is not directly possible. However, TI provides a set of intrinsic functions for this purpose in the header files in 430.h and msp430g2553.h. We provide the intrinsic functions used for the MSP430G2553 in the Appendix.

We provide the sample code in Listing 6.9 to give examples of the topics considered in this section. We provide the Expressions window in Fig. 6.10(a). To observe how the define statements are handled during the compiling process, we provide the first Disassembly window in Fig. 6.11(a). As can be seen in this

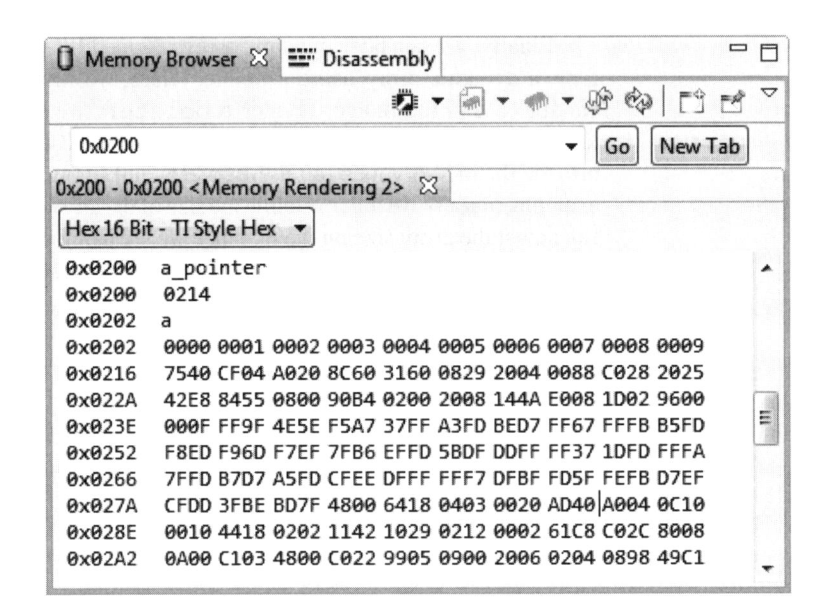

Figure 6.9 The memory map for the dynamic array using a pointer.

figure, for the C code lines `a = 2*CONST` and `b = 4*CONST` the assembly codes are given directly. In other words, the values `2*CONST` and `4*CONST` are calculated during the compiling process. The end result is directly assigned. We provide the second Disassembly window in Fig. 6.11(b) to show that the global const int variable `c` is in fact saved in the code memory. Finally, in Fig. 6.10(b) we provide the bits of the status register. This is to show that the intrinsic function `_get_SR_register()` actually assigned the bit values of the status register to the integer variable `SR_bits`.

Listing 6.9 The sample code for miscellaneous issues.

```
#include <msp430.h>

#define CONST 4

 int a = 0,b = 0;
 const int c[ ] = {1,2,3,4};
 int d = 32767;

void main(void)
{
 WDTCTL = WDTPW|WDTHOLD;

 int SR_bits;

 a = 2*CONST;
 b = 4*CONST;
 d = d+c[1];
 SR_bits = _get_SR_register();

 while(1);
}
```

Figure 6.10

The Expressions and Registers windows for the miscellaneous C concepts.

(x)= Variables | Expressions | Registers

Expression	Type	Value	Address
(x)= a	int	8	0x0200
(x)= b	int	16	0x0202
(x)= d	int	-32767	0x0204
(x)= SR_bits	int	0x0104 (Hex)	0x03FC

(a) The Expressions window

(x)= Variables | Expressions | Registers

Name	Value
SR	0x0104
V	1
SCG1	0
SCG0	0
OSCOFF	0
CPUOFF	0
GIE	0
N	1
Z	0
C	0

(b) The Registers window

The header file math.h can be used for advanced mathematical operations in the MSP430 in the C language. In Listing 6.10, we use the `sin()` function, defined under this header file, to fill the `sine_arr` array with one period of the sine wave. The Expressions window can be used to see the `sine_arr` array entries after the program is run.

Listing 6.10 The C code for the usage of the math.h header file.

```
#include <msp430.h>
#include "math.h"

#define M 20
#define PI 3.1415

float sine_arr[M];

void main(void)
{
 WDTCTL = WDTPW|WDTHOLD;

 int count;
 for(count = 0;count<M;count++)
  sine_arr[count] = sin(2*PI*count/M);

  while(1);
}
```

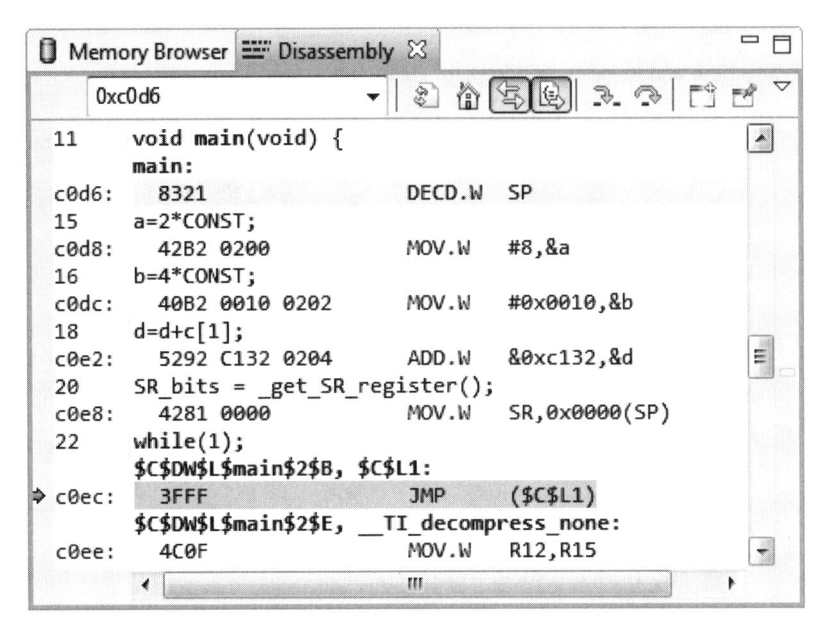

(a) The first Disassembly window

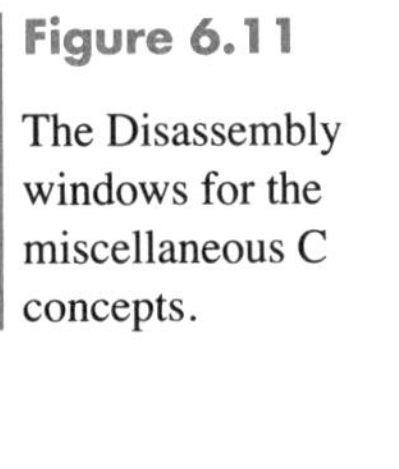

Figure 6.11

The Disassembly windows for the miscellaneous C concepts.

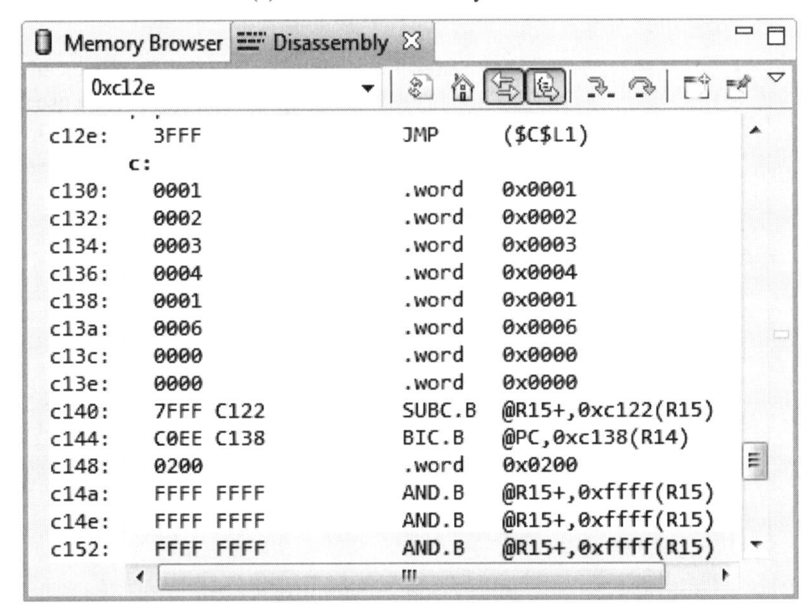

(b) The second Disassembly window

In Chap. 9, we will use the `pragma` keyword extensively. This will force the compiler to include the following code block in the compiling process. We will see why this option is crucial in Chap. 9.

6.7 Summary

The MSP430 can be programmed in both assembly and C languages. In this chapter, we considered the latter approach. Although we focused on the C programming of the MSP430, we assumed that the user has a basic knowledge of C. Here we extend this basic knowledge on the MSP430. Therefore, we first explored the memory management issues and C data types. Since CCS allows us to observe the memory map of the MSP430, we were able to see how local and global variables are handled. Then we looked at the result of division by zero in the C language. We also practiced on array and pointer operations. Finally, we provided an example of using the math.h header file for advanced mathematical operations.

6.8 Problems

6.1 Write a program in the C language such that:

a. It contains a function which calculates the third power of a given integer.
b. Your program should calculate the third powers of numbers between 1 and 10 using your function. The results should be saved in an array.
c. Take the overflow possibility into account.

6.2 Write a C program to calculate the first 10 elements of the Fibonacci series. The user only provides the first two entries of the series. The rest will be calculated by the program. The result will be saved in an array.

6.3 What is the calculated value `y` in Listing 6.11?

Listing 6.11 The C code for Prob. 6.3.

```
#include <msp430.h>

void main (void)
{
 WDTCTL = WDTPW|WDTHOLD;

 int count;
 int y[5];
 int x[] = {1,1,0,1,1};
 int h[] = {1,-1,0,0,0};

 for(count = 0; count<5; count++)
 {
 y[count] = x[count]*h[4-count];
 }

 while(1);
}
```

Note: This program is similar to the discrete convolution operation in signal processing.

6.4 Write a complete C program which calculates the difference between a given number (`out` = 10.3, in this problem) and the reference value (`ref` = 8.2, in this problem). The difference will be assigned to the variable `err`. Your program should also produce the control signal (`cont`, in this problem) as +1 if the variable `err` is less than zero and −1 if it is greater than zero.

Note: This program is similar to a basic feedback application in digital control.

6.5 Find the first four values of the variable `y` as the code in Listing 6.12 is run.

Listing 6.12 The C code for Prob. 6.5.

```
#include <msp430.h>

void main (void)
{
 WDTCTL = WDTPW|WDTHOLD;

 int a = 4;
 int mask = 0x0003;
 int y = 0xFFFF;

 while(a)
 {
 a -= 1;
 y = (y^mask)&a;
 }

 while(1);
}
```

6.6 What will be the entries of the array `arr` as the code in Listing 6.13 is run?

Listing 6.13 The C code for Prob. 6.6.

```
#include <msp430.h>

void main()
{
 WDTCTL = WDTPW|WDTHOLD;

 float arr[] = {2.56,4.88,6.93,0.0,0.0};
 float *ptr = arr,sum = 0;
 short i;

 for(i = 0; i<3; i++)
 {
 sum += *ptr++;
 }
 *ptr++ = sum;
 *ptr = sum/3;

 while(1);
}
```

6.7 Using a suitable intrinsic function, change the contents of the status register. Observe the result in the Registers window.

7 MSP430 Instruction Set

Chapter Outline

We explored the architecture of the MSP430 microcontroller in Chap. 4. In fact, the architecture also contains the instruction set of the microcontroller. This chapter is about the instruction set and the addressing modes of the MSP430. Related to these, we will also consider the stack here. We have devoted a separate chapter to the instruction set due to its importance.

7.1 Introduction

The MSP430 has 27 instructions based on its reduced instruction set computing (RISC) architecture. We will explore these in groups as double operand, single operand, and jump instructions. As you know, every operation to be performed on the microcontroller should be represented in binary form (ones and zeros). The instruction set is no exception to this. This is called machine language. We will consider machine language in detail in Sec. 7.2. Since reading and decoding patterns of zeros and ones is nearly impossible for the user, the assembly language is used. Here each instruction is represented by a mnemonic. In the following sections, we will describe the MSP430 instruction set in terms of mnemonics.

7.1.1 *Double-Operand Instructions*

There are 12 double-operand instructions. These can be separated into arithmetic, logical and register control, and data instructions. In these, the operation is done by two operands called source (src) and destination (dst). The result is written to the destination.

Double-operand instructions may work on both word and byte levels. This is set by attaching a suffix to the mnemonic, either .w (word) or .b (byte). The default is word-level processing. We provide the mnemonic, operation, and a brief description of the operation for double-operand instructions in Table 7.1.

There are several issues to be considered in Table 7.1. As can be seen, there is neither a multiplication nor a division operation for the MSP430. Also, the subtraction operation is done in two's complement form. In this table *.not, .and, .or, .xor* words stand for binary logical operations. In the **cmp** instruction, the result is not written to the destination. Only the appropriate status bits are affected by this operation. Although the **mov** command is almost always called move, it only copies the source to the destination. This may cause confusion. However, this usage originates from historical roots.

In Table 7.2, we provide the effect of the instructions on the status register (SR) bits. In this table, the signs have the following meanings: +, the corresponding bit is affected; -, the corresponding bit is not affected; 0, the corresponding bit is reset; 1, the corresponding bit is set. We will use the same notation in the following sections.

7.1.2 *Single-Operand Instructions*

The MSP430 has seven single-operand instructions. These may also work on both word and byte levels. This is set by, suffix to the mnemonic either .w (word) or .b (byte). The default is adding a word-level processing. The only exception to this dual operation is **sxt** and **swpb**. These instructions can only work on word level. Single-operand instructions are described in Table 7.3.

Table 7.1

Description of double-operand instructions.

Mnemonic	Operation	Brief Description
Arithmetic instructions		
add src, dst	$src + dst \rightarrow dst$	Add source to destination
addc src, dst	$src + dst + C \rightarrow dst$	Add source and carry to destination
dadd src, dst	$src + dst \rightarrow dst$ (dec)	Decimal add source and carry to destination
sub src, dst	$dst + .not.src + 1 \rightarrow dst$	Subtract source from destination
subc src, dst	$dst + .not.src + C \rightarrow dst$	Subtract source and not carry from destination
Logical and register control instructions		
and src, dst	$src.and.dst \rightarrow dst$	And source with destination
bic src, dst	$.not.src.and.dst \rightarrow dst$	Clear bits in destination
bis src, dst	$src.or.dst \rightarrow dst$	Set bits in destination
bit src, dst	$src.and.dst$	Test bits in destination
xor src, dst	$src.xor.dst \rightarrow dst$	Xor source with destination
Data instructions		
cmp src, dst	$dst - src$	Compare source to destination
mov src, dst	$src \rightarrow dst$	Copy source to destination

Table 7.2

The effect of the double-operand instructions on the status register bits.

	SR Bit			
Mnemonic	V	N	Z	C
add src,dst	+	+	+	+
addc src,dst	+	+	+	+
dadd src,dst	+	+	+	+
sub src,dst	+	+	+	+
subc src,dst	+	+	+	+
and src,dst	0	+	+	+
bic src,dst	–	–	–	–
bis src,dst	–	–	–	–
bit src,dst	0	+	+	+
xor src,dst	+	+	+	+
cmp src,dst	+	+	+	+
mov src,dst	–	–	–	–

Table 7.3

Description of the single-operand instructions.

Mnemonic	Operation	Brief Description
Logical and register control instructions		
rra dst	$MSB \rightarrow MSB \rightarrow \cdots LSB \rightarrow C$	Roll destination right
rrc dst	$C \rightarrow MSB \rightarrow \cdots LSB \rightarrow C$	Roll destination right through carry
swpb dst	Swap bytes	Swap bytes in destination
sxt dst	$bit7 \rightarrow bit8 \cdots bit15$	Sign extend destination
push dst	$SP - 2 \rightarrow SP, src \rightarrow @SP$	Push source on stack
Program flow control instructions		
call dst	$SP - 2 \rightarrow SP, PC + 2 \rightarrow @SP,$ $dst \rightarrow PC$	Subroutine call to destination
reti	$@SP+ \rightarrow SR, @SP+ \rightarrow @SP$	Return from interrupt

In Table 7.3, SP stands for the stack pointer and SR for the status register. As in the previous section, we show the effect of single-operand instructions on the status register bits in Table 7.4.

7.1.3 *Jump Instructions*

Jump instructions are given in Table 7.5. In this table, C, Z, and N represent status register bits or flags (as mentioned in Sec. 4.2.2). Jump instructions redirect the program execution flow based on these bits. In other words, jump instructions alter the program counter.

The jump instruction **jmp** $ deserves special consideration. Using $ as a label indicates that the program will jump to the current address. Hence, an infinite loop will be formed. Therefore, the execution of the program will not end. We will use this structure in our assembly code samples for this purpose.

Table 7.4

The effect of the single-operand instructions on the status register bits.

	SR Bit			
Mnemonic	V	N	Z	C
rra dst	0	+	+	+
rrc dst	+	+	+	+
swpb dst	–	–	–	–
sxt dst	0	+	+	+
push dst	–	–	–	–
call dst	–	–	–	–
reti	+	+	+	+

Table 7.5

Description of the jump instructions.

Mnemonic	Condition	Brief Description
jc/jhs label	C = 1	Jump to label if the carry bit is set
		Jump if higher or same
jge label	(N .xor V) = 0	Jump to label if greater than or equal
jl label	(N .xor V) = 1	Jump to label if less than
jmp label	NONE	Jump to label unconditionally
jn label	N = 1	Jump to label if the negative bit is set
jnc/jlo label	C = 0	Jump to label if the carry bit is reset / Jump if lower
jnz/jne label	Z = 0	Jump to label if the zero bit is reset / Jump if not equal
jz/jeq label	Z = 1	Jump to label if the zero bit is set / Jump if equal

7.1.4 *Emulated Instructions*

The MSP430 has 24 emulated instructions in addition to the 27 instructions mentioned above. These are given in Table 7.6. To note here, the emulated instructions only help the readability of the assembly code written. Besides, they are automatically replaced by the original instructions (or their pairs) during the compiling step.

There are some important points on the usage of emulated instructions. In the **sbc** instruction, the constant FFFFh will be replaced by FFh for the byte-level operation. This is also the case for the **inv** instruction. We provide the effect of emulated instructions on the status register bits in Table 7.7.

7.2 Anatomy of an Instruction

As we have mentioned in the previous sections, the information in a microcontroller is represented in binary form (as ones and zeros). Instructions are no exception. In this section, we provide the format for double, single, and jump instructions in Table 7.8 in machine language.

In Table 7.8, **As** represents the addressing bits used to define the addressing mode used by the source operand. **S-reg** represents the register used by the source operand. **Ad** represents the addressing bits used to define the addressing mode used by the destination operand. **D-reg** represents the register used by the destination operand. Finally, **b/w** represents word- or byte-level selection bit.

We provide three instructions from three groups (double operand, single operand, and jump) in Listing 7.1. We obtain the Disassembly window as given in Fig. 7.1 after executing the assembly code. As can be seen in this figure, the assembly code line `mov.w R5,R4` (double-operand instruction) has the hexadecimal representation 4504h. The assembly code line `rrc R5` (single-operand instruction) has the hexadecimal representation 1005h. Finally, the assembly code line `jmp Mainloop` (jump instruction) has the hexadecimal representation 3FFCh. These are the machine language representations of the sample code lines given. For more detailed information on this issue, please see [2].

Table 7.6
Emulated instructions.

Mnemonic	Operation	Emulation	Brief Description
Arithmetic instructions			
adc dst	$dst + C \rightarrow dst$	**addc** #0, dst	Add carry to destination
dadc dst	$dst + C \rightarrow dst$ (decimal)	**dadd** #0, dst	Decimal add carry to destination
dec dst	$dst - 1 \rightarrow dst$	**sub** #1, dst	Decrement destination
decd dst	$dst - 2 \rightarrow dst$	**sub** #2, dst	Decrement destination twice
inc dst	$dst + 1 \rightarrow dst$	**add** #1, dst	Increment destination
incd dst	$dst + 2 \rightarrow dst$	**add** #2, dst	Increment destination twice
sbc dst	$dst + FFFFh \rightarrow dst$	**subc** #0, dst	Subtract source and borrow
Logical and register control instructions			
inv dst	$.not.dst \rightarrow dst$	**xor** #FFFFh, dst	Invert bits in destination
rla dst	$C \leftarrow MSB \leftarrow \cdots LSB \leftarrow 0$	**add** dst, dst	Roll left arithmetically
rlc dst	$C \leftarrow MSB \leftarrow \cdots LSB \leftarrow C$	**addc** dst, dst	Roll left through carry
Program flow control instructions			
br dst	$dst \rightarrow PC$	**mov** dst, PC	Branch to destination
dint	$0 \rightarrow GIE$	**bic** #8, SR	Disable (general) interrupts
eint	$1 \rightarrow GIE$	**bis** #8, SR	Enable (general) interrupts
nop	None	**mov** R3, R3	No operation
ret	$@SP \rightarrow PC, SP + 2 \rightarrow SP$	**mov** $@SP+$, PC	Return from subroutine
Data instructions			
clr dst	$0 \rightarrow dst$	**mov** #0, dst	Clear destination
clrc	$0 \rightarrow C$	**bic** #1, SR	Clear the carry flag
clrn	$0 \rightarrow N$	**bic** #4, SR	Clear the negative flag
clrz	$0 \rightarrow Z$	**bic** #2, SR	Clear the zero flag
pop dst	$@SP \rightarrow temp, SP + 2 \rightarrow SP$ $temp \rightarrow dst$	**mov** @SP+, dst	Pop word/byte from destination
setc	$1 \rightarrow C$	**bis** #1, SR	Set the carry flag
setn	$1 \rightarrow N$	**bis** #4, SR	Set the negative flag
setz	$1 \rightarrow Z$	**bis** #2, SR	Set the zero flag
tst dst	dst+FFFFh+1	**cmp** #0, dst	Test destination

7.3 MSP430 Addressing Modes

The MSP430 has seven addressing modes. These can be used by all instructions given in the previous section. We briefly list these addressing modes in Table 7.9. In this table, ADDR1 and ADDR2 represent symbols for the memory locations. Each addressing mode is explained below with examples.

Table 7.7

The effect of the emulated instructions on the status register bits.

Mnemonic	SR Bit V	N	Z	C
adc dst	+	+	+	+
dadc dst	+	+	+	+
dec dst	+	+	+	+
decd dst	+	+	+	+
inc dst	+	+	+	+
incd dst	+	+	+	+
sbc dst	+	+	+	+
inv dst	+	+	+	+
rla dst	+	+	+	+
rlc dst	+	+	+	+
br dst	–	–	–	–
dint	–	–	–	–
eint	–	–	–	–
nop	–	–	–	–
ret	–	–	–	–
clr dst	–	–	–	–
clrc	–	–	–	0
clrn	–	0	–	–
clrz	–	–	0	–
pop dst	–	–	–	–
setc	–	–	–	1
setn	–	1	–	–
setz	–	–	1	–
tst dst	0	+	+	1

RISC / CISC machine
- 27 orthogonal instructions
  - 8 Jump instructions
  - 7 single operand instructions
  - 12 double operand instructions
- 4 basic 6 addressing modes
- 8/16-bit instruction addressing

• Memory architecture
- 16 16-bit registers
- 16-bit arithmetic logic unit (ALU)
- 16-bit address bus (64 KB address)
- 16-bit data bus (8-bit addressability)
- 8/16-bit peripherals

- addressability: number of bits stored in each memory location.

Table 7.8

Format for instructions.

Bits	15	12	11	9	7	6	5	3	0
Double operand	Op-code		S-reg		Ad	b/w	As	D-reg	
Single operand			Op-code			b/w	Ad	D/S-reg	
Jump	Op-code	Condition			10-bit two's complement PC offset				

7.3.1 *Immediate Mode*

The immediate mode can be used to assign numbers directly to a register or a memory location. In assigning numbers, Code Composer Studio (CCS) allows binary, octal, decimal, and hexadecimal values with the format given in Chap. 3. The # sign is used before the number to represent a constant value. We illustrate the usage of the immediate mode in Listing 7.2.

Listing 7.1 Anatomy of an instruction example.

```
 .cdecls C,LIST,"msp430.h"

 .text
 .retain
 .retainrefs

RESET
 mov.w #WDTPW|WDTHOLD,WDTCTL
 mov.w #__STACK_END,SP

Mainloop:
 mov.w #01h,R5
 mov.w R5,R4
 rrc R5
 jmp Mainloop

;------------------------
;Stack Pointer definition
;------------------------
 .global __STACK_END
 .sect .stack

;-------------------
;Interrupt Vectors
;-------------------
 .sect RESET_VECTOR
 .short RESET
 .end
```

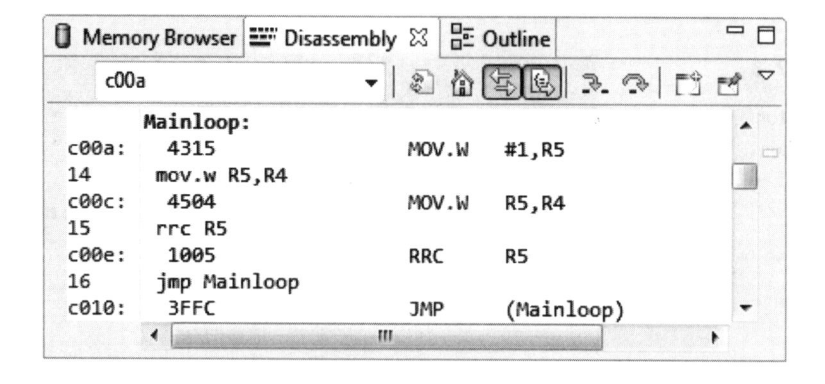

Figure 7.1 Disassembly window for the anatomy of an instruction example.

Table 7.9 Addressing modes.

Addressing Mode	Property	Example
Immediate	Constant values	`mov #45h,&ADDR1`
Register	Fast	`mov R10,R11`
Absolute	Direct access to a memory location	`mov &ADDR1,&ADDR2`
Symbolic	Easy to read code, relative	`mov ADDR1,ADDR2`
Indexed	Table processing	`mov 2(R10),6(R10)`
Indirect register	Access memory with pointers	`mov @R10,R11`
Indirect autoincrement	Table processing	`mov @R10+,R11`

The code in Listing 7.2 assigns hexadecimal numbers 0000h to 0003h to successive memory locations 0200h to 0206h. It also assigns the hexadecimal number 0A0Ah to the general-purpose register R5.

Listing 7.2 Usage of the immediate mode.

```
    .cdecls C,LIST,"msp430.h"

    .text
    .retain
    .retainrefs

RESET
    mov.w #WDTPW|WDTHOLD,WDTCTL
    mov.w #__STACK_END,SP

    mov.w #0000h,&0200h
    mov.w #0001h,&0202h
    mov.w #0002h,&0204h
    mov.w #0003h,&0206h
    mov.w #0A0Ah,R5

    jmp $

;------------------------
;Stack Pointer definition
;------------------------
    .global __STACK_END
    .sect .stack

;-------------------
;Interrupt Vectors
;-------------------
    .sect RESET_VECTOR
    .short RESET
    .end
```

7.3.2 *Register Mode*

The operations are performed on registers in the register mode. Since the registers are on the CPU, they are easy to access. Therefore, their processing speed is fast. The code in Listing 7.3 copies the content of register R5 to register R6.

Listing 7.3 Usage of the register mode.

```
    .cdecls C,LIST,"msp430.h"

    .text
    .retain
    .retainrefs

RESET
    mov.w #WDTPW|WDTHOLD,WDTCTL
    mov.w #__STACK_END,SP

    mov.w R5,R6

    jmp $

;------------------------
;Stack Pointer definition
;------------------------
    .global __STACK_END
    .sect .stack

;-------------------
;Interrupt Vectors
;-------------------
    .sect RESET_VECTOR
    .short RESET
    .end
```

7.3.3 *Absolute Mode*

The absolute mode is used to reach a memory address directly. In the MSP430, peripherals are also taken as addresses as explained in Sec. 4.3.1. Therefore, the absolute mode may be used to reach and alter them. The memory address is indicated by the & sign. We illustrate the usage of the absolute mode in Listing 7.4. The eight line in Listing 7.4 copies the contents of memory address 0200h to register R9. The ninth line copies the contents of the memory address 0200h to the memory address 0206h.

Listing 7.4 Usage of the absolute mode.

```
    .cdecls C,LIST,"msp430.h"

    .text
    .retain
    .retainrefs

RESET
    mov.w #WDTPW|WDTHOLD,WDTCTL
    mov.w #__STACK_END,SP

    mov.w &0200h,R9
```

```
 mov.w &0200h,&0206h

 jmp $

;------------------------
;Stack Pointer definition
;------------------------
 .global __STACK_END
 .sect .stack

;-------------------
;Interrupt Vectors
;-------------------
 .sect RESET_VECTOR
 .short RESET
 .end
```

7.3.4 *Symbolic Mode*

Memory locations and variables are represented by words in symbolic mode. This makes the assembly program easy to understand. This also allows us to transfer the code to another MSP430 microcontroller family member easily (the relativity property).

We provide Listing 7.5 to illustrate the usage of symbolic mode. Here, the program counter (PC) has been defined beforehand (in the msp430g2553.h header file). We can copy the value of the PC to register R5 by `mov.w PC,R5`. As can be seen here, the usage of the symbol PC is straightforward in assembly language. The second line in Listing 7.5 copies the port P1 input values to register R8 (we will see this in detail in Chap. 8).

Listing 7.5 Usage of the symbolic mode.

```
 .cdecls C,LIST,"msp430.h"

 .text
 .retain
 .retainrefs

RESET
 mov.w #WDTPW|WDTHOLD,WDTCTL
 mov.w #__STACK_END,SP

 mov.w PC,R5
 mov.b P1IN,R6

 jmp $

;------------------------
;Stack Pointer definition
;------------------------
 .global __STACK_END
 .sect .stack

;-------------------
```

```
;Interrupt Vectors
;-------------------
 .sect RESET_VECTOR
 .short RESET
 .end
```

7.3.5 Indexed Mode

The indexed mode is used to process a table in memory. A register (memory) holds the base address of the table in this mode. This value is incremented to reach successive memory locations. Meanwhile, the base address value is not changed.

We provide Listing 7.6 to illustrate indexed mode usage. Here, four hexadecimal numbers are assigned to successive memory locations first. Then, the value at each location is assigned to register R6 step by step. To do so, register R5 is taken as the base address. The increments are done by two since we are using the `mov` instruction at word level (`mov.w`) in reaching memory. If we want to reach the memory at byte level, we should use `mov.b`. The second part in Listing 7.6 provides such an example.

Listing 7.6 Usage of the indexed mode.

```
 .cdecls C,LIST,"msp430.h"

 .text
 .retain
 .retainrefs

RESET
 mov.w #WDTPW|WDTHOLD,WDTCTL
 mov.w #__STACK_END,SP

;word level operations
 mov.w #0000h,&0200h
 mov.w #0002h,&0202h
 mov.w #0004h,&0204h
 mov.w #0006h,&0206h

 mov.w #0200h,R5
 mov.w 0(R5),R6
 mov.w 2(R5),R6
 mov.w 4(R5),R6
 mov.w 6(R5),R6

;byte level operations
 mov.b #00h,&0200h
 mov.b #01h,&0201h
 mov.b #02h,&0202h
 mov.b #03h,&0203h

 mov.w #0200h,R5
 mov.b 0(R5),R6
 mov.b 1(R5),R6
```

```
 mov.b 2(R5),R6
 mov.b 3(R5),R6

 jmp $

;------------------------
;Stack Pointer definition
;------------------------
 .global __STACK_END
 .sect .stack

;-------------------
;Interrupt Vectors
;-------------------
 .sect RESET_VECTOR
 .short RESET
 .end
```

7.3.6 *Indirect Register Mode*

The indirect register mode performs a pointer-based operation. In this mode, the memory address saved in the register is reached. We illustrate the usage of the indirect register mode in Listing 7.7. Here, the @ sign is used to represent the memory address processing (pointer). In the ninth line of Listing 7.7, the memory address 0202h is saved in register R10. In the tenth line, the value in this memory address is copied to register R11.

Listing 7.7 Usage of the indirect register mode.

```
 .cdecls C,LIST,"msp430.h"

 .text
 .retain
 .retainrefs

RESET
 mov.w #WDTPW|WDTHOLD,WDTCTL
 mov.w #__STACK_END,SP

 mov.w #0CCCCh,&0202h
 mov.w #0202h,R10
 mov.w @R10,R11

 jmp $

;------------------------
;Stack Pointer definition
;------------------------
 .global __STACK_END
 .sect .stack

;-------------------
```

```
;Interrupt Vectors
;-------------------
 .sect RESET_VECTOR
 .short RESET
 .end
```

7.3.7 Indirect Autoincrement Mode

The indirect autoincrement mode can also be used for table processing. This mode uses the indirect register mode. The register value is incremented after each operation automatically without any extra code line.

In Listing 7.8, we repeat the operations in Listing 7.6 using the indirect autoincrement mode. In the first part of Listing 7.8, increments are done in 2 bytes since we have a word operation. If we had byte operations, the increments would be in 1 byte. We provide such an example in the second part of Listing 7.8.

Listing 7.8 Usage of the indirect autoincrement mode.

```
 .cdecls C,LIST,"msp430.h"

 .text
 .retain
 .retainrefs

RESET
 mov.w #WDTPW|WDTHOLD,WDTCTL
 mov.w #__STACK_END,SP

;word level operations
 mov.w #0000h,&0200h
 mov.w #0002h,&0202h
 mov.w #0004h,&0204h
 mov.w #0006h,&0206h
 mov.w #0200h,R5

 mov.w @R5+,R6
 mov.w @R5+,R6
 mov.w @R5+,R6
 mov.w @R5+,R6

;byte level operations
 mov.b #00h,&0200h
 mov.b #01h,&0201h
 mov.b #02h,&0202h
 mov.b #03h,&0203h
 mov.w #0200h,R5

 mov.b @R5+,R6
 mov.b @R5+,R6
 mov.b @R5+,R6
```

```
 mov.b @R5+,R6

 jmp $

;------------------------
;Stack Pointer definition
;------------------------
 .global __STACK_END
 .sect .stack

;-------------------
;Interrupt Vectors
;-------------------
 .sect RESET_VECTOR
 .short RESET
 .end
```

7.4 The Stack

The stack is a last-in, First-out (LIFO) address list. Local variables are saved in the stack automatically for C programs. The stack is also used during function and interrupt calls by the CPU. As mentioned in Sec. 4.2.2, the special-purpose register R1 is assigned as the stack pointer for the MSP430. At the beginning of an assembly program, this pointer should be initialized. For the MSP430G2553 microcontroller, the memory address for this initialization is 0400h. This is the highest data memory location as given in Table 4.2.

There are two assembly instructions related to the stack, **push** and **pop**. Push adds the number (as the last entry) to the stack. Pop gets the last number from the stack. Therefore, these two commands perform the LIFO operation. In Listing 7.9, we provide an example on how to initialize the stack and the usage of the push and pop operations. This program is similar to the previous table-based processing. The reader should pay attention to the inverse ordering of the table entries due to the LIFO structure.

Listing 7.9 Usage of the stack.

```
 .cdecls C,LIST,"msp430.h"

 .text
 .retain
 .retainrefs

RESET
 mov.w #WDTPW|WDTHOLD,WDTCTL
 mov.w #__STACK_END,SP

 push.w #0006h
 push.w #0004h
 push.w #0002h
 push.w #0000h

 pop.w R6
```

```
 pop.w R6
 pop.w R6
 pop.w R6

 jmp $

;------------------------
;Stack Pointer definition
;------------------------
 .global __STACK_END
 .sect .stack

;-------------------
;Interrupt Vectors
;-------------------
 .sect RESET_VECTOR
 .short RESET
 .end
```

7.5 Assembly Program Structure

An assembly program has a specific structure. It needs extra directives and adjustments during linking and compiling steps. They are explained in detail in [13, 11]. In this section, we briefly explain the directives used in assembly programs throughout the book. We pick Listing 7.10 to demonstrate the assembly program structure and the usage of directives. The directives in Listing 7.10 are explained below.

Listing 7.10 Structure of an assembly program.

```
1      .cdecls C,LIST,"msp430.h"

2
3      .text
4      .retain
5      .retainrefs

6
7     RESET
8      mov.w #WDTPW|WDTHOLD,WDTCTL
9      mov.w #__STACK_END,SP

10
11     mov.b #11h,R4
12     mov.w #00AAh,R5
13     and.w R4,R5

14
15     jmp $

16
17    ;------------------------
18    ;Stack Pointer definition
```

```
19    ;------------------------
20     .global __STACK_END
21     .sect .stack

22
23    ;-------------------
24    ;Interrupt Vectors
25    ;-------------------
26     .sect RESET_VECTOR
27     .short RESET
28     .end
```

- **.cdecls**: This directive is used to include the C header files to the assembly program. As a result, C header file definitions can be used directly.
- **.text**: This directive indicates the beginning of the executable code block.
- **.retain**: This directive is used to disable the removal of nonreferenced code blocks (like interrupt service routines) during the linking process.
- **.retainrefs**: This directive expands the retain operation to code blocks from other sections.
- **.global**: This directive defines a global symbol such that it can be reached from any part of the program.
- **.sect**: This directive defines a section (a contiguous block of code or data in the memory) that can hold code or data.
- **.short**: This directive initializes one or more 16-bit integers.
- **.end**: This directive indicates the end of the assembly program.

Based on the preceding definitions, we can read Listing 7.10 as follows. In *line 1*, we include the C header file msp430.h using the `.cdecls` directive. In *line 3*, we indicate the starting point of the program by the `.text` directive. In *lines 4* and *5*, we adjust the linker properties by the `.retain` and `.retainref` directives. In *lines 20* and *21*, we associate the constant `__STACK_END` with the stack address (defined by `.stack`) using directives `.sect` and `.short` in a joint manner. Similarly, in *lines 26* and *27*, we associate the `RESET` label with the `RESET_VECTOR` using directives `.sect` and `.short`. We will explain the reason for this operation in Chap. 9. In *line 28*, we indicate the endpoint of the program by the `.end` directive.

7.6 Sample Programs on Instruction Set Usage

In this section, we provide several assembly programs. The reader should execute each program step by step to observe the result of each operation. Therefore, the Assembly Step Into option in the Debug view should be used.

We should emphasize two important topics before starting. First, comments can be added to an assembly program by the ';' sign. We have been using the ';' sign to add comments. Second, CCS needs an extra 0 in front of a hexadecimal number starting with characters A–F. Therefore, the user should write 0FFFFh instead of FFFFh in the assembly code. Otherwise the CCS compiler gives an error message.

We demonstrate the usage of arithmetic, logical and register control, data, and jump instructions in Listing 7.11. We emphasized the difference between the binary and decimal addition operations in arithmetic instructions. The result of the subtraction operation will be a negative number in Listing 7.11. Therefore, the reader can observe the two's complement representation here. We defined the constants in binary form in logical and register control instructions. The reader should adjust the number format for each register to see the result of each operation clearly. We have used the data instructions in the previous steps. Therefore, they are not new. Here, the reader should observe the effect of the **cmp** instruction on the status register. Finally, we provide one jump instruction in Listing 7.11 to form an infinite loop. We will extensively use jump instructions next.

Listing 7.11 Usage of arithmetic, logical and register control, data, and jump instructions.

```
 .cdecls C,LIST,"msp430.h"

 .text
 .retain
 .retainrefs

RESET
 mov.w #WDTPW|WDTHOLD,WDTCTL
 mov.w #__STACK_END,SP

;Arithmetic Instructions
 mov.w #0009h,R5
 mov.w #0006h,R6
 add.w R5,R6
 mov.w #0006h,R6
 dadd.w R5,R6
 mov.w #0006h,R6
 sub.w R5,R6

;Logical and Register Control Instructions
 mov.b #00001111b,R5
 mov.b #00000011b,R6
 and.b R5,R6
 mov.b #00000011b,R6
 xor.b R5,R6
 rra.b R6
 swpb  R5

;Data Instructions
 mov.w #0006h,R5
 mov.w #0009h,R6
 cmp.w R5,R6

;Jump Instructions
 jmp $

;-----------------------
;Stack Pointer definition
```

```
;-----------------------
 .global __STACK_END
 .sect .stack

;-------------------
;Interrupt Vectors
;-------------------
 .sect RESET_VECTOR
 .short RESET
 .end
```

Control structures are not explicitly defined in assembly language. Therefore, we provided several examples on forming C-like control structures using assembly instructions in Listing 7.12. In all these examples, jump operations are mandatory. To note here, these are not the only control structures in assembly language. The reader can form his or her structure using different jump instructions.

Listing 7.12 Control structures in assembly language.

```
 .cdecls C,LIST,"msp430.h"

 .text
 .retain
 .retainrefs

RESET
 mov.w #WDTPW|WDTHOLD,WDTCTL
 mov.w #__STACK_END,SP

;if then
 mov.w #0005h,R4
 cmp.w #0004h,R4
 jge Greater
 dec R4
 jmp done1
Greater:
 inc R4
done1:

;if then
 mov.w #000Ah,R4
 mov.w #0009h,R5
 sub.w R4,R5
 jn Less
 dec R5
 jmp done2
Less:
 inc R4
done2:

;for
 mov.w #000Ah,R4
Loop1:
```

```
 dec R4
 jne Loop1

;while
 mov.w #0006h,R4
 mov.w #0002h,R5
Loop2:
 dec R4
 cmp.w R5,R4
 jge Loop2

 jmp $

;------------------------
;Stack Pointer definition
;------------------------
 .global __STACK_END
 .sect .stack

;-------------------
;Interrupt Vectors
;-------------------
 .sect RESET_VECTOR
 .short RESET
 .end
```

Finally, we provide an example on the usage of subroutines (functions) in Listing 7.13. Here, the contents of registers R5 and R6 are swapped through register R7. This operation is done in the user-defined subroutine.

Listing 7.13 Usage of subroutines in assembly language.

```
 .cdecls C,LIST,"msp430.h"

 .text
 .retain
 .retainrefs

RESET
 mov.w #WDTPW|WDTHOLD,WDTCTL
 mov.w #__STACK_END,SP

 mov.w #0005h,R5
 mov.w #0006h,R6
 mov.w #0007h,R7

 call #Replace

 jmp $

Replace:
 mov.w R6,R7
```

```
mov.w R5,R6
mov.w R7,R5
ret

;------------------------
;Stack Pointer definition
;------------------------
.global __STACK_END
.sect .stack

;-------------------
;Interrupt Vectors
;-------------------
.sect RESET_VECTOR
.short RESET
.end
```

7.7 Summary

Although the C language can be used in most operations under the MSP430, the assembly language gives insight into the basics of the microcontroller. Therefore, in this chapter we considered the instruction set and the addressing modes of the MSP430. We started with the double-operand, single-operand, and jump instructions. We provided sample codes on their usage. Then we briefly described the anatomy of an instruction. As a separate section, we looked at the addressing modes of the MSP430. This was followed by a discussion of the stack and its usage. Finally, we provided an overview of assembly program structure. Since the assembly language is important for understanding the microcontroller, we provide the assembly language codes for all applications besides C in the following chapters.

7.8 Problems

7.1 Assume that your register R4 holds the hexadecimal number 4001h and the memory location 02F0h holds the hexadecimal number 0F18h. Write a program in assembly language such that:

a. These two numbers are added.
b. One's complement of the sum is saved in register R6.
c. The least significant two hexadecimal digits of the sum are swapped with its most significant two hexadecimal digits. The output is written to register R12.

7.2 Write a program in assembly language such that:

a. Two binary numbers are saved in two separate memory locations.
b. As the numbers are added, an overflow will occur.
c. As the numbers are added, no overflow occurs.

Check whether an overflow occurred or not from the status register.

7.3 Repeat Prob. 7.2 using the subtraction operation.

7.4 Extend Prob. 7.2 such that, if the result is greater than (less than) zero, the program will jump to the label `greater` (`less`).

7.5 Write an assembly program such that:

a. It reads the input from port P1 and assigns it to register R9.

b. If the second most significant bit of this value is greater than zero, assign FFFFh to register R10.

7.6 Write an assembly program to change the little endian ordering of the MSP430 to big endian for selected memory locations.

7.7 Write a program in assembly language such that:

a. It contains a subroutine which performs the *and* operation with the entry of register R6 and hexadecimal number 0001h. The result will be saved in the same register.

b. Repeat part a by applying the same operation to five successive memory locations. Use an appropriate addressing mode.

7.8 Write a program in assembly language to calculate the first 10 elements of the Fibonacci series.

a. The user only provides the first two entries of the series.

b. The rest will be calculated by the program.

c. Use appropriate memory locations.

d. The numbers can be represented in hexadecimal form.

7.9 Write an assembly program such that:

a. When the least significant bit of registers R4 and R5 have the value 1, the register R9 gets the value 0FF0h.

b. When only one of the least significant bits of either register R4 or R5 has the value 1, the one's complement of the value in register R9 will be saved in register R10.

7.10 What will be the values at memory locations 02F0h, 02F2h, 02F4h, 02F6h, and 02F8h when the program given in Listing 7.14 is run?

Listing 7.14 The assembly code for Prob. 7.10.

```
 .cdecls C,LIST,"msp430.h"

 .text
 .retain
 .retainrefs

RESET
 mov.w #WDTPW|WDTHOLD,WDTCTL
 mov.w #__STACK_END,SP

Mainloop:
 mov.w #0006h,&02F0h
 mov.w #0009h,&02F2h
 clr.w &02F6h
 clr.w &02F8h
 mov.w &02F2h,&02F4h
 add.w &02F0h,&02F4h
 cmp.w #000Ah,&02F4h
 jhs Greater
```

```
 jlo Less

Greater:
 mov.w &02F0h,&02F8h
 jmp Mainloop

Less:
 mov.w &02F2h,&02F6h
 jmp Mainloop

;------------------------
;Stack Pointer definition
;------------------------
 .global __STACK_END
 .sect .stack

;-------------------
;Interrupt Vectors
;-------------------
 .sect RESET_VECTOR
 .short RESET
 .end
```

7.11 Write an assembly program with the following specifications.

a. In the main block, you should have two registers R4 and R5. They should be checked in an infinite loop. If R4 is greater than R5, then the `greater` subroutine will be called. If R4 is less than R5, then the `less` subroutine will be called. If R4 equals R5, then no operations will be done.

b. In the *greater* subroutine, your code will fill the decimal numbers 1, 2, 3, 4, 5 in hexadecimal form to five successive memory locations. After this operation, the value in R4 will be decreased by one.

c. In the *less* subroutine, your code will fill the decimal numbers 10, 9, 8, 7, 6 in hexadecimal form to five successive memory locations. After this operation, the value in R4 will be decreased by one.

7.12 Write an assembly program to calculate the division of the hexadecimal number 00FFh by 00A0h. Use only available registers to save your variables. Write the result to register R7.

7.13 What will the register values R4, R5, and R6 be as the program in Listing 7.15 is run?

Listing 7.15 The assembly code for Prob. 7.13.

```
 .cdecls C,LIST,"msp430.h"

 .text
 .retain
 .retainrefs

RESET
 mov.w #WDTPW|WDTHOLD,WDTCTL
 mov.w #__STACK_END,SP

 mov.w #0004h,R5
```

```
 mov.w R5,R4
 clr R6
 dec R5
 inc R4
 sub.w R4,R5
 sub.w #9FFFh,R4
 add.w #000Bh,R6
 and.w R5,R6

 jmp $

;------------------------
;Stack Pointer definition
;------------------------
 .global __STACK_END
 .sect .stack

;-------------------
;Interrupt Vectors
;-------------------
 .sect RESET_VECTOR
 .short RESET
 .end
```

7.14 Write down the values assigned to register R14 in four steps (labeled in Listing 7.16).

Listing 7.16 The assembly code for Prob. 7.14.

```
 .cdecls C,LIST,"msp430.h"

 .text
 .retain
 .retainrefs

RESET
 mov.w #WDTPW|WDTHOLD,WDTCTL
 mov.w #__STACK_END,SP

 mov.w #0006h,&0200h
 mov.w #000Ah,&0202h
 mov.w #0014h,&0204h
 mov.w #008Dh,&0206h
 mov.w #0200h,R13

 mov.w 2(R13),R14 ;step 1
 sub.w 0(R13),R14 ;step 2
 add.w 4(R13),R14 ;step 3
 add.w 6(R13),R14 ;step 4

 jmp $

;------------------------
;Stack Pointer definition
;------------------------
```

```
 .global __STACK_END
 .sect .stack

;-------------------
;Interrupt Vectors
;-------------------
 .sect RESET_VECTOR
 .short RESET
 .end
```

7.15 Write an assembly program to divide a hexadecimal number by 2. The number to be divided should be kept in a suitable memory location of the MSP430G2553. The division result will be kept in register R5 and the remainder will be kept in register R6.

7.16 Three numbers are written in the designated memory locations with the code given in Listing 7.17.

Listing 7.17 The assembly code for Prob. 7.16.

```
 .cdecls C,LIST,"msp430.h"

 .text
 .retain
 .retainrefs

RESET
 mov.w #WDTPW|WDTHOLD,WDTCTL
 mov.w #__STACK_END,SP

 mov.w #007Dh,&0200h ;the first number
 mov.w #00B5h,&0202h ;the second number
 mov.w #00E8h,&0204h ;the third number

 jmp $

;------------------------
;Stack Pointer definition
;------------------------
 .global __STACK_END
 .sect .stack

;-------------------
;Interrupt Vectors
;-------------------
 .sect RESET_VECTOR
 .short RESET
 .end
```

Write an assembly program such that:

a. It contains a subroutine which performs the *and* operation between the first and second numbers. Then, it performs the *or* operation between the result of the *and* operation and the third number.
b. The final result must be written to the memory address 020Dh.

c. You cannot change the numbers in the addresses given above.

7.17 Two numbers are written in the designated memory locations with the code given in Listing 7.18.

Listing 7.18 The assembly code for Prob. 7.17.

```
 .cdecls C,LIST,"msp430.h"

 .text
 .retain
 .retainrefs

RESET
 mov.w #WDTPW|WDTHOLD,WDTCTL
 mov.w #__STACK_END,SP

 mov.w #2D97h,&0220h ;the first number
 mov.w #6239h,&0222h ;the second number

 jmp $

;------------------------
;Stack Pointer definition
;------------------------
 .global __STACK_END
 .sect .stack

;-------------------
;Interrupt Vectors
;-------------------
 .sect RESET_VECTOR
 .short RESET
 .end
```

Write an assembly program such that:

a. It contains a subroutine which performs the *xor* operation between the first and second numbers. Then, it performs the *not* operation on the result of the *or* operation.
b. The final result must be written to the memory address 023Ch.
c. You cannot change the numbers in the addresses given above.

7.18 Write an assembly program to calculate the sum of the four numbers given in Prob. 7.14.

7.19 Add a subroutine to your assembly program in Prob. 7.18 to calculate the average of the four numbers.

7.20 Assume that there are four numbers represented in single floating-point format. These are saved in successive memory locations starting from 0200h. Write an assembly program to convert them to the fixed-point UQ16.16 format. The converted numbers should be saved in successive memory locations starting from 0300h.

7.21 Analyze the assembly code given in Listing 7.19. Form a table for registers and fill their values as the code is run.

Listing 7.19 The assembly code for Prob. 7.21.

```
 .cdecls C,LIST,"msp430.h"

 .text
 .retain
 .retainrefs

RESET
 mov.w #WDTPW|WDTHOLD,WDTCTL
 mov.w #__STACK_END,SP

 mov.w #0200h,R5
 mov.w #0001h,R6

Loop1:
 mov.w R6,0(R5)
 incd.w R5
 inc.w R6
 cmp.w #0214h,R5
 jlo Loop1

 mov.w #0000h,R7
 mov.w #0200h,R5

Loop2:
 cmp.w #020Ah,R5
 jlo Less
 sub.w @R5,R7
 jmp Incr

Less:
 add.w @R5,R7
Incr:
 incd.w R5
 cmp.w #0214h,R5
 jlo Loop2

 jmp $

;------------------------
;Stack Pointer definition
;------------------------
 .global __STACK_END
 .sect .stack

;-------------------
;Interrupt Vectors
;-------------------
 .sect RESET_VECTOR
 .short RESET
 .end
```

8 Digital Input and Output

Chapter Outline

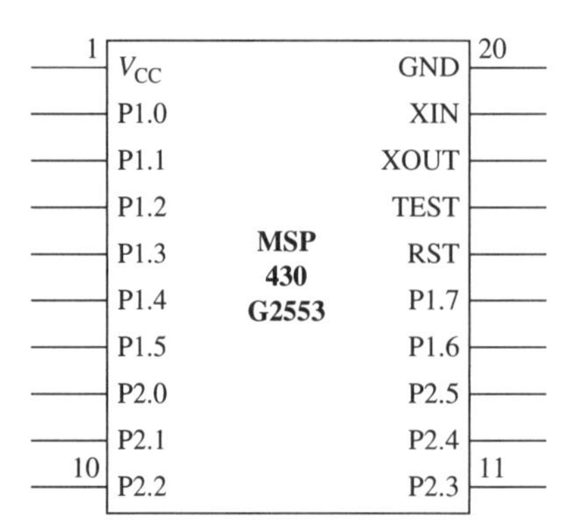

Figure 8.1 Pin layout of the MSP430G2553.

A microcontroller interacts with the outside world through its input and output ports. The interaction can be in either analog or digital form. In this chapter, we focus on the digital input and output (digital I/O) characteristics of the MSP430 microcontroller. We will develop methods to use them. We will also use Grace to configure the digital I/O.

8.1 Pin Layout for Digital I/O

Digital input and output (I/O) is the simplest form of communication between the microcontroller and the outside world. The input or the output is either 0 or 1 in this form. In other terms, the input or the output is either 0 V or V_{CC}. There are two ports called P1 and P2 in the MSP430G2553. These are generally called Px. Each port has eight pins associated with it. These are called P1.0–P1.7 and P2.0–P2.7. The general pin layout of the MSP430G2553 is given in Fig. 8.1. The usage table for these pins is given in Table 8.1. As can be seen in this table, all pins in ports P1 and P2 can be used for digital I/O.

We diagram the basic hardware for the pins in Fig. 8.2. As can be seen, all pins can be used for multipurpose besides being used for digital I/O. They are labeled *Other* in this figure. In this chapter, we only consider the digital I/O characteristics of the pins. Next, we will explore how to reach a specific pin in a given port.

8.2 Digital I/O Registers

Each pin in a port can be set either as input or output. This is done by the register **PxDIR**. To set a specific pin as input, the corresponding bit in this register should be reset (to 0). In a similar manner, to set a specific pin as output, the corresponding bit in the PxDIR register should be set (to 1). Due to the byte-based operation of the MSP430, all the port pins should be taken into account in this operation. In

Table 8.1

The pin usage table for digital I/O in the MSP430G2553.

Pin	Port Name	Usage Area
1	V_{CC}	Source voltage
2	P1.0	General-purpose digital I/O
3	P1.1	General-purpose digital I/O
4	P1.2	General-purpose digital I/O
5	P1.3	General-purpose digital I/O
6	P1.4	General-purpose digital I/O
7	P1.5	General-purpose digital I/O
8	P2.0	General-purpose digital I/O
9	P2.1	General-purpose digital I/O
10	P2.2	General-purpose digital I/O
11	P2.3	General-purpose digital I/O
12	P2.4	General-purpose digital I/O
13	P2.5	General-purpose digital I/O
14	P1.6	General-purpose digital I/O
15	P1.7	General-purpose digital I/O
16	RST	Reset
17		
18	P2.7	General-purpose digital I/O
19	P2.6	General-purpose digital I/O
20	V_{SS}	Ground voltage

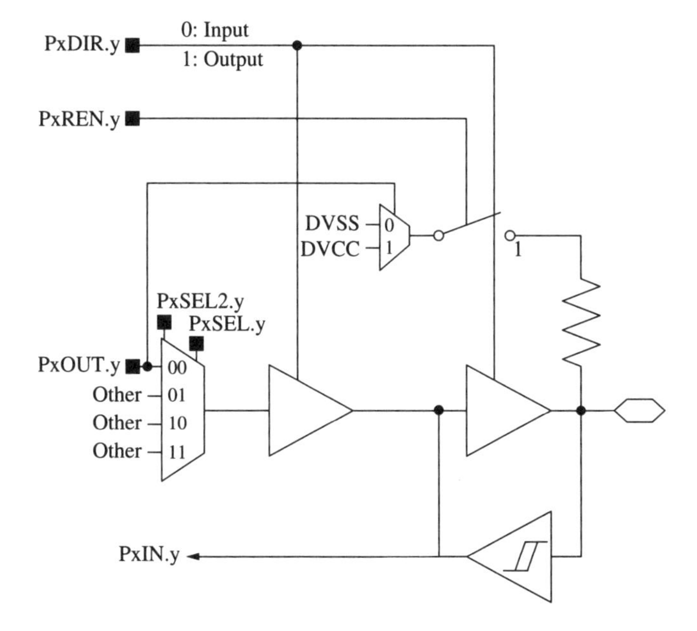

Figure 8.2

Basic hardware for the pins.

fact, this is applicable for all port registers. As an example, let's say that we want to assign the first and the seventh pins of the first port (P1.0, P1.6) as output. We would also like to assign the fourth pin of the same port (P1.3) as input. To do so, we should assign the binary number 01000001b (41h) to P1DIR. Therefore, the code line `P1DIR=0x41` in C language does the job. This corresponds to `mov.b #41h,P1DIR` in assembly language. Now, the input can be connected to pin P1.3 and outputs can be connected to pins P1.0 and P1.6.

Based on the preceding definition, P1.3 is set as input. The P1IN register should be checked to read values from this pin. In general, we will call this register **PxIN** (for P1 or P2). The digital input fed to the microcontroller is directly observed from this register. As we have mentioned previously, we cannot observe a specific bit in this register. To do so, we need to apply a binary mask to extract the desired input value. We only need the value of the pin P1.3 for our example. Therefore, we should apply an *and* operation between the P1IN register and the binary mask 00001000b (08h). This will be done by the code line `P1IN&0x08` in C language. This corresponds to `bit #08h,P1IN` in assembly language. While processing the digital input values, please also take into account the active high and low settings (to be discussed in the next section).

To feed output values to pins P1.0 or P1.6, the P1OUT register is used. In general, we will call this register **PxOUT** (for P1 or P2). If we want to feed a 0 V to output from a specific pin, the corresponding bit value in PxOUT should be reset (to 0). We should set the pin (to 1) to feed V_{CC} to output. As an example, we should assign 01000000b (40h) to P1OUT to feed V_{CC} to output from pin P1.6. This corresponds to the code line `P1OUT = 0x40` in C language. This corresponds to the code line `bis #40h,P1OUT` in assembly language.

Each pin in ports P1 and P2 has a pull-up/down resistor. These are controlled by the **PxREN** register. By default, these resistors are disabled. To enable a resistor connected to a specific pin, the corresponding bit in PxREN should be set (to 1). As an example, the code line `P2REN = 0x02` should be used to enable the pull-up/down resistor of the pin P2.1 in C language. This corresponds to the code line `bis #02h,P2REN` in assembly language. After the pull-up/down resistor is enabled, the selection between the pull-up or down option is done with the PxOUT register. If the related bit of the PxOUT register is set, the internal resistor will be used as pull up. Otherwise, the internal resistor will be used as pull down.

As mentioned in the previous section, MSP430 pins are used for more than one purpose. The registers **PxSEL** and **PxSEL2** are used to select the usage area of the pins. If a specific pin will be used for digital I/O, the corresponding bits in PxSEL and PxSEL2 should be reset. If the pin will be used for a specific purpose other than the digital I/O, then PxSEL and PxSEL2 should be set accordingly. In Table 8.2, we provide these settings for different application types.

The only exception to Table 8.2 is in the comparator usage described in Chap. 11. When the comparator output (CAOUT) is given from P1.3, PxSEL=1 and PxSEL2 = 1. When the CAOUT is given from P1.7, PxSEL=1 and PxSEL2=0.

Table 8.2

Application type based on PxSEL and PxSEL2 function settings.

Application	PxSEL	PxSEL2
Digital I/O	0	0
Timer and clock usage	1	0
Capacitive sensing	0	1
Digital communication	1	1

8.3 Digital I/O Hardware Issues

There are two major hardware issues to be dealt with when using digital I/O. The first is the definition and setup of active high/low input. The second is switch bouncing. We will explain them next.

8.3.1 Active High/Low Input

There are two setup options to use a push button in a digital circuit. In the first setup, the microcontroller gets V_{CC} volts (logic 1) on its pin when the button is pressed. This is called the *active high* input. In the second setup, the microcontroller gets 0 V (logic 0) on its pin when the button is pressed. This is called the *active low* input. To note here, active high or low inputs are not related to the microcontroller. They are based on the connection type to the digital I/O pin. The circuit diagrams for the active high/low input setups are given in Fig. 8.3.

For the MSP430 LaunchPad, the preferred setup is active low. Therefore, when the button connected to pin P1.3 is pressed, it will generate logic 0. When it is not pressed, it will give logic 1. This should be taken into account while reading the value from the P1IN register using a binary mask.

8.3.2 Switch Bouncing

The second hardware-based issue in digital I/O is switch (button) bouncing. If the button is pressed once, it may generate output more than once, depending on its physical characteristics. Therefore, the microcontroller may see one input

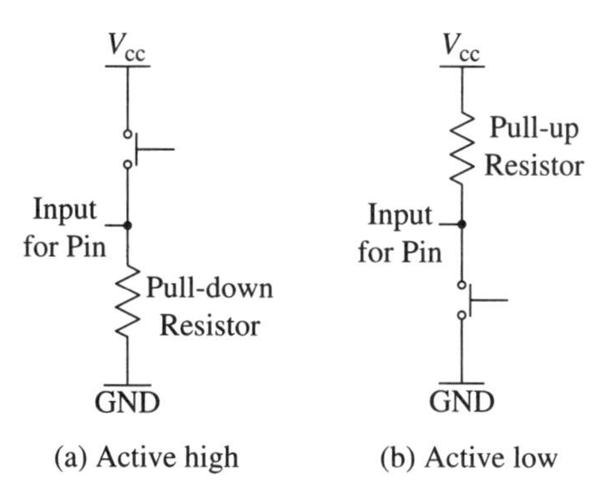

Figure 8.3

Active high and low input circuit diagrams for the push button.

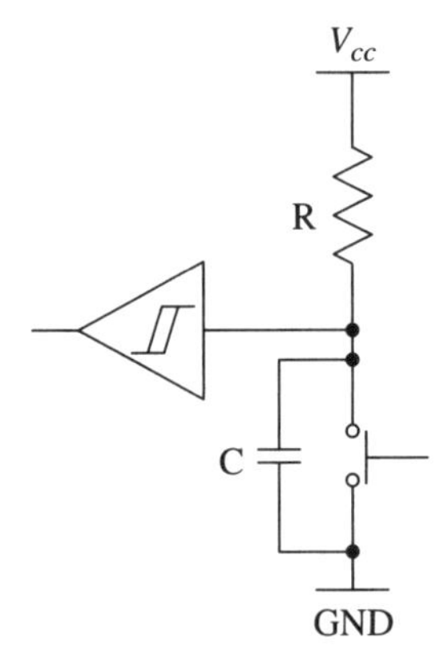

Figure 8.4

Hardware solution of the switch bouncing problem.

and its successive shadow versions. To eliminate this effect, either a software- or a hardware-based solution can be used. In the software-based solution, a delay should be added to the button input reading part of the code. The input will not be observed, and possible shadow inputs will be eliminated during this delay. Although the software-based solution is easier to implement, actual inputs will also be eliminated during delay. Therefore, it should be used with caution.

There are also hardware-based solutions for switch bouncing. The most feasible circuitry is a low-pass RC filter (composed of a resistor and a capacitor) followed by a Schmitt trigger for the MSP430. This setup is shown in Fig. 8.4. Each digital I/O pin of the MSP430G2553 has a Schmitt trigger [16]. There is also an internal pull-up resistor and a capacitor connected parallel to the push button (connected to P1.3) in MSP430 LaunchPad Rev.1.4. Therefore, the circuitry in Fig. 8.4 is ready for this LaunchPad version. However, the parallel capacitor is discarded in MSP430 LaunchPad Rev.1.5. Only the internal pull-up resistor (if enabled) can be used to form the circuit. The user should also connect an external capacitor to solve the switch bouncing problem by hardware in this LaunchPad version. In the next section, we provide sample codes to address this issue.

In Fig. 8.4, the low-pass RC filter is used to eliminate the high-frequency shadow inputs coming from the button. Then remaining glitches are eliminated by the Schmitt trigger. In this circuitry, the time constant of the filter ($\tau = R \times C$) must be larger than the switch bouncing time to eliminate all shadow inputs. The internal pull-up resistor values may vary between 20 and 50 kΩ. Throughout the book, we will assume that a 2- to 5-ms time constant is enough to eliminate switch bouncing. Therefore, a 100-ηF external capacitor should be used. The time constant of the filter can be adjusted by this capacitance value depending on other constraints.

8.4 Coding Practices for Digital I/O

In this section, we provide sample C and assembly codes for digital I/O. Since our setup is based on the MSP430 LaunchPad, we will use the push button and LEDs (red, green) on it. Therefore, ports used in the codes will be the same.

8.4.1 *Digital I/O in C*

We first consider an application where the red LED (connected to P1.0 on the MSP430 LaunchPad) turns on when the push button (connected to P1.3 on the MSP430 LaunchPad) is pressed. When the button is released, the red LED turns off. We provide the C code in Listing 8.1 for MSP430 LaunchPad Rev.1.4. Here, the location of the red LED and the push button is defined first. The port direction is set accordingly. Initially, the red LED is turned off. Then the input from the push button is checked in an infinite loop. Here there are three important issues. First, a masking operation should be done to check for a specific pin since port P1 can be observed at a byte level. Second, the active low setup of the MSP430 LaunchPad should be taken into account. Therefore, the control within the if statement is checked for 0 not 1. Third, the switch bouncing is not taken into account here since the LED turns on only during the button press.

Listing 8.1 Turning on the red LED when the push button is pressed, for MSP430 LaunchPad Rev.1.4.

```
#include <msp430.h>

#define LED BIT0
#define BUTTON BIT3

void main(void)
{
 WDTCTL = WDTPW|WDTHOLD;

 P1DIR = LED;
 P1OUT = 0x00;

 while(1){
 if((P1IN & BUTTON) == 0x00) // Active low input
 P1OUT = LED;    // Turn on the LED
 else
 P1OUT = 0x00;  // Turn off the LED
 }
}
```

We repeat the previous application given in Listing 8.1 for MSP430 LaunchPad Rev.1.5. We provide the C code in Listing 8.2. As can be seen here, the code is slightly modified. First, the internal resistor connected to the push button (connected to P1.3 on the MSP430 LaunchPad) is enabled by the code line `P1REN = BUTTON;`. This resistor is also set as pull-up by the code line `P1OUT = BUTTON;`. The push button check conditions are also modified accordingly so that this setup is not changed during operation.

Listing 8.2 Turning on the red LED when the push button is pressed, for MSP430 LaunchPad Rev.1.5.

```
#include <msp430.h>

#define LED BIT0
```

```
#define BUTTON BIT3

void main(void)
{
 WDTCTL = WDTPW|WDTHOLD;

 P1DIR = LED;
 P1REN = BUTTON;
 P1OUT = BUTTON;

 while(1){
 if((P1IN & BUTTON) == 0x00) // Active low input
 P1OUT |= LED;    // Turn on the LED
 else
 P1OUT &= ~LED; // Turn off the LED
 }
}
```

We next provide a more complex example. Here, the green LED (connected to P1.6 on the MSP430 LaunchPad) is toggled every time the push button (connected to P1.3 on the MSP430 LaunchPad) is pressed. We provide the C code for MSP430 LaunchPad Rev.1.4 in Listing 8.3. Here, the location of the red LED and the push button is defined first. The port direction is set accordingly. Initially, the green LED is turned off. Then, the input from the push button is checked in an infinite loop. When the button is pressed, the code waits for 5 ms by the intrinsic function `__delay_cycles(5000);`. The code checks the push button condition after this delay again. If the push button is still pressed, then the green LED toggles. The code waits in an infinite loop for the release of the push button. This step ensures that the green LED cannot be toggled again unless the push button is released.

Listing 8.3 Toggling the green LED when the push button is pressed, for MSP430 LaunchPad Rev.1.4.

```
#include <msp430.h>

#define LED BIT6
#define BUTTON BIT3

void main(void)
{
 WDTCTL = WDTPW|WDTHOLD;
 P1DIR = LED;
 P1OUT = 0x00;

 while(1){
 if((P1IN & BUTTON) == 0x00){
 __delay_cycles(5000);
 if((P1IN & BUTTON) == 0x00){
```

```
 P1OUT ^= LED;
 while((P1IN & BUTTON) == 0x00);
 }}}
}
```

We modify the C code given in Listing 8.3 for MSP430 LaunchPad Rev.1.5 next. We provide the C code in Listing 8.4. Here, the internal resistor connected to the push button (connected to P1.3 on the MSP430 LaunchPad) is enabled by the code line `P1REN = BUTTON;`. This resistor is also set as pull-up by the code line `P1OUT = BUTTON;`. This was also the case in Listing 8.2.

Listing 8.4 Toggling the green LED when the push button is pressed, for MSP430 LaunchPad Rev.1.5.

```
#include <msp430.h>

#define LED BIT6
#define BUTTON BIT3

void main(void)
{
 WDTCTL = WDTPW|WDTHOLD;
 P1DIR = LED;
 P1REN = BUTTON;
 P1OUT = BUTTON;

 while(1){
 if((P1IN & BUTTON) == 0x00){
 __delay_cycles(5000);
 if((P1IN & BUTTON) == 0x00){
 P1OUT ^= LED;
 while((P1IN & BUTTON) == 0x00);
 }}}
}
```

8.4.2 *Digital I/O in Assembly*

We provide assembly code for digital I/O in Listing 8.5. Here, the microcontroller waits in an infinite loop. It turns the red and green LEDs on and off based on the status of the button pressed.

Listing 8.5 Digital I/O in assembly.

```
 .cdecls C,LIST,"msp430.h"

 .text
 .retain
 .retainrefs

RESET
 mov.w #WDTPW|WDTHOLD,WDTCTL
```

```
 mov.w #__STACK_END,SP

 bis.b #01000001b,P1DIR

Mainloop:
 bit.b #00001000b,P1IN
 jc Off

On:
 bic.b #00000001b,P1OUT
 bis.b #01000000b,P1OUT
 jmp Mainloop

Off:
 bis.b #00000001b,P1OUT
 bic.b #01000000b,P1OUT
 jmp Mainloop

;------------------------
;Stack Pointer definition
;------------------------
 .global __STACK_END
 .sect .stack

;-------------------
;Interrupt Vectors
;-------------------
 .sect RESET_VECTOR
 .short RESET
 .end
```

8.5 Digital I/O in Grace

Grace can be used to configure the input and output ports of the MSP430. Let's start with a new Grace project (generated in accordance with Sec. 5.7). We can configure the pin properties by clicking on the blue port (P1, P2, P3) block. As we click on the block, a new tab named GPIO appears in main.cfg. This new tab is named GPIO-Overview. It has several buttons. For our application (having the 20-pin version of the MSP430G2553 microcontroller) the overview, Pinout20-TSSOP/20-PDIP, power user, and P1/P2 views are useful. As we mentioned in Sec. 5.7, Overview provides basic info and sample code blocks on the GPIO.

8.5.1 *The Pinout20-TSSOP/20-PDIP Mode*

The Pinout20-TSSOP/20-PDIP mode provides the active microcontroller block diagram as shown in Fig. 8.5. In this mode, the property of each pin can be changed from the drop-down list which appears when the blue pointer by the pin is clicked. This automatically changes the PxSEL, PxSEL2, and PxDIR registers.

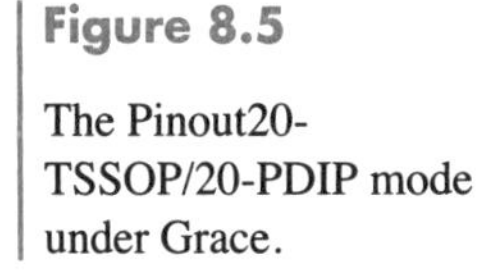

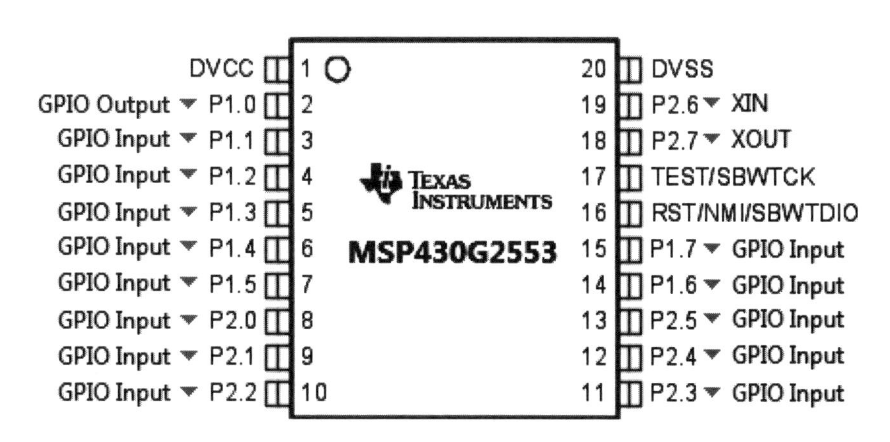

Figure 8.5

The Pinout20-TSSOP/20-PDIP mode under Grace.

8.5.2 *The Power User Mode*

In the power user mode, shown in Fig. 8.6, the following properties of each pin can be configured: GPIO function, output state, pull-up/down resistor enable, and interrupt enable. The GPIO function option is the same as the Pinout20-TSSOP/20-PDIP mode. The output state allows us to set the initial value (low or high) of the pin when used in the GPIO output mode. The pull-up/down resistor enable option allows the user to disable or enable the pull-up/down resistors connected to the pin when used in the GPIO input mode. Finally, the interrupt enable option can be used to configure the interrupt properties related to the pin. We will see how to use this property in Chap. 9.

8.5.3 *The P1/P2 Mode*

This mode provides the port registers (PxOUT, PxDIR, PxSEL, PxSEL2, and PxREN) as given in Fig. 8.7. All port registers can be set and reset in this mode by clicking on the appropriate check button. Changes made here also affect the other modes automatically.

8.5.4 *Coding Practices*

In this section, we redo the digital I/O application given in Listing 8.1 using Grace. As a reminder, this application turns on the red LED (connected to P1.0 on the MSP430 LaunchPad) when the button (connected to P1.3 on the MSP430 LaunchPad) is pressed. When the button is not pressed, the red LED turns off. We start by generating a Grace project. Then, we configure the pins P1.0 and P1.3 under Grace. The pin P1.0 should be set as GPIO output. The pin P1.3 should be set as input. These settings can be done by any of the three GPIO views. When we add the code block given in Listing 8.6 to the main.c file of the Grace project, we are done. After compiling the project, we can run our application.

Figure 8.6

The power user mode under Grace.

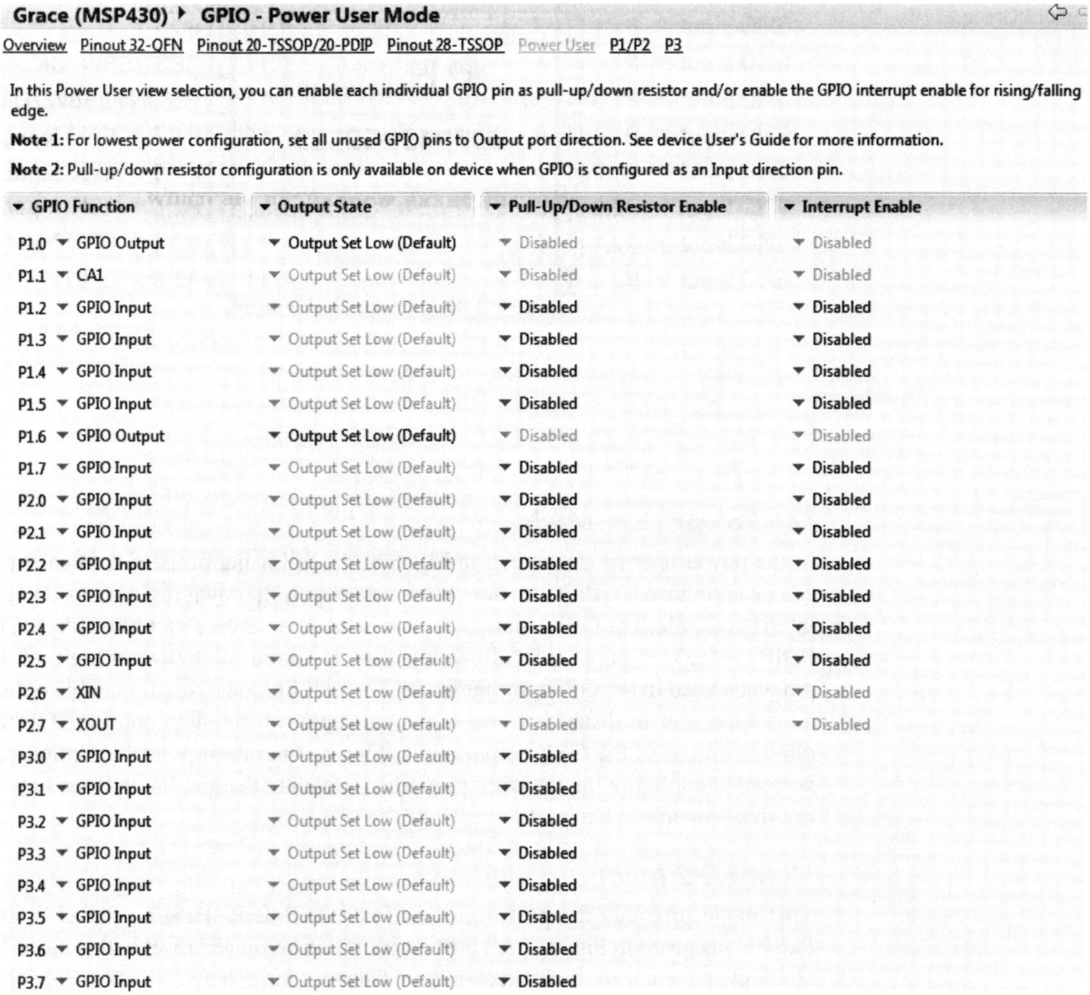

Grace (MSP430) ▸ GPIO - Power User Mode

Overview Pinout 32-QFN Pinout 20-TSSOP/20-PDIP Pinout 28-TSSOP Power User P1/P2 P3

In this Power User view selection, you can enable each individual GPIO pin as pull-up/down resistor and/or enable the GPIO interrupt enable for rising/falling edge.

Note 1: For lowest power configuration, set all unused GPIO pins to Output port direction. See device User's Guide for more information.

Note 2: Pull-up/down resistor configuration is only available on device when GPIO is configured as an Input direction pin.

Pin	GPIO Function	Output State	Pull-Up/Down Resistor Enable	Interrupt Enable
P1.0	GPIO Output	Output Set Low (Default)	Disabled	Disabled
P1.1	CA1	Output Set Low (Default)	Disabled	Disabled
P1.2	GPIO Input	Output Set Low (Default)	Disabled	Disabled
P1.3	GPIO Input	Output Set Low (Default)	Disabled	Disabled
P1.4	GPIO Input	Output Set Low (Default)	Disabled	Disabled
P1.5	GPIO Input	Output Set Low (Default)	Disabled	Disabled
P1.6	GPIO Output	Output Set Low (Default)	Disabled	Disabled
P1.7	GPIO Input	Output Set Low (Default)	Disabled	Disabled
P2.0	GPIO Input	Output Set Low (Default)	Disabled	Disabled
P2.1	GPIO Input	Output Set Low (Default)	Disabled	Disabled
P2.2	GPIO Input	Output Set Low (Default)	Disabled	Disabled
P2.3	GPIO Input	Output Set Low (Default)	Disabled	Disabled
P2.4	GPIO Input	Output Set Low (Default)	Disabled	Disabled
P2.5	GPIO Input	Output Set Low (Default)	Disabled	Disabled
P2.6	XIN	Output Set Low (Default)	Disabled	Disabled
P2.7	XOUT	Output Set Low (Default)	Disabled	Disabled
P3.0	GPIO Input	Output Set Low (Default)	Disabled	
P3.1	GPIO Input	Output Set Low (Default)	Disabled	
P3.2	GPIO Input	Output Set Low (Default)	Disabled	
P3.3	GPIO Input	Output Set Low (Default)	Disabled	
P3.4	GPIO Input	Output Set Low (Default)	Disabled	
P3.5	GPIO Input	Output Set Low (Default)	Disabled	
P3.6	GPIO Input	Output Set Low (Default)	Disabled	
P3.7	GPIO Input	Output Set Low (Default)	Disabled	

8.6 Digital Safe Application

The purpose of this application is to learn how to use the digital I/O pins of the MSP430 microcontroller. As a real-world application, we will design a digital safe system. In this section, we provide the equipment list, the layout of the circuit, the procedure, and the system design specifications.

Figure 8.7

The P1/P2 mode under Grace.

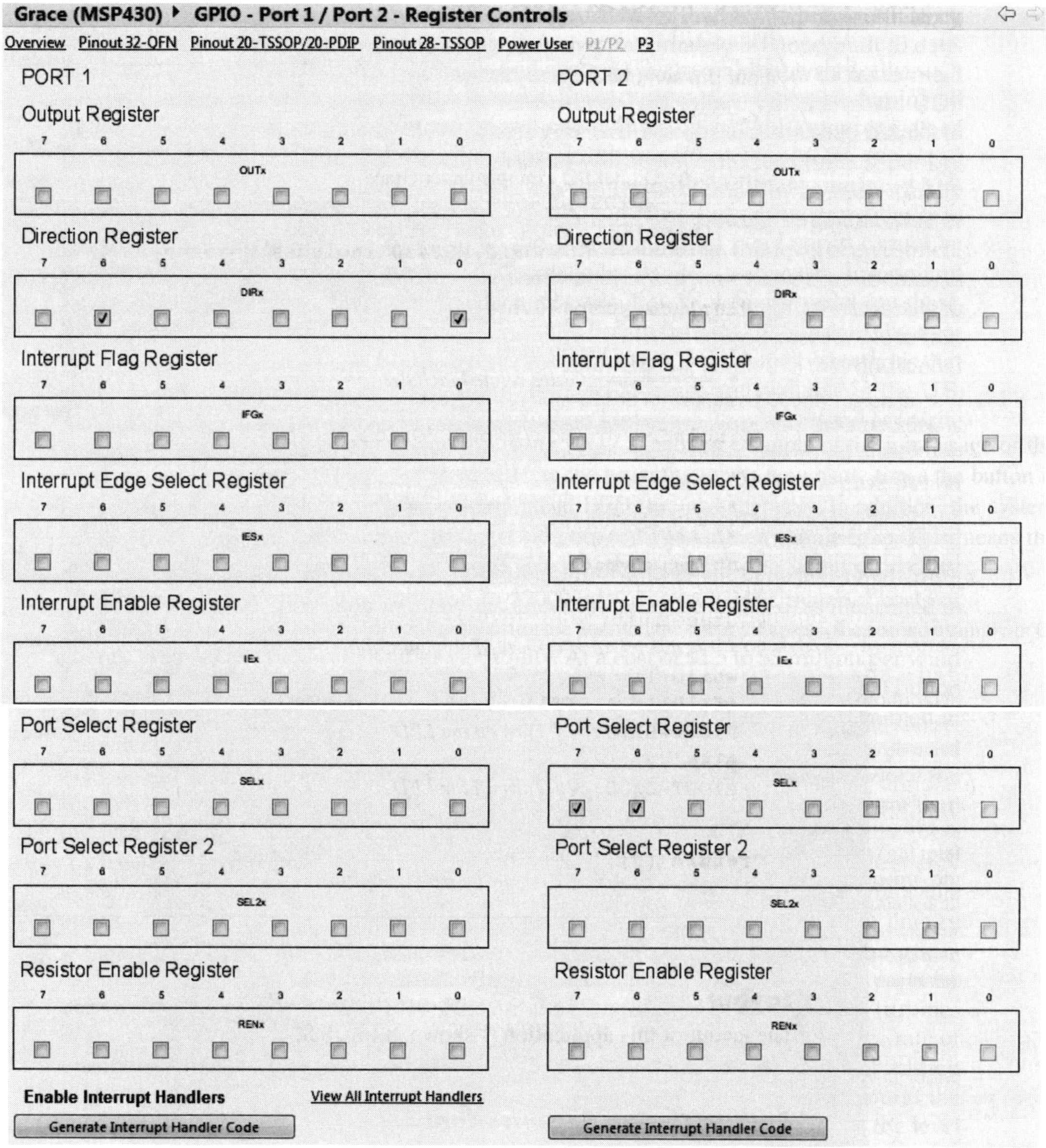

Grace (MSP430) ▸ GPIO - Port 1 / Port 2 - Register Controls

Overview Pinout 32-QFN Pinout 20-TSSOP/20-PDIP Pinout 28-TSSOP Power User P1/P2 P3

PORT 1

Output Register (OUTx)

Direction Register (DIRx)

Interrupt Flag Register (IFGx)

Interrupt Edge Select Register (IESx)

Interrupt Enable Register (IEx)

Port Select Register (SELx)

Port Select Register 2 (SEL2x)

Resistor Enable Register (RENx)

PORT 2

Output Register (OUTx)

Direction Register (DIRx)

Interrupt Flag Register (IFGx)

Interrupt Edge Select Register (IESx)

Interrupt Enable Register (IEx)

Port Select Register (SELx)

Port Select Register 2 (SEL2x)

Resistor Enable Register (RENx)

Enable Interrupt Handlers

View All Interrupt Handlers

Generate Interrupt Handler Code

Generate Interrupt Handler Code

8.6.1 *Equipment List*

Below, we provide the equipment list to be used in this application.

- Three LEDs (green, yellow, red)
- Three 220 Ω resistors
- Four dip switches
- Four push buttons

Listing 8.6 Digital I/O example under Grace.

```
/*
 * ======== Standard MSP430 includes ========
 */
#include <msp430.h>

/*
 * ======== Grace related includes ========
 */
#include <ti/mcu/msp430/Grace.h>

/*
 *  ======== main ========
 */
int main(void)
{
 Grace_init();
//  Activate Grace-generated configuration
 while(1){
 if((P1IN & BIT3)==0x00) // Active low input
 P1OUT=BIT0; // Turn on the LED
 else
 P1OUT=0x00; // Turn off the LED
}

return (0);
}
```

8.6.2 *Layout*

The layout of this application is shown in Fig. 8.8.

8.6.3 *System Design Specifications*

The design of the digital safe will have three main blocks. These are listed below.

- **Block 1**: At startup, the user will press the *enter new password* push button. Then he or she will enter a password using *password switches*. Afterwards, he or she will lock the system using the *verify new password* push button. The yellow LED will turn on. The system will wait for an input.

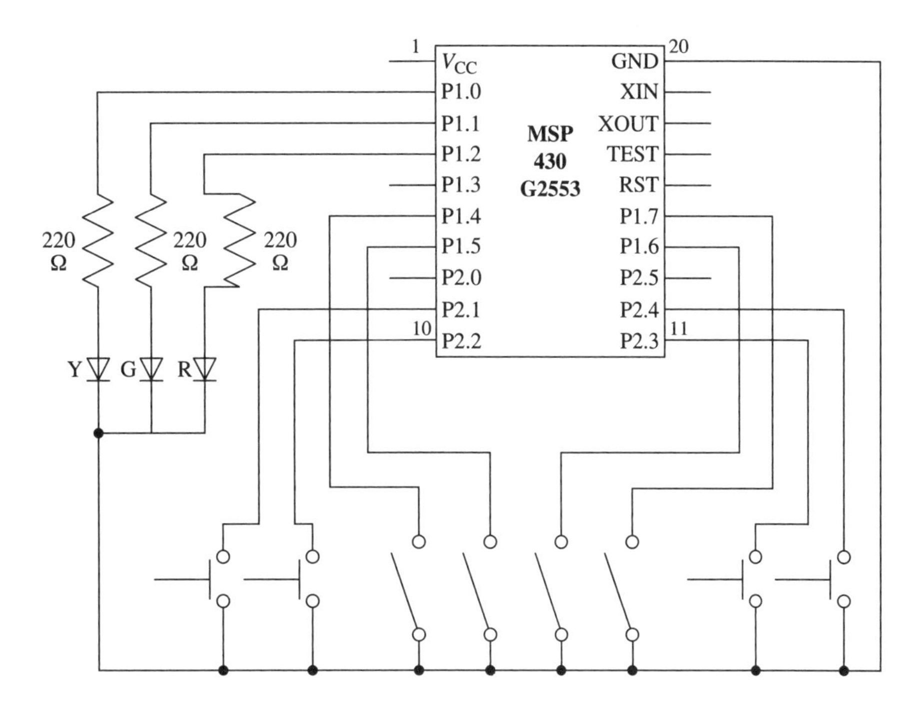

Figure 8.8

Layout of the digital safe application.

- **Block 2**: If the user wants to unlock the system, first he or she should press the *enter your password* push button and enter the password using *password switches*. Then, the *verify your password* push button should be pressed to unlock the safe. If the entered password is correct, the green LED will turn on. Otherwise, the red LED will turn on and the system will wait for the correct password.
- **Block 3**: After the first entry, the password can only be changed if the previous password is entered correctly.

8.6.4 *The C Code for the System*

In the first part of the code, given in Listing 8.7, constants are defined. This is done to make the code more readable. In Listing 8.7, `B1`, `B2`, `B3`, and `B4` are used for input from switches. `BUTTON1`, `BUTTON2`, `BUTTON3`, and `BUTTON4` are used for input from push buttons. `YellowLedOn`, `GreenLedOn`, and `RedLedOn` are used for turning on corresponding LEDs. `AllLedsOff` is used for turning off all LEDs at the same time.

Listing 8.7 Digital safe, the C code part I.

```
#define BUTTON1 ((P2IN & 0x02) == 0x00)
#define BUTTON2 ((P2IN & 0x04) == 0x00)
#define BUTTON3 ((P2IN & 0x08) == 0x00)
#define BUTTON4 ((P2IN & 0x10) == 0x00)
#define B1 (P1IN & 0x10)
```

```
#define B2 (P1IN & 0x20)
#define B3 (P1IN & 0x40)
#define B4 (P1IN & 0x80)
#define YellowLedOn (P1OUT |= 0x01)
#define GreenLedOn  (P1OUT |= 0x02)
#define RedLedOn    (P1OUT |= 0x04)
#define AllLedsOff  (P1OUT &= ~0x07)
```

In the second part of the code, given in Listing 8.8, local variables to be used in the code are defined. They must be defined at the beginning of the `main` function to prevent some compiling errors in CCS. In Listing 8.8, the `NewPassword` array is used for determining the new password. The `YourPassword` array is used for entering the password to unlock the system. The integers `EnterNewPassword` and `EnterYourPassword` are used for changing and entering the password. The requirement of entering the correct password for changing the old password is checked with the `Control` integer.

Listing 8.8 Digital safe, the C code part II.

```
int Control = 0;
int NewPassword[4];
int YourPassword[4];
int EnterNewPassword = 0;
int EnterYourPassword = 0;
```

In the third part of the code, given in Listing 8.9, the hardware setup is done. In the first line of Listing 8.9, the watchdog timer is disabled. The reason for this step will be explained in Sec. 10.5. In the second line, the port P2 is configured as digital I/O completely. In the third and fourth lines, pin directions are assigned. For port P1, `P1DIR = 0x0F` is used because three LEDs are connected to P1.0, P1.1, and P1.2 and four dip switches are connected to P1.4, P1.5, P1.6, and P1.7. The unused pin P1.3 is set as output to prevent accidental input changes. For port P2, `P2DIR = 0xE1` is used because four push buttons are connected to pins P2.1, P2.2, P2.3, and P2.4. Unused pins P2.0, P2.5, P2.6, and P2.7 are set as output. In the fifth and sixth lines, pull-up/down resistors are enabled for pins in which a button or a switch is connected. In the seventh and eighth lines, output registers are set as `P1OUT=0xF0` and `P2OUT=0x1E`. Low bits of these registers are used for turning off LEDs initially. Unnecessary power consumption is prevented for unused output pins by this procedure. On the other hand, high bits of these registers are used for choosing pull-up resistors for input pins.

Listing 8.9 Digital safe, the C code part III.

```
WDTCTL = WDTPW|WDTHOLD;

P2SEL = 0x00;
P1DIR = 0x0F;
P2DIR = 0xE1;
P1REN = 0xF0;
```

```
P2REN = 0x1E;
P1OUT = 0xF0;
P2OUT = 0x1E;
```

Finally, the C code for the system (with all its components) is given in Listing 8.10. The code block doing the operation is put in an infinite loop. Therefore, the system will wait for an input, checking for the password all the time. Initially, the `Control` variable is zero and the system is in Block 1. When `BUTTON1` is pressed, the `EnterNewPassword` variable is changed to 1 and values read from switches are assigned to the `NewPassword` array. Unless `BUTTON2` is pressed, the `EnterNewPassword` variable is kept at 1 and switches can be changed. But when `BUTTON2` is pressed, this variable is toggled to zero and the new password is determined. The yellow LED will turn on to indicate that this step is done. The `Control` variable is also changed to 1 to get the system in Block 2. When `BUTTON3` is pressed, the `EnterYourPassword` variable is changed to 1 and values coming from switches are assigned to the `YourPassword` array. When `BUTTON4` is pressed, `YourPassword` and `NewPassword` arrays are compared. If they do not match, the password is decided as wrong and the red LED turns on for a warning. If they match, the entered password is decided as true and the green LED will turn on to indicate that the safe is unlocked. The `Control` variable is also changed to zero to get the system in Block 1 to determine a new password. It can be seen that the `Control` variable is used alone to accomplish Block 3.

Listing 8.10 Digital safe, the complete C code.

```
#include <msp430.h>

#define BUTTON1 ((P2IN & 0x02) == 0x00)
#define BUTTON2 ((P2IN & 0x04) == 0x00)
#define BUTTON3 ((P2IN & 0x08) == 0x00)
#define BUTTON4 ((P2IN & 0x10) == 0x00)
#define B1 (P1IN & 0x10)
#define B2 (P1IN & 0x20)
#define B3 (P1IN & 0x40)
#define B4 (P1IN & 0x80)
#define YellowLedOn (P1OUT |= 0x01)
#define GreenLedOn  (P1OUT |= 0x02)
#define RedLedOn    (P1OUT |= 0x04)
#define AllLedsOff  (P1OUT &= ~0x07)

void main(void)
{
 WDTCTL = WDTPW|WDTHOLD;

 int Control = 0;
 int NewPassword[4];
 int YourPassword[4];
 int EnterNewPassword = 0;
 int EnterYourPassword = 0;

 P2SEL = 0x00;
```

```
P1DIR = 0x0F;
P2DIR = 0xE1;
P1REN = 0xF0;
P2REN = 0x1E;
P1OUT = 0xF0;
P2OUT = 0x1E;

while(1)
{
if (Control == 0)
{
if (BUTTON1)
{
EnterNewPassword = 1;
}
if (EnterNewPassword == 1)
{
AllLedsOff;
NewPassword[0] = B1;
NewPassword[1] = B2;
NewPassword[2] = B3;
NewPassword[3] = B4;
if (BUTTON2)
{
EnterNewPassword = 0;
YellowLedOn;
Control = 1;
}
}
}
if (Control == 1)
{
if (BUTTON3)
{
EnterYourPassword = 1;
}
if (EnterYourPassword == 1)
{
YourPassword[0] = B1;
YourPassword[1] = B2;
YourPassword[2] = B3;
YourPassword[3] = B4;
if (BUTTON4)
{
EnterYourPassword = 0;
if ((YourPassword[0] == NewPassword[0]) &&
(YourPassword[1] == NewPassword[1]) &&
(YourPassword[2] == NewPassword[2]) &&
(YourPassword[3] == NewPassword[3]))
{
AllLedsOff;
```

```
GreenLedOn;
Control = 0;
}
else
{
AllLedsOff;
RedLedOn;
}}}}}
}
```

8.7 Summary

A microcontroller interacts with other devices through its ports. In this chapter, we focused on the digital I/O in the MSP430. We reviewed specific registers to set up the digital I/O properties. Then we considered the two important hardware issues related to push buttons. Finally, we designed a real-life application (a digital safe) using digital I/O. We provided all the hardware and software design information related to the digital safe. We hope that this information will encourage the reader to develop new projects using digital I/O.

8.8 Problems

8.1 Write a C program for the MSP430 to calculate the number of zeros and ones in an array. If the number of zeros is more than the number of ones, the red LED (connected to P1.0 on the MSP430 LaunchPad) will turn on. Otherwise, the green LED (connected to P1.6 on the MSP430 LaunchPad) will turn on.

8.2 Repeat Prob. 8.1 in assembly language.

8.3 Repeat Prob. 8.1 using Grace.

8.4 Write a C program for the MSP430 to multiply numbers (except zeros) in an array. Then divide the result by the length of the array. If the result is less than the first predefined value, the red LED (connected to P1.0 on the MSP430 LaunchPad) will turn on. If it is between the first and second predefined values, the green LED (connected to P1.6 on the MSP430 LaunchPad) will turn on. If it is more than the second predefined value, both LEDs will turn on.

8.5 Repeat Prob. 8.4 using Grace.

8.6 Write a C program for the MSP430 to compute the average of 10 floating-point numbers. If the average is greater than zero, the red LED (connected to P1.0 on the MSP430 LaunchPad) will turn on. Otherwise, the green LED (connected to P1.6 on the MSP430 LaunchPad) will turn on. Initially both LEDs are turned off.

8.7 Repeat Prob. 8.6 using pointers and pointer arithmetic only.

8.8 Repeat Prob. 8.7 using Grace.

8.9 Write a C program for the MSP430 with the following specifications: When the push button (connected to P1.3 on the MSP430 LaunchPad) is pressed, the red LED (connected to P1.0 on the MSP430 LaunchPad) will turn on and wait for a certain time. Then, both the red and

green LEDs (connected to P1.0 and P1.6 on the MSP430 LaunchPad) will turn on and wait for a certain time. Afterwards, the red LED (connected to P1.0 on the MSP430 LaunchPad) will turn off and the green LED (connected to P1.6) will turn on and wait for a certain time. Finally, both LEDs will turn off. This procedure is repeated indefinitely.
Hint: Use loop operations to generate waiting times.

8.10 Repeat Prob. 8.9 in assembly language.

8.11 Repeat Prob. 8.9 using Grace.

8.12 Write a C program for the MSP430 with the following specifications. When the push button (connected to P1.3 on the MSP430 LaunchPad) is pressed four times, the red LED (connected to P1.0 on the MSP430 LaunchPad) will turn on and the green LED (connected to P1.6 on the MSP430 LaunchPad) will turn off. When the push button is pressed two more times, the red LED will turn off and the green LED will turn on. This procedure is repeated indefinitely.

8.13 Repeat Prob. 8.12 in assembly language.

8.14 Repeat Prob. 8.12 using Grace.

Interrupts

Chapter Outline

The interrupt is the main tool for event-driven programming in a microprocessor. If the user wants to write a program to react to predefined actions, the only solution is using interrupts. The most important (and confusing) property of interrupts is their unpredictability. Since interrupts are generated by hardware, it is not possible to predict when they will occur. This chapter is about interrupts on the MSP430. We start by explaining what happens when an interrupt occurs. Then, step-by-step we explore the interrupt concept.

9.1 What Happens When an Interrupt Occurs?

We experience interrupts in our daily lives. Let's assume that the fire alarm is activated for a fire drill during the class hour. This is an interrupt. The instructor halts the lesson and everyone leaves the class. This is what we do after the interrupt. After the fire drill is done, the lesson resumes from where it was left. This is returning from the interrupt. The pattern is the same for the CPU. Let's analyze what happens when an interrupt occurs.

First of all, the interrupt must come from an external source (such as a button or a switch) or an internal source (like a timer or an analog-to-digital conversion [ADC] signal). The user should enable the interrupt option for the desired source to process the interrupt. As the interrupt comes, the CPU stops what it is doing. If it is executing an instruction, this is done. Then, without executing the next instruction the CPU saves the program counter (PC), status register (SR), and variables to the stack. The PC is set to the *interrupt vector*, which is a predefined memory address for that specific interrupt. Therefore, the execution continues from that address. As a matter of fact, the interrupt vector address holds another address to be branched to. This branched address holds a subroutine to be processed in response to the interrupt. This is called the *interrupt service routine* (ISR). The user is responsible for the code block to be written in the ISR. The CPU recalls the saved PC, SR, and variables from the stack as the ISR is executed. Then it turns back to the main program and continues executing the next instruction.

Let's analyze a simple example to clarify what happens when an interrupt occurs. We want to turn on the red LED connected to port P1.0 by the push button connected to pin P1.3 of the MSP430 LaunchPad. In this case, the interrupt source will be pin P1.3. Here we assume that the port settings are done as given in Chap. 8. The interrupt from pin P1.3 is enabled (to be explained in Sec. 9.6 in detail). If we only want to turn on the red LED when the button is pressed, the main program will just have an infinite loop. In other words, the CPU will wait in an infinite loop doing nothing. When the user pushes the button, an interrupt will be generated. The CPU will stop the infinite loop. The CPU will set the PC to the address of the port interrupt vector. As the CPU reaches this address, it will check what is written there. The interrupt vector holds another address pointing to the ISR as we mentioned previously. Now, the CPU sets the PC to this address. Therefore, the CPU reaches the ISR in the next step. As programmers, it is our responsibility to write the code block in the ISR. Since our task is just to turn on the red LED, the code in the ISR will just set pin P1.0 to V_{CC}. As this code block is executed, the CPU will turn back from the ISR and continue to wait in the infinite loop. In the following sections, we will explore all these steps in detail.

Table 9.1

Interrupt sources, flags, and vectors.

Interrupt Source	Interrupt Flag	Type	Interrupt Address	Priority
Power-Up	PORIFG			
External Reset	RSTIFG			
Watchdog Timer+	WDTIFG	reset	FFFEh	31
Flash key violation	KEYV			highest
PC out of range				
NMI	NMIIFG			
Oscillator fault	OFIFG	(non)-maskable	FFFCh	30
Flash memory access violation	ACCVIFG			
Timer1_A3	TA1CCR0, CCIFG	maskable	FFFAh	29
Timer1_A3	TA1CCR2, TA1CCR1 CCIFG,TAIFG	maskable	FFF8h	28
Comparator_A+	CAIFG	maskable	FFF6h	27
Watchdog Timer+	WDTIFG	maskable	FFF4h	26
Timer0_A3	TA0CCR0, CCIFG	maskable	FFF2h	25
Timer0_A3	TA0CCR2, TA0CCR1 CCIFG,TAIFG	maskable	FFF0h	24
USCI_A0/USCI_B0 receive USCI_B0 I^{2}C status	UCA0RXIFG,UCB0RXIFG	maskable	FFEEh	23
USCI_A0/USCI_B0 transmit USCI_B0 I^{2}C receive/transmit	UCA0TXIFG,UCB0TXIFG	maskable	FFECh	22
ADC10	ADC10IFG	maskable	FFEAh	21
I/O Port P2 (up to eight flags)	P2IFG.0 to P2IPG.7	maskable	FFE6h	19
I/O Port P1 (up to eight flags)	P1IFG.0 to P1IPG.7	maskable	FFE4h	18

9.2 Types of Interrupts

The MSP430 has different kinds of interrupt sources as listed in Table 9.1. These are divided into three groups: reset, non-maskable, and maskable. The reset interrupt has the highest priority. Maskable and non-maskable interrupts (NMIs) are enabled by individual interrupt enable bits. The main difference between them is that maskable interrupts are also controlled by the global interrupt enable (GIE) bit in the SR. NMIs do not have such control. There are other NMIs aside from those listed in Table 9.1. Detailed information on these can be found in [16].

There is a possibility that more than one interrupt will occur at the same time. Hence, there must be an order between the interrupt sources. This is called the priority order among interrupts. This order indicates that when two interrupts occur at the same time, the one with higher priority will take precedence over the lower priority one. The priority order for the MSP430 is provided in Table 9.1.

9.3 Interrupt Flags

Interrupt flags are actually register bits. When an interrupt occurs, its specific flag is set. Therefore, the CPU becomes aware of the interrupt. The list of interrupt flags for the MSP430 appears in Table 9.1. While reset, non-maskable, and universal serial communications interface (USCI) interrupt flags are placed in interrupt flag registers, maskable interrupt flags (except USCI interrupt flags) are located in the related module's register. There is also a CPU interrupt flag controlled by the GIE. When a maskable interrupt occurs, the CPU interrupt flag and the related flag in the module register are set.

The CPU should also know whether to process the incoming interrupt or not. The interrupt enable bit is used for this purpose. Each interrupt has a specific enable bit in a different register address. This bit must be set to request an interrupt. It is sufficient to set the related enable bit for NMIs. But the GIE bit must also be set for maskable interrupts. The GIE bit warns the CPU about the interrupt process for these. The individual interrupt enable bit gives information about the interrupt source.

For some microcontrollers, an interrupt with higher priority may occur while an interrupt with a lower priority is in progress. This is called a nested interrupt. But this may cause a stack overflow. The GIE bit is cleared when the ISR is called in the MSP430. This prevents calling any other maskable interrupts. In other words, nested interrupts are not allowed in the MSP430. To enable the interrupt from the same source again, the related interrupt flag must be reset in the ISR. The GIE flag is cleared automatically, so there is no need to reset it.

9.4 Interrupt Vectors

As we mentioned in Sec. 9.1, after an interrupt occurs and the necessary information about the main process is saved, the CPU needs to go to the ISR. The memory address for the ISR is kept in an interrupt vector. Generally, each interrupt source has a specific interrupt vector. But some of them share the same interrupt vector. For each interrupt source, the interrupt vector address is given in Table 9.1. To simplify coding, these addresses are defined as constants in the MSP430 header file given in the Appendix. In Table 9.2, they are shown as interrupt vector definitions.

Table 9.2 Interrupt vector definitions.

Interrupt Source	Interrupt Vector Definition
I/O Port 1	PORT1_VECTOR
I/O Port 2	PORT2_VECTOR
ADC10	ADC10_VECTOR
USCI A0/B0 Transmit	USCIAB0TX_VECTOR
USCI A0/B0 Receive	USCIAB0RX_VECTOR
Timer0_A CC1	TIMER0_A1_VECTOR
Timer0_A CC0	TIMER0_A0_VECTOR
Watchdog Timer	WDT_VECTOR
Comparator A	COMPARATORA_VECTOR
Timer1_A CC1	TIMER1_A1_VECTOR
Timer1_A CC0	TIMER1_A0_VECTOR
Non-Maskable Interrupt	NMI_VECTOR
Reset	RESET_VECTOR

9.5 Interrupt Service Routines

The interrupt service routine (ISR), also known as the interrupt handler, is the code block which is executed when an interrupt occurs. The ISR is very similar to a function (subroutine). But unlike a function, ISR is called by the interrupt. As discussed before, the MSP430 doesn't allow nested interrupts. Therefore, the ISR must be kept short. It must perform its actions as quickly as possible and return to the main code to allow other interrupts. To do so, the programmer should also reset the related interrupt flag at the end of the ISR.

9.5.1 ISR in C

The `pragma` keyword should be used in defining an ISR in C. Since the ISR will not be related to the main code (remember, their only connection is through hardware), the C compiler may not include it in the compilation process. To avoid this, we define the ISR with the `# pragma` keyword in front of it. To define to which interrupt vector this ISR is related, we should also put it before the ISR. The ISR should also be distinguished from a function. Therefore, the `__interrupt` keyword should be added before its name. In Listing 9.1, these definitions are given on a sample port interrupt.

Listing 9.1 ISR definitions in C.

```
// Main Code Block to be Added Here

#pragma vector=PORT1_VECTOR
// define the interrupt vector
__interrupt void PORT1_ISR(void){
// Interrupt Service Routine

// ISR Code to be Added Here

 P1IFG = 0x00; // clear the interrupt flag
}
```

9.5.2 ISR in Assembly

Defining the ISR in the assembly language is easier than in C. First of all, there is no `pragma` keyword in the assembly code. The interrupt vector should be defined here also. This is necessary to associate the ISR with the interrupt. In fact, we applied this procedure in Sec. 7.5. There, we associated the `RESET` label with the `RESET_VECTOR`. Next, we provide a sample code block in Listing 9.2 for the port-based interrupt in assembly language. Here, the interrupt vectors and the ISR are associated at the bottom of the code block. The ISR is just like a subroutine. The only exception is the instruction `reti`, which is used instead of `ret` to return from the ISR.

Listing 9.2 ISR definitions in assembly language.

```
  .cdecls C,LIST,"msp430.h"

  .text
```

```
 .retain
 .retainrefs

RESET
 mov.w #WDTPW|WDTHOLD,WDTCTL
 mov.w #__STACK_END,SP

;The Main Code Block to be Added Here

;---------------------
P1_ISR ;ISR definition
;---------------------

;ISR Code to be Added Here

 bic.b #08h,P1IFG ;Clear the P1.3 interrupt flag
 reti ;Return from the ISR

;------------------------
;Stack Pointer definition
;------------------------
 .global __STACK_END
 .sect .stack

;-------------------
;Interrupt Vectors
;-------------------
 .sect RESET_VECTOR
 .short RESET
 .sect  PORT1_VECTOR
 .short P1_ISR
 .end
```

9.6 Port Interrupts

The interrupt may be generated from several sources as mentioned earlier. Since digital input and output (I/O) was considered in Chap. 8, here we focus on the interrupts from ports. We will also focus on interrupts from other sources in the following chapters.

Ports P1 and P2 can be used as interrupt sources for the MSP430. Each pin of these ports can be used for a different interrupt. Unfortunately, these share the same interrupt vector, hence the same ISR. Interrupt flags can be used to overcome this issue. Using them, the task of each pin interrupt can be controlled separately by the condition of the related interrupt flag. There are three special registers to control port interrupts: interrupt enable register (**PxIE**), interrupt edge select register (**PxIES**), and interrupt flag register (**PxIFG**).

The PxIE register is used to enable the interrupt for the associated pin. To enable the port interrupt from a specific pin, the corresponding bit in PxIE should be set. To disable the interrupt from the same pin, the corresponding bit should be reset. Initially all the port interrupts are disabled.

The PxIES register is used to select the signal edge in which the interrupt occurs (on a specific pin). The interrupt occurs when the input goes from low to high, if the bit corresponding to the pin is reset. The interrupt occurs when the input goes from high to low, if the bit is set. PxIE and GIE bits must be set beforehand to enable the interrupt from that pin.

The PxIFG register is used to check the interrupt condition. When an interrupt occurs from a pin, the related interrupt flag is set. PxIFG must be cleared at the end of the ISR to allow a new interrupt. In the same manner, all the interrupt flags should be reset at the beginning of the program to avoid any confusion.

9.7 Coding Practices for Interrupts

In this section, we provide several C and assembly codes related to port interrupts under the MSP430. These provide basic examples on how to use interrupts on the MSP430 LaunchPad.

9.7.1 *Interrupts in C*

The first C code on interrupts, given in Listing 9.3, toggles the red LED when the button is pressed. Here, it is important to note that all the interrupts are enabled by an intrinsic function `_enable_interrupts()`.

Listing 9.3 Toggle the red LED by an interrupt, in C language.

```
#include <msp430.h>

#define LED BIT0
#define BUTTON BIT3

void main(void)
{
 WDTCTL = WDTPW|WDTHOLD;

 P1DIR = LED;
 P1OUT = LED;

 P1IE = BUTTON;// enable interrupt from port P1
 P1IES = BUTTON;// interrupt edge select from high to low
 P1IFG = 0x00;// clear the interrupt flag

 _enable_interrupts();// enable all interrupts

 while(1); // wait for an interrupt
}

#pragma vector=PORT1_VECTOR
// define the interrupt vector
__interrupt void PORT1_ISR(void){
//  Interrupt Service Routine
 P1OUT ^= 0x01; // toggle LED
 P1IFG = 0x00; // clear the interrupt flag
}
```

In the second C code, given in Listing 9.4, the number of button presses is counted by the ISR. Here, the variable `count` is specifically defined as global. Therefore, it can be kept between successive interrupts.

Listing 9.4 Count the number of button presses by interrupts, in C language.

```
#include <msp430.h>

#define BUTTON BIT3

 int count = 0;

void main(void)
{
 WDTCTL = WDTPW|WDTHOLD;

 P1IE = BUTTON;// Enable interrupt from port1
 P1IES = BUTTON;// Interrupt edge select from high to low
 P1IFG = 0x00;// Clear the interrupt flag

 _enable_interrupts();// Enable all interrupts

 while(1);
}

#pragma vector=PORT1_VECTOR
// define the interrupt vector
__interrupt void PORT1_ISR(void){
//  Interrupt Service Routine for Port 1
 count += 1;
 P1IFG = 0x00;// Clear the interrupt flag
}
```

In the third C code, given in Listing 9.5, the red and green LEDs are turned on based on the total number of button presses. Again, the variable `count` is specifically defined as global for the same reason given in Listing 9.4.

Listing 9.5 Turn on and off LEDs by the total number of interrupts, in C language.

```
#include <msp430.h>

#define REDLED BIT0
#define GREENLED BIT6
#define BUTTON BIT3
 int count = 0;

void main(void)
{
 WDTCTL = WDTPW|WDTHOLD;

 P1DIR = REDLED|GREENLED;
 P1OUT = 0x00;

 P1IE = BUTTON;// enable interrupt from port1
```

```
 P1IES = BUTTON;// interrupt edge select from high to low
 P1IFG = 0x00;// clear interrupt flag

 _enable_interrupts();// enable all interrupts

 while(1);
}

#pragma vector=PORT1_VECTOR
// define interrupt vector
__interrupt void PORT1_ISR(void){
//  Interrupt Service Routine
 count++;
 if (count == 4)P1OUT = REDLED;// Turn on the RED LED
 if (count == 6){
 P1OUT = 0x00;
 P1OUT = GREENLED;
 count = 0;
 }// Turn on the GREEN LED
 P1IFG = 0x00;// clear the interrupt flag
}
```

9.7.2 Interrupts in Assembly

In the first assembly-based port interrupts application, we redo the code given in Listing 9.3. Here, we toggle red and green LEDs when the button is pressed instead of toggling the red LED alone. The assembly code for this operation is given in Listing 9.6.

Listing 9.6 Toggle red and green LEDs when the button is pressed, in assembly language.

```
 .cdecls C,LIST,"msp430.h"

 .text
 .retain
 .retainrefs

RESET
 mov.w #WDTPW|WDTHOLD,WDTCTL
 mov.w #__STACK_END,SP

 mov.b #41h,P1DIR ;P1.0 and P1.6 output, else input
 mov.b #01h,P1OUT ;P1.0 set, else reset
 bis.b #08h,P1IE  ;P1.3 Interrupt enabled
 bis.b #08h,P1IES ;P1.3 high/low edge
 bic.b #08h,P1IFG ;P1.3 IFG cleared

 bis.w #GIE,SR ;Enable interrupts

 jmp $

;--------------------------
P1_ISR  ;Toggle P1.0 Output
;--------------------------
 xor.b #41h,P1OUT ;P1.0 and P1.6 toggle
```

```
 bic.b #08h,P1IFG ;P1.3 IFG Cleared
 reti ;Return from ISR

;-------------------------
;Stack Pointer definition
;-------------------------
 .global __STACK_END
 .sect .stack

;-------------------
;Interrupt Vectors
;-------------------
 .sect RESET_VECTOR
 .short RESET
 .sect  PORT1_VECTOR
 .short P1_ISR
 .end
```

In the second assembly-based port interrupt application, we redo the code given in Listing 9.5. The final assembly code is given in Listing 9.7.

Listing 9.7 Turn on and off LEDs by the total number of interrupts, in assembly language.

```
 .cdecls C,LIST,"msp430.h"

 .text
 .retain
 .retainrefs

RESET
 mov.w #WDTPW|WDTHOLD,WDTCTL
 mov.w #__STACK_END,SP

 mov.b #41h,P1DIR ;P1.0 and P1.6 output, else input
 bic.b #41h,P1OUT ;P1.0 and P1.6 reset
 bis.b #08h,P1IE  ;P1.3 Interrupt enabled
 bis.b #08h,P1IES ;P1.3 high/low edge
 bic.b #08h,P1IFG ;P1.3 IFG cleared
 clr R5

 bis.w #GIE,SR ;Enable all interrupts

 jmp $

;--------------------------
P1_ISR  ;Toggle P1.0 Output
;--------------------------
 inc R5
 cmp.w #4d,R5
 jeq RedLED
 cmp.w #6d,R5
 jeq GreenLED
```

```
 bic.b #08h,P1IFG ;P1.3 IFG Cleared
 reti ;Return from ISR

RedLED:
 bis.b #01h,P1OUT
 bic.b #40h,P1OUT
 bic.b #08h,P1IFG ;P1.3 IFG Cleared
 reti ;Return from ISR

GreenLED:
 bic.b #01h,P1OUT
 bis.b #40h,P1OUT
 clr R5
 bic.b #08h,P1IFG ;P1.3 IFG Cleared
 reti ;Return from ISR

;------------------------
;Stack Pointer definition
;------------------------
 .global __STACK_END
 .sect .stack

;-------------------
;Interrupt Vectors
;-------------------
 .sect RESET_VECTOR
 .short RESET
 .sect  PORT1_VECTOR
 .short P1_ISR
 .end
```

9.8 Interrupts in Grace

We can use Grace to handle interrupts. There are several interrupt sources for the MSP430 as tabulated in Table 9.1. Up to now, we only considered port interrupts. Therefore, we will focus on them in this section.

9.8.1 *Port Interrupts*

The port power user or P1/P2 mode of general purpose input and output (GPIO) has a View All Interrupt Handlers link. As we press it, a new tab named Interrupt Vectors appears. This is given in Fig. 9.1. In this tab, all interrupt sources are listed under the All Interrupts list. We should select either Port 1 or Port 2 in this list to generate a port interrupt. As we select one of these options, the select buttons for specific pins appear on the right as shown in Fig. 9.1.

The Generate Interrupt Handler Code button generates the prototype ISR under the file InterruptVectors_init.c. This file can be opened by pressing the link Open Interrupt Vector File on the right top corner of Fig. 9.1. As a reminder, this file is automatically generated. Unfortunately, it is reset whenever a hardware option is changed under Grace. Therefore, it should be used with caution.

Figure 9.1

Interrupt vectors under Grace.

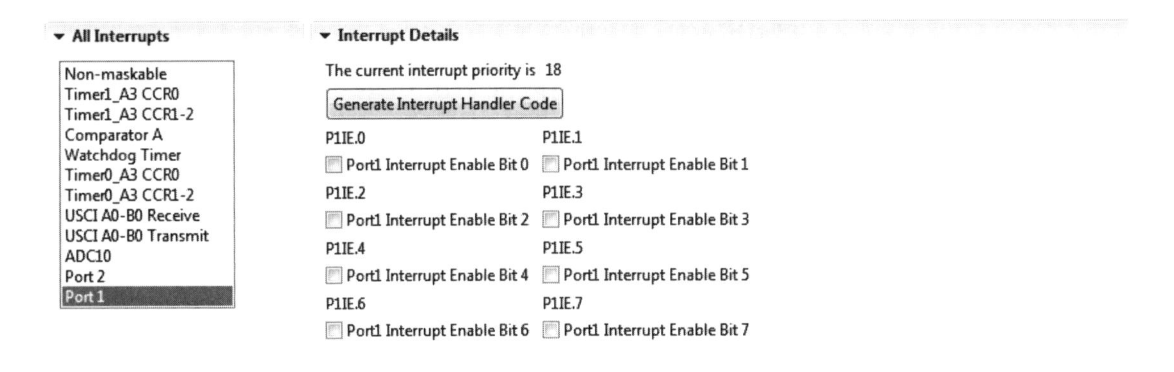

9.8.2 *Coding Practices*

In this section, we redo the port interrupt application given in Listing 9.3 using Grace. As a reminder, this application toggles the red LED (connected to P1.0 on the MSP430 LaunchPad) when the button (connected to P1.3 on the MSP430 LaunchPad) is pressed. We start by generating a Grace project. Then, we configure pins P1.0 and P1.3 under Grace. The pin P1.0 should be set as GPIO output. The pin P1.3 should be set as input. These settings can be done by any of the three GPIO views given in Sec. 8.5. The port interrupt property of P1.3 should also be done either from the power user or the P1/P2 mode of the GPIO. In this application, we do not add any code lines to the main.c file. Since we are using the ISR, we generate an ISR prototype using the Generate Interrupt Handler Code button. We fill the prototype Port 1 ISR block under InterruptVectors_init.c as given in Listing 9.8. After compiling the project, we can run our application.

Listing 9.8 Generating a port interrupt under Grace.

```
#pragma vector=PORT1_VECTOR
__interrupt void PORT1_ISR_HOOK(void)
{
 P1OUT ^= BIT0;
 P1IFG = 0x00;
}
```

9.9 Washing Machine Application

The aim in this application is to learn how to set and use port interrupts of the MSP430 microcontroller. As a real-world application, we will design a washing machine system using a stepper motor. In this section, we provide the equipment list, layout of the circuit, procedure, and system design specifications.

9.9.1 *Equipment List*

Following is the equipment list to be used in this application.

- One 12-V dc adaptor
- One LM7805 voltage regulator
- One 330-ηF capacitor
- One 10-μF electrolytic capacitor
- One stepper motor
- One ULN2003 motor driver
- Five push buttons
- Three 100-ηF capacitors
- Two LEDs (yellow and red)
- Two 220-Ω resistors

The **stepper motor** is a device that rotates in steps, rather than turning smoothly as a dc motor does. The rotation step size can be 0.9 (half-stepping) or 1.8 (full-stepping) degrees. Therefore, a full rotation needs 400 and 200 steps respectively. The speed of the motor is determined by the time delay between each step. In this application, we use a four-phase stepper motor.

We should feed a binary sequence to rotate the stepper motor. This sequence will hold the states. For our four-phase motor, the binary sequence for half-stepping is given in Table 9.3. For full-stepping, this sequence will be as in Table 9.4. We should feed one of these sequences in a periodic manner in order to rotate the stepper motor continuously. We should also add a time delay between each state in the sequence for the motor to work properly.

Table 9.3 Half-step control sequence.

States	Output1	Output2	Output3	Output4
State1	1	0	0	0
State2	1	1	0	0
State3	0	1	0	0
State4	0	1	1	0
State5	0	0	1	0
State6	0	0	1	1
State7	0	0	0	1
State8	1	0	0	1

Table 9.4

Full-step control sequence.

States	Output1	Output2	Output3	Output4
State1	1	0	0	0
State2	0	1	0	0
State3	0	0	1	0
State4	0	0	0	1

9.9.2 *Layout*

The layout of this application is given in Fig. 9.2. The **voltage supply** block will be used in future applications also. Therefore, it is given in Fig. 9.3.

9.9.3 *System Design Specifications*

The washing machine will be controlled by five push buttons. Two of them are *main on/off* and *rotation speed*. The remaining three buttons are for program selection as follows:

- *Prewash*: 30 rotations in one direction, then 30 rotations in the other direction.
- *Normal wash*: 100 rotations in one direction, then 100 rotations in the other direction.

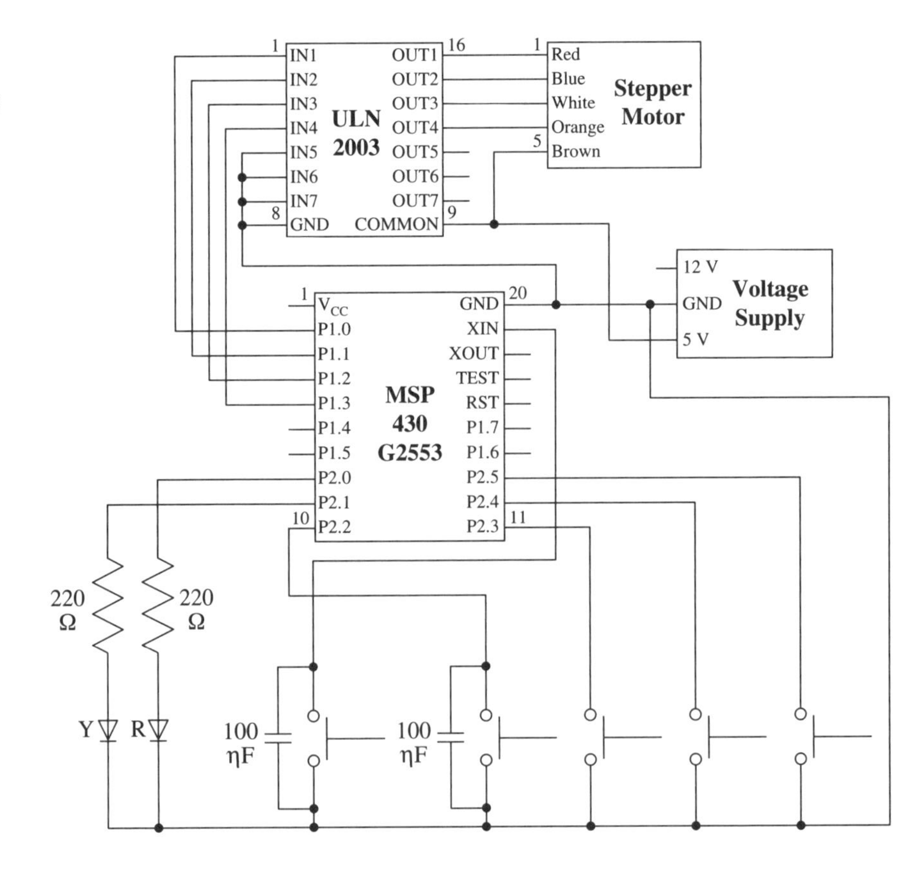

Figure 9.2 Layout of the washing machine application.

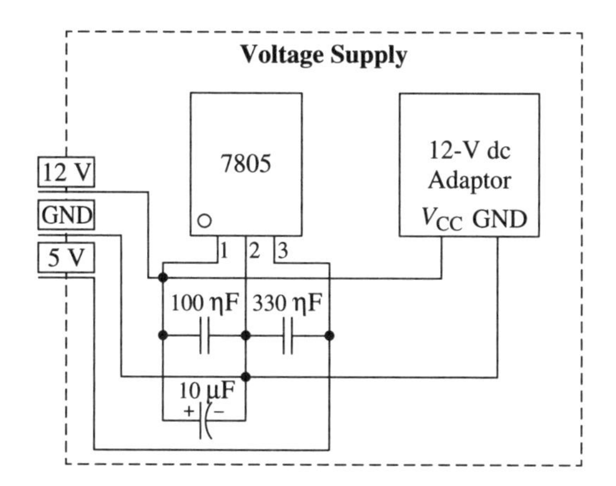

Figure 9.3

The voltage supply block.

- *Final spin*: 50 rotations in one direction, but faster than prewash and normal wash.

When the main on/off button is pressed, the system will be activated. To indicate this, the red LED will turn on. In this state, all programs (prewash, normal wash, and final spin) can be performed. Each program can be selected by a specific button. There is an extra button for adjusting the rotation speed to slow or fast. Depending on the selection, the yellow LED will be either on or off. When the main on/off button is pressed again, the system will be deactivated. To indicate this, the red LED will turn off.

9.9.4 The C Code for the System

In the first part of the code, constants for interrupts and output pins are defined. This is done to make the code more readable. Here, `ONOFF`, `RSPEED`, `NWASH`, `PWASH`, and `FSPIN` are used for interrupts from push buttons. `YellowLedToggle` and `RedLedToggle` are used for toggling the LEDs. Also `NormalWash`, `PreWash`, and `FinalSpin` are defined as constants with appropriate values. The code block for this part is given in Listing 9.9.

Listing 9.9 Washing machine, the C code part I.

```
#define ONOFF ((P2IFG & 0x40) == 0x40)
#define RSPEED ((P2IFG & 0x04) == 0x04)
#define NWASH ((P2IFG & 0x08) == 0x08)
#define PWASH ((P2IFG & 0x10) == 0x10)
#define FSPIN ((P2IFG & 0x20) == 0x20)
#define RedLedToggle (P2OUT ^= 0x01)
#define YellowLedToggle (P2OUT ^= 0x02)
#define NormalWash 1
#define PreWash 2
#define FinalSpin 3
```

In the second part of the code, global variables are defined. The code block for this part is given in Listing 9.10. Here, the `Program` variable is used for determining the wash cycle. The `RotationSpeed` variable is used for choosing the rotation speed option for the system. The `open` variable is used for controlling the on/off property of the main button. These are defined as global variables since they are used by the ISR.

Listing 9.10 Washing machine, the C code part II.

```
int Program;
int RotationSpeed = 0;
int open = 0;
```

In the third part of the code, given in Listing 9.11, the hardware setup is done. In the first line of Listing 9.11, the watchdog timer is disabled. The reason for this step will be explained in Sec. 10.5. In the second line, port P2 is configured as digital I/O completely. In the third and fourth lines, pin directions are assigned. For port P1, `P1DIR=0xFF` is used because the stepper motor is connected to pins P1.0, P1.1, P1.2, and P1.3. Unused pins P1.4, P1.5, P1.6, and P1.7 are set as output. For port P2, `P2DIR=0x83` is used since five push buttons are connected to pins P2.2, P2.3, P2.4, P2.5, and P2.6. Two LEDs are connected to P2.0 and P2.1. Unused pin P2.7 is again set as output. In the fifth line, pull-up/down resistors for button-connected pins of port P2 are enabled. In the sixth and seventh lines, output registers are set as `P1OUT=0x00` and `P2OUT=0x7C`. Unnecessary power consumption is prevented for unused output pins by this procedure. On the other hand, high bits of the P2OUT register are used for choosing pull-up resistors for input pins. In the next three lines, interrupt configurations for port P2 are done. Interrupt is enabled for pins P2.2, P2.3, P2.4, P2.5, and P2.6 by `P2IE=0x7C` (since a push button is connected to each). All of these five interrupts are triggered by a high-to-low transition. Therefore, we set `P2IES=0x7C`. Also, all interrupt flags are cleared at the beginning of the code by `P2IFG=0x00`. Finally, in the last line the GIE bit is set to enable maskable interrupts by the intrinsic function `_enable_interrupts()`.

Listing 9.11 Washing machine, the C code part III.

```
WDTCTL = WDTPW|WDTHOLD;

P2SEL = 0x00;
P1DIR = 0xFF;
P2DIR = 0x83;
P2REN = 0x7C;
P1OUT = 0x00;
P2OUT = 0x7C;
P2IE = 0x7C;
P2IES = 0x7C;
P2IFG = 0x00;

_enable_interrupts();
```

ISR settings for port P–2 based interrupts are given in Listing 9.12. There are five interrupt sources coming from five different buttons. The main on/off button toggles the variable `open`. It also toggles the red LED to inform the user whether the system is on or off. The rotation speed button toggles the variable `RotationSpeed`. It also warns the user about the selected rotation speed by toggling the yellow LED. The other three buttons are used for choosing the wash program (prewash, normal wash, and final spin). Each button assigns a different number to the variable `program`. The related interrupt flag is cleared at the end of the ISR to get a new interrupt.

Listing 9.12 Washing machine, the C code part IV.

```
#pragma vector =PORT2_VECTOR
__interrupt void PORT2_ISR(void){
 if(ONOFF){
 open ^= 1;
 RedLedToggle;
 P2IFG &= ~0x40;
 }
 if ((RSPEED) && (open == 1)){
 RotationSpeed ^= 1;
 YellowLedToggle;
 P2IFG &= ~0x04;
 }
 if ((NWASH) && (open == 1)){
 Program = NormalWash;
 P2IFG &= ~0x08;
 }
 if ((PWASH) && (open == 1)){
 Program = PreWash;
 P2IFG &= ~0x10;
 }
 if ((FSPIN) && (open == 1)){
 Program = FinalSpin;
 P2IFG &= ~0x20;
 }
 P2IFG=0x00;
}
```

Finally, the C code for the system (with all its components) is given in Listing 9.13. The code block doing the operation is put in an infinite loop. Therefore, the system will wait for an input and check for the buttons all the time.

Listing 9.13 Washing machine, the C code.

```
#include <msp430.h>

#define ONOFF ((P2IFG & 0x40) == 0x40)
#define RSPEED ((P2IFG & 0x04) == 0x04)
#define NWASH ((P2IFG & 0x08) == 0x08)
#define PWASH ((P2IFG & 0x10) == 0x10)
```

```
#define FSPIN ((P2IFG & 0x20) == 0x20)
#define RedLedToggle (P2OUT ^= 0x01)
#define YellowLedToggle (P2OUT ^= 0x02)
#define NormalWash 1
#define PreWash 2
#define FinalSpin 3

 int Program=0;
 int RotationSpeed = 0;
 int open = 0;

void delay_ms(int);
void Wash();

void main(void)
{
 WDTCTL = WDTPW|WDTHOLD;

 P2SEL = 0x00;
 P1DIR = 0xFF;
 P2DIR = 0x83;
 P2REN = 0x7C;
 P1OUT = 0x00;
 P2OUT = 0x7C;
 P2IE = 0x7C;
 P2IES = 0x7C;
 P2IFG = 0x00;

 _enable_interrupts();

 while(1){
 if(Program!=0) Wash();
 }
}

void delay_ms(int a){
 while(a != 0)
 {
 _delay_cycles(1000);
 a--;
 }
}

void Wash(){
 int speed, turn, fast, slow, Rturn, Lturn, pos;
  volatile unsigned int seqr[8] = {0x08,0x0C,0x04,0x06,
 0x02,0x03,0x01,0x09};
  volatile unsigned int seql[8] = {0x09,0x01,0x03,0x02,
 0x06,0x04,0x0C,0x08};

 if (Program == 1){
 slow = 20, fast = 10, Rturn = 100, Lturn = 100;
 }
 if (Program == 2){
 slow = 20, fast = 10, Rturn = 30, Lturn = 30;
```

```
    }
    if (Program == 3){
    slow = 10, fast = 5, Rturn = 50, Lturn = 0;
    }

    if(RotationSpeed == 0)
    speed = fast;
    else
    speed = slow;

    for (turn=0; turn<Rturn; turn++){
    if(open == 1){
    pos = 0;
    while(pos<8){
    P1OUT = seqr[pos];
    pos++;
    delay_ms(speed);
    }
    P1OUT = 0x00;
    }
    else break;
    }
    for (turn=0; turn<Lturn; turn++){
    if(open == 1){
    pos = 0;
    while(pos<8){
    P1OUT = seql[pos];
    pos++;
    delay_ms(speed);
    }
    P1OUT = 0x00;
    }
    else break;
    }
    Program = 0;
}

#pragma vector =PORT2_VECTOR
__interrupt void PORT2_ISR(void){
    if(ONOFF){
    open ^= 1;
    RedLedToggle;
    P2IFG &= ~0x40;
    }
    if ((RSPEED) && (open == 1)){
    RotationSpeed ^= 1;
    YellowLedToggle;
    P2IFG &= ~0x04;
    }
    if ((NWASH) && (open == 1)){
    Program = NormalWash;
    P2IFG &= ~0x08;
```

```
    }
    if ((PWASH) && (open == 1)){
    Program = PreWash;
    P2IFG &= ~0x10;
    }
    if ((FSPIN) && (open == 1)){
    Program = FinalSpin;
    P2IFG &= ~0x20;
    }
    P2IFG=0x00;
  }
```

There are two functions used in this code: `delay_ms` and `Wash`. Delay times between the stepper motor states are obtained by the `delay_ms` function. This function calls the intrinsic `_delay_cycles` function with a value of 1 ms. This intrinsic function is called within a loop (by the time coming from the input variable `a`) to obtain the desired delay time. The `Wash` function is used to rotate the stepper motor according to the value chosen by the rotation speed and one of the three program selection buttons. Program selection buttons determine the number of right turns, number of left turns, and the delay time coefficient (for slow or fast rotation speed). The rotation speed button is used to select the fast or slow rotation speed. To rotate the stepper motor, a `while` loop is added to send the states given in Table 9.3. The iteration number of these sequences is controlled by `for` loops. Each `for` loop has a condition `open==1`. This is used for breaking the `for` loop when the main on/off button is pressed.

9.10 Summary

This chapter is about the interrupt-based programming of the MSP430. We first considered what happens when an interrupt occurs. Then, step-by-step we explored the interrupt concept. Therefore, we analyzed the interrupt types, interrupt vectors, and ISR. We only considered port-based interrupts here, since we discussed digital I/O concepts previously. In later chapters, we will also see timer, ADC, and digital communication-based interrupts. We provided sample C and assembly codes using interrupts. We also considered the interrupt concept under Grace. Finally, we provided a real-life application using port interrupts.

9.11 Problems

9.1 What is the difference between the interrupt service routine and the interrupt vector?

9.2 Write a C program for the MSP430 that will count the number of times the push button (connected to P1.3 on the MSP430 LaunchPad) is pressed.

a. The button-pressing operation should be defined in an ISR.

b. Observe the count value from the Watch window.

9.3 Repeat Prob. 9.2 in assembly language.

9.4 Expand Prob. 9.2 such that

a. At the beginning of the program, the green LED (connected to P1.6 on the MSP430 LaunchPad) will turn on.
b. When the count reaches multiples of five, the green LED (connected to P1.6 on the MSP430 LaunchPad) will toggle.

9.5 Repeat Prob. 9.4 in assembly language.

9.6 Repeat Prob. 9.4 using Grace.

9.7 Repeat Probs. 8.9 and 8.10 using interrupts.

9.8 Repeat Prob. 8.9 using interrupts under Grace.

9.9 Repeat Probs. 8.12 and 8.13 using interrupts.

9.10 Repeat Prob. 8.12 using interrupts under Grace.

9.11 Write a C program for the MSP430 such that the global integer array `x` with 10 elements will be filled initially. For this problem, fill it at the beginning of the code. When an interrupt comes from the push button (connected to P1.3 on the MSP430 LaunchPad), the ISR will be called. The ISR will calculate the global integer array `y` defined as `y[n] = 2*x[n] - x[n-1]` where `n` is the index for the array. In fact, this is a simple filtering operation working with interrupts. In the actual application, the interrupt should come from some other source. The array `y` should also be filled by an other module or a peripheral (such as ADC).

9.12 Repeat Prob. 9.11 in assembly language.

9.13 Repeat Prob. 9.11 using Grace.

10 Oscillators, Clocks, and Timers

Chapter Outline

In earlier microcontrollers, there was just one clock (supplied by one oscillator) to handle all time-based operations. In modern microcontrollers this approach has been abandoned, and different time-based operations are handled by different clocks (supplied by different oscillators). As a modern microcontroller, the MSP430 also has this property. Therefore, it has three oscillators and three clocks. In this chapter, we will start with the oscillators. The clocks of the MPS430 are managed by the basic clock module+ (BCM+). Next, we will focus on it. The CPU also depends on a clock signal to operate. Therefore, to halt the operation of the CPU, we should disable its clock. This is the main idea behind low-power modes. We will also introduce them in this chapter. Then we will consider the watchdog timer. Finally, we will focus on the Timer_A (TA) module of the MSP430.

10.1 Oscillators

The oscillator is the basic building block of the clock. The MSP430 has three oscillators: the digitally controlled oscillator (**DCO**), very low power oscillator (**VLO**), and low-frequency/high-frequency external oscillator (**LFXT1**). Their properties are briefly listed in Table 10.1.

As can be seen in Table 10.1, the DCO and VLO are based on an internal resistor capacitor (RC)–based circuitry. Therefore the DCO and VLO are cheap and quick to start. Unfortunately, they have poor accuracy. On the other hand, LFXT1 is based on an external crystal, which is expensive and needs a longer time to start. However, crystal-based oscillators are accurate and stable. Internal RC oscillators are sufficient if accurate timing is not required. The oscillators are briefly summarized in the following subsections. How to handle their faulty operations is also explained in a separate subsection.

10.1.1 *Digitally Controlled Oscillator*

The digitally controlled oscillator (DCO) is a high-frequency integrated oscillator. Its frequency can be as high as 16 MHz. The DCO is based on the internal RC circuitry. Hence, it is not accurate as crystal oscillators. On the other hand, the DCO draws less current. Hence, it consumes less energy. The frequency accuracy of the DCO varies in a 2% range.

10.1.2 *Very Low Power Oscillator*

The very low power oscillator (VLO) is a low-frequency internal RC oscillator. Its frequency can be around 12 kHz. The VLO is generally used for periodically waking up the device from low-power modes. The frequency accuracy of the VLO varies in a 5% range. Therefore, it is not as accurate as the DCO.

Table 10.1

Oscillators in the MSP430.

Oscillator	Brief Description	Frequency Range
DCO	Internal RC oscillator	1 MHz, 8 MHz, 12 MHz, 16 MHz
VLO	Internal RC oscillator	12 kHz
LFXT1	External crystal-based oscillator	

10.1.3 *Low-Frequency External Oscillator*

The low-frequency external oscillator (LFXT1) can be used by connecting an external crystal between XIN and XOUT pins. Capacitors should also be connected between these pins and the ground. The MSP430 has internal capacitors for this purpose. External crystal oscillators are more accurate and stable than internal RC-based oscillators. However, they are expensive, consume more energy, and need more time to reach their stable state. Therefore, they should not be preferred unless accurate timing is required.

10.1.4 *Oscillator Faults*

External crystal oscillators may cause errors due to their startup stabilization time or to a failure during operation. When an oscillator fault occurs, system clocks sourced from it also malfunction. In such a case, only the master clock source is switched to the DCO for clocking the CPU. But this process may also cause problems. Fortunately, the CPU can detect these faults through the individual oscillator fault bits such as the oscillator fault interrupt flag (**OFIFG**). Then the oscillator fault can be fixed by software.

The MSP430G2553 has only LFXT1, which uses an external crystal. Therefore, checking the OFIFG is sufficient. The best way to fix an oscillator fault is to use the NMI. The OFIFG calls the NMI handler if the oscillator fault interrupt enable (**OFIE**) bit is set. Then the OFIFG is cleared repeatedly (until it stays cleared) in the associated interrupt service routine (ISR). The OFIE bit is cleared automatically when the NMI is handled. Therefore, it must be set again. For more information on this issue, please see [17].

10.2 Clocks

The MSP430 has three clocks sourced by the oscillators explained in the previous section. These are the master clock (**MCLK**), sub-main clock (**SMCLK**), and auxiliary clock (**ACLK**). Their properties are listed in Table 10.2.

10.2.1 *The Basic Clock Module+*

The clocks of the MSP430 are handled by the basic clock module+ (BCM+). A block diagram of this module is given in Fig. 10.1. As can be seen in this figure, the clock source, type, frequency division ratio, and other properties can be configured. This is done by the dedicated BCM+ registers to be explained next.

Table 10.2

Clocks in the MSP430.

Clock	Used by	Sourced from	Initial Frequency
MCLK	CPU	DCO, LFXT1, VLO	1 MHz
SMCLK	Peripherals	DCO, LFXT1, VLO	1 MHz
ACLK	Peripherals	LFXT1, VLO	

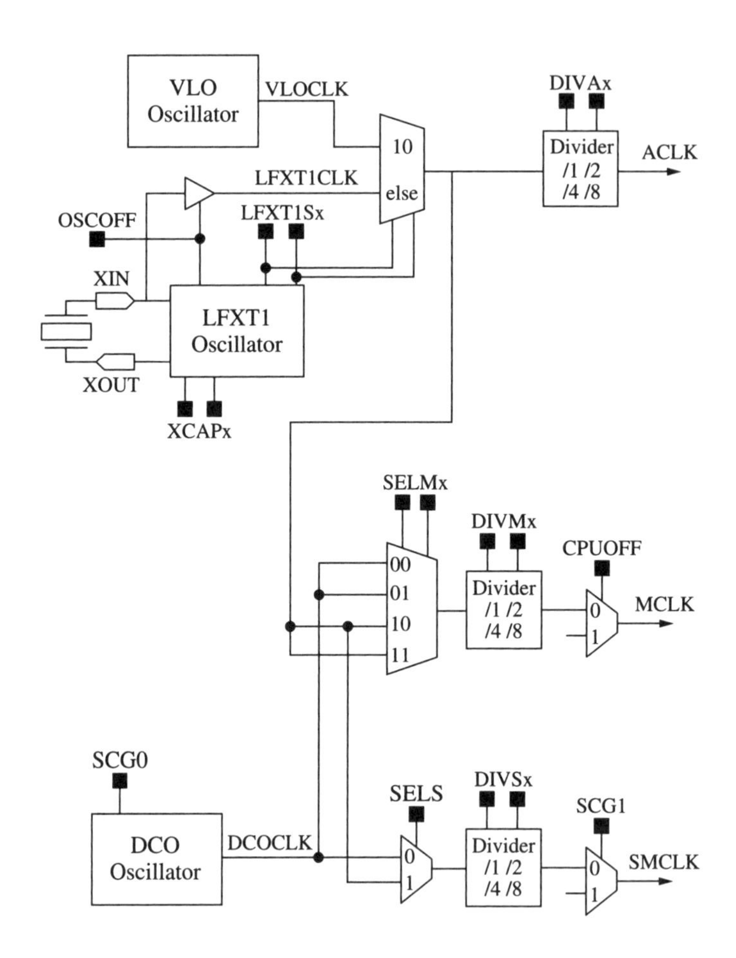

Figure 10.1

Block diagram of the BCM+.

10.2.2 *BCM+ Registers*

There are six dedicated registers to configure the BCM+. These are the DCO control register (**DCOCTL**), basic clock system control register 1 (**BCSCTL1**), basic clock system control register 2 (**BCSCTL2**), basic clock system control register 3 (**BCSCTL3**), interrupt enable register (**IE1**), and interrupt flag register (**IFG1**).

In this book, we will not consider DCOCTL. More information on it can be found in [17]. Instead, we will adjust the frequency of the DCO by predefined values. The MSP430 has four calibrated DCO frequency values: 1 MHz, 8 MHz, 12 MHz, and 16 MHz. These are represented by the constants given in Table 10.3.

The entries of the BCSCTL1 register are shown in Table 10.4. This register is mainly responsible for the auxiliary clock. In Table 10.4, the **XTS** bit is used for LFXT1 mode selection. When it is reset, low-frequency mode is selected. When it is set, high-frequency mode is selected. To note here, setting XTS is not supported for the MSP430G2553. **DIVAx** bits are used for frequency division by 1, 2, 4, and 8 for the ACLK. Constants for these values are DIVA_0, DIVA_1, DIVA_2,

Table 10.3

Calibration codes for the DCO and BCS.

Constant	Register	Frequency (MHz)
CALDCO_1MHZ	DCOCTL	1
CALDCO_8MHZ	DCOCTL	8
CALDCO_12MHZ	DCOCTL	12
CALDCO_16MHZ	DCOCTL	16
CALBC1_1MHZ	BCSCTL1	1
CALBC1_8MHZ	BCSCTL1	8
CALBC1_12MHZ	BCSCTL1	12
CALBC1_16MHZ	BCSCTL1	16

Table 10.4

BCM+ control register 1 (BCSCTL1).

Bits	7	6	5 - 4	3 - 0
		XTS	DIVAx	RSELx

Table 10.5

BCM+ control register 2 (BCSCTL2).

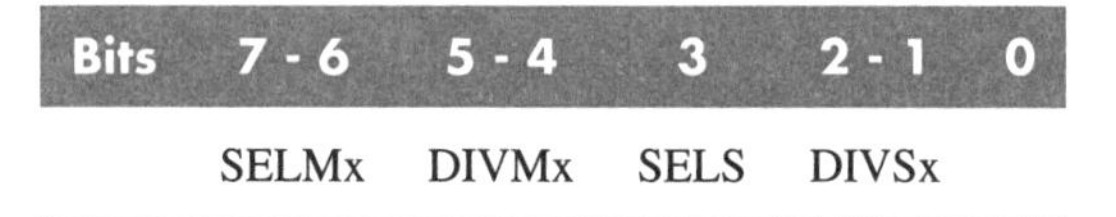

Bits	7 - 6	5 - 4	3	2 - 1	0
	SELMx	DIVMx	SELS	DIVSx	

and DIVA_3 respectively. **RSELx** bits are used for the DCO. Hence, we will not explain them here.

The entries of the BCSCTL2 register are shown in Table 10.5. This register is mainly responsible for the master and sub-main clock. In Table 10.5, **SELMx** bits select the MCLK oscillator. Constants for these bits are SELM_0 and SELM_1 (for DCOCLK), SELM_2 and SELM_3 (for VLOCLK or LFXTCLK). **DIVMx** bits are used for frequency division by 1, 2, 4, and 8 for the MCLK. Constants for these values are DIVM_0, DIVM_1, DIVM_2, and DIVM_3 respectively. The **SELS** bit is used for the SMCLK oscillator. When this bit is reset, DCOCLK is used. When it is set, LFXT1CLK or VLOCLK is used. **DIVSx** bits are used for frequency division by 1, 2, 4, and 8 for the SMCLK. Constants for these values are DIVS_0, DIVS_1, DIVS_2, and DIVS_3 respectively.

The entries of the BCSCTL3 register are shown in Table 10.6. This register is mainly responsible for the external oscillator and clock. In Table 10.6, **LFXT1Sx** bits are used for low-frequency clock and range select. Constants for these bits are LFXT1S_0 (for external crystal), LFXT1S_1 (reserved), LFXT1S_2 (for VLO),

Table 10.6

BCM+ control register 3 (BCSCTL3).

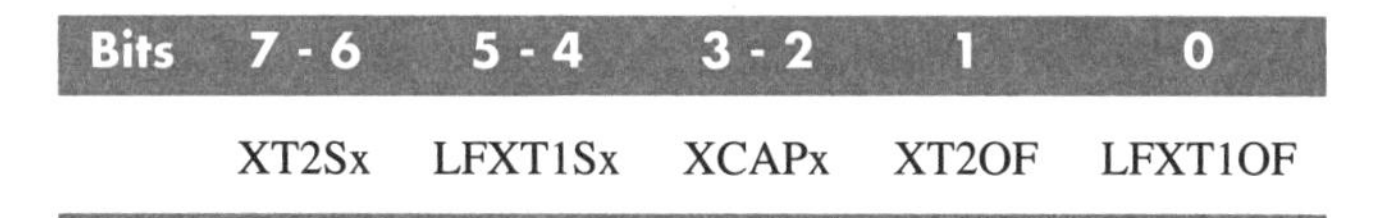

Bits	7 - 6	5 - 4	3 - 2	1	0
	XT2Sx	LFXT1Sx	XCAPx	XT2OF	LFXT1OF

and LFXT1S_3 (for digital input signal). The MSP430G2553 does not have an XT2 oscillator. Therefore, **XT2Sx** and **XT2OF** bits are not used. **XCAPx** bits are used for selecting internal capacitors for external crystal oscillator. Here, 1-pF, 6-pF, 10-pF, or 12.5-pF capacitor can be chosen. If an external clock source will be used for the system, these bits must be reset. **LFXT1OF** bit represents whether an oscillator fault is present or not.

10.2.3 Coding Practices for the BCM+ Module

In Listing 10.1, we provide a code block for adjusting clocks. In the first two lines of the code, the VLO is set to produce a 3-kHz ACLK clock signal. In the third line of the code, LFTX1 with a 10-pF internal capacitor is used to produce a 32-kHz ACLK clock signal.

Listing 10.1 The C code block for adjusting clocks.

```
BCSCTL3 |= LFXT1S_2; // ACLK from VLO
BCSCTL1 |= DIVA_2; // Divide ACLK frequency by four

BCSCTL3 |= LFXT1S_0|XCAP_3;
// ACLK from crystal oscillator,
// with 10 pF internal capacitors
```

10.3 BCM+ in Grace

In the Device Overview window of Grace (given in Fig. 5.11), the **Oscillators Basic Clock System+** block is used to configure the BCM+. The "Enable Clock in my configuration" box should be checked first to use the BCM+ under Grace. As a note, the BCM+ is called Basic Clock System+ in Grace. As in the previous blocks, there are three options for this configuration. We will consider each next.

10.3.1 The Basic User Mode

The basic user mode is shown in Fig. 10.2 for the BCM+. In this mode, high- and low-speed clock frequencies for the CPU and other peripherals can be set separately. Calibrated clock frequencies (1 MHz, 8 MHz, 12 MHz, and 16MHz) can be used for the high-speed clock source. The other option is entering the desired frequency into the "manually configure" box. The closest producible frequency value will be generated. A 12-kHz VLO or 32-kHz crystal frequency values can be chosen for the low-speed clock source.

10.3.2 The Power User Mode

In the power user mode of the BCM+ (shown in Fig. 10.3), system clocks can also be configured. The high-speed clock source is specified as the DCO. It can be disabled from the "Disable DCO" check box if it is not used. For the low-speed clock source, 12-kHz VLO, 32-kHz crystal or an external digital source (by marking the related check box) can be selected from the Select Clock Source drop-down list. If the external crystal is used for low-speed clock source, internal capacitor values should also be determined from the Int. Load Eff. Capacitance

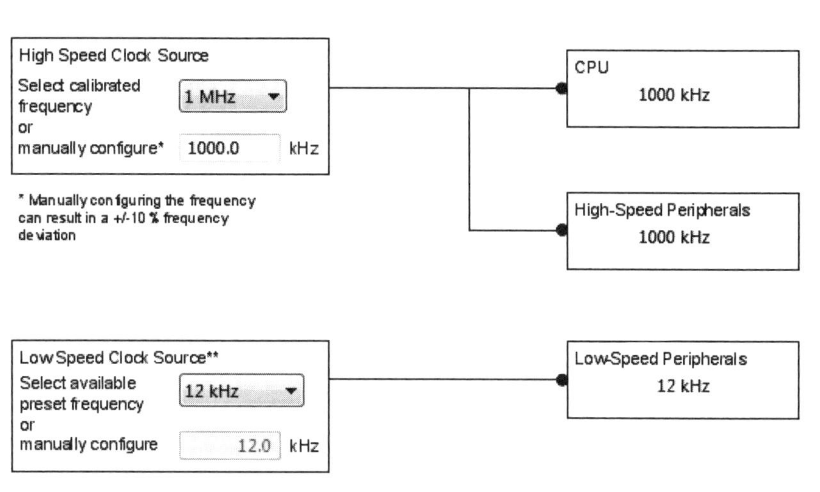

Figure 10.2

Basic user mode for the BCM+.

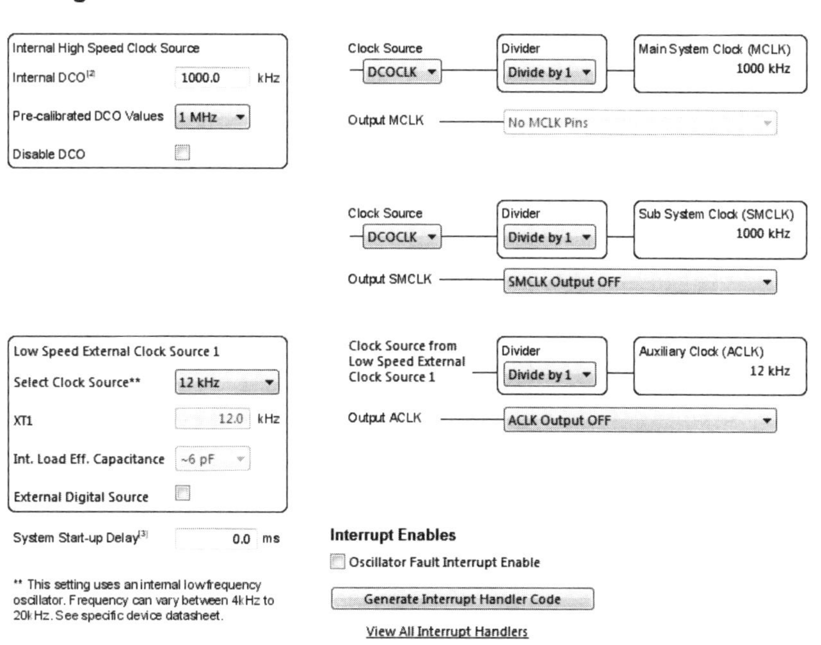

Figure 10.3

Power user mode for the BCM+.

drop-down list. An initial delay, in terms of milliseconds, can be added to the configured clock from the System Start-up Delay box. Then, the configured clock source can be used as the system clock as MCLK or SMCLK. ACLK is always sourced from the Low-Speed External Clock Source 1. Frequency division can also be applied to the clocks in this mode by the associated "Divider" drop-down list. The SMCLK and ACLK can also be fed to the related output pins by the Output SMCLK and Output ACLK drop-down lists. The oscillator fault interrupt can also be set by checking the Oscillator Fault Interrupt Enable box. Then, the prototype ISR can be generated under the InterruptVectors_init.c file by pressing the Generate Interrupt Handler Code button.

10.3.3 *The Register Controls Mode*

Finally, the register controls mode can be used to configure the BCM+ registers. The register controls mode is shown in Fig. 10.4. In these registers, some bits are disabled since the user cannot change them. There are also drop-down lists to adjust some register entries.

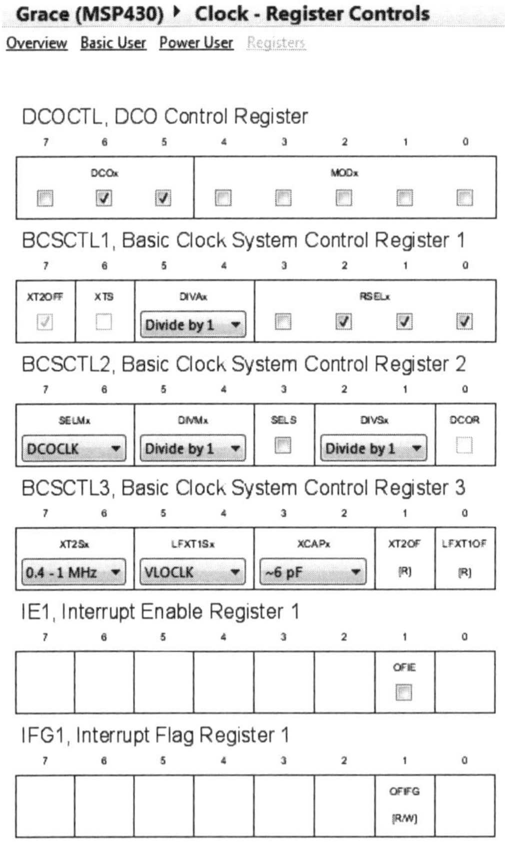

Figure 10.4 The register controls mode for the BCM+.

Grace (MSP430) ▸ Clock - Register Controls
Overview Basic User Power User Registers

DCOCTL, DCO Control Register
DCOx | MODx

BCSCTL1, Basic Clock System Control Register 1
XT2OFF | XTS | DIVAx: Divide by 1 | RSELx

BCSCTL2, Basic Clock System Control Register 2
SELMx: DCOCLK | DIVMx: Divide by 1 | SELS | DIVSx: Divide by 1 | DCOR

BCSCTL3, Basic Clock System Control Register 3
XT2Sx: 0.4 - 1 MHz | LFXT1Sx: VLOCLK | XCAPx: ~6 pF | XT2OF (R) | LFXT1OF (R)

IE1, Interrupt Enable Register 1
OFIE

IFG1, Interrupt Flag Register 1
OFIFG (R/W)

10.4 Low-Power Modes

Power consumption is a critical feature for modern battery-controlled devices. Therefore, modern microcontrollers are designed to work in low-power modes (LPMs). The MSP430 has one active and five low-power modes, as listed below.

- **Active mode (AM)**: CPU, all clocks and enabled peripheral modules are active. Draws about 230-μA current.
- **Low-power mode 0 (LPM0)**: CPU and MCLK are disabled. SMCLK and ACLK are still active. Draws about 56-μA current.
- **Low-power mode 1 (LPM1)**: CPU and MCLK are disabled. SMCLK and ACLK are still active. DCO is disabled if it is not used.
- **Low-power mode 2 (LPM2)**: CPU, MCLK, and SMCLK are disabled. ACLK and DCO remain active. Draws about 22-μA current.
- **Low-power mode 3 (LPM3)**: CPU, MCLK, SMCLK, and DCO are disabled. ACLK remains active. Draws about 0.5-μA current. This is also called the *standby mode*.
- **Low-power mode 4 (LPM4)**: CPU, all clocks, and the crystal oscillator are disabled. Only RAM is retained. Draws about 0.1-μA current. This is also called the *off mode*.

In each mode, only necessary modules (peripherals and the CPU) are active. This is achieved by disabling and enabling clocks feeding the modules. We provide the effect of each low-power mode on the SR bits in Table 10.7.

Since various operations should be done to enter or exit a low-power mode, there are predefined constants in the MSP430 header file. In Table 10.8, constants for entering low-power modes are tabulated.

Similarly, in Table 10.9, constants for exiting low-power modes are tabulated. These can be directly used in C and assembly programs.

Some points should be taken into account when using low-power modes with interrupts. First, if the LPM_EXIT command is not entered in the ISR, the CPU turns back to the code line where the interrupt is generated. This property can be used to form an infinite loop in time-based operations. We will provide examples

Table 10.7

The effect of the low-power modes on the SR bits.

	SR Bit			
Mode	SCG1	SCG0	OSCOFF	CPUOFF
Active	0	0	0	0
LPM0	0	0	0	1
LPM1	0	1	0	1
LPM2	1	0	0	1
LPM3	1	1	0	1
LPM4	1	1	1	1

Table 10.8

Predefined constants for entering LPM.

Constant	Description
LPM0	Enter Low-power mode 0
LPM1	Enter Low-power mode 1
LPM2	Enter Low-power mode 2
LPM3	Enter Low-power mode 3
LPM4	Enter Low-power mode 4

Table 10.9

Predefined constants for exiting LPM.

Constant	Description
LPM0_EXIT	Exit Low-power mode 0
LPM1_EXIT	Exit Low-power mode 1
LPM2_EXIT	Exit Low-power mode 2
LPM3_EXIT	Exit Low-power mode 3
LPM4_EXIT	Exit Low-power mode 4

in Sec. 10.7.4 on this issue. Second, some problems may occur while exiting from low-power modes in an ISR in assembly programming. This is mainly because of the operation of the CPU. Since the CPU saves all the data in the stack while handling the ISR, this point should be taken into account. For example, we should use `bic.w #LPM0,0(SP)` to exit from LPM0 in an ISR. We will provide examples in Sec. 10.7.4 on this issue.

10.5 The Watchdog Timer

The watchdog timer resets the system periodically unless disabled before generating the reset signal. This operation aims to eliminate any undesired infinite loops in operation due to software failure. The watchdog timer can also be used as a timer that can generate periodic interrupts. The watchdog timer module is specifically called the Watchdog Timer+ (WDT+) in the MSP430. The layout of the WDT+ module is given in Fig. 10.5.

The WDT+ is controlled by a 16-bit register called the **WDTCTL**. The entries for this register are given in Table 10.10. Here, the **WDTPW** bits are used for entering the password. To stop the reset signal (power up clear, PUC), 05Ah should be written to it. When the **WDTHOLD** bit is set, the WDT+ is stopped. The **WDTNMIES** bit sets the WDT+ non-maskable interrupt edge select. When this bit is reset, the interrupt is generated on the rising edge. When it is set, the interrupt is generated on the falling edge. The **WDTNMI** bit is used to select a reset or a non-maskable interrupt. Since the default work of the WDT+ is to periodically reset the CPU, this bit is initially reset. If this bit is set (for non-maskable

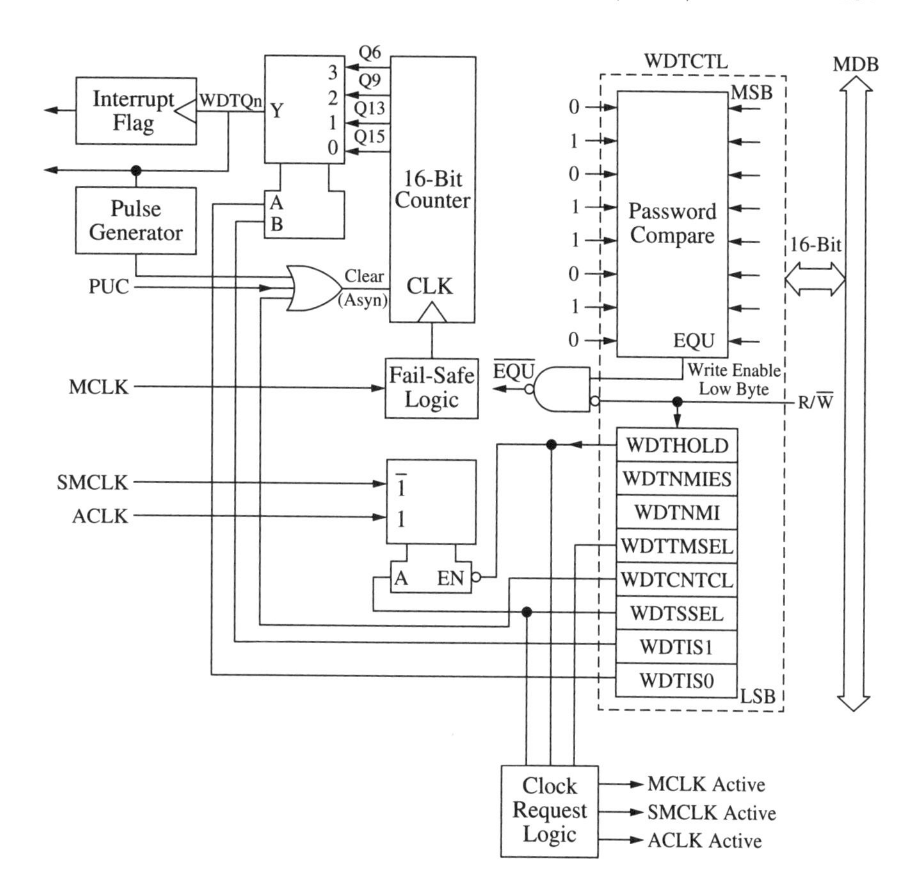

Figure 10.5

Block diagram of the WDT+ module.

interrupt generation), the NMIIE bit inside the IE1 register should be set at the same time. The **WDTTMSEL** is used as the mode selection bit. When this bit is set, the WDT+ can be used as an interval timer (without any watchdog operation). When it is reset, the WDT+ is used as a watchdog. The **WDTCNTCL** bit clears the watchdog counter to 0000h. The **WDTSSEL** bit selects the WDT+ clock source. When this bit is reset, SMCLK is used. When it is set, ACLK is used. The **WDTISx** bits are used for WDT+ interval select (both for watchdog and timer operations). Assigning binary values 00, 01, 10, and 11 to these will lead to the division of the WDT+ clock source by 2^{15} (32,768), 2^{13} (8192), 2^9 (512), and 2^6 (64) respectively. In a way, they work as frequency dividers. The watchdog timer has a specific counter called **WDTCNT**. It cannot be reached by software.

Table 10.10

WDT+ control register (WDTCTL).

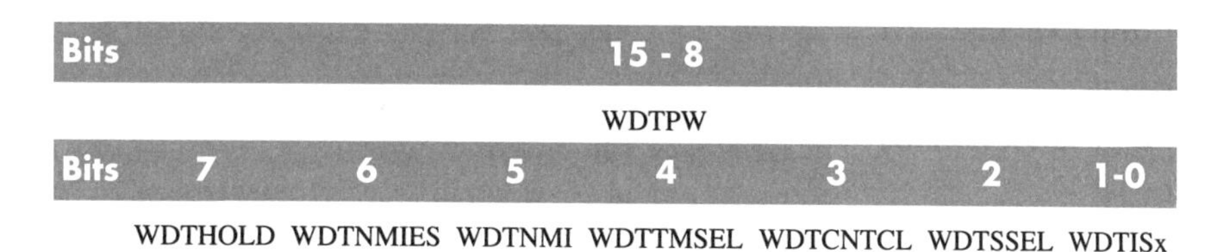

Bits	15 - 8						
	WDTPW						
Bits	**7**	**6**	**5**	**4**	**3**	**2**	**1-0**
	WDTHOLD	WDTNMIES	WDTNMI	WDTTMSEL	WDTCNTCL	WDTSSEL	WDTISx

Table 10.11

WDT+ constants when SMCLK (1 MHz) is used.

Shortcut	Setting	Time (ms)
WDT_MRST_32	WDTPW+WDTCNTCL	32.000
WDT_MRST_8	WDTPW+WDTCNTCL+WDTIS0	8.000
WDT_MRST_0_5	WDTPW+WDTCNTCL+WDTIS1	0.500
WDT_MRST_0_064	WDTPW+WDTCNTCL+WDTIS1+WDTIS0	0.064

10.5.1 *WDT+ Used as a Watchdog*

The WDT+ is activated when the system is powered up or reset. The WDT+ should be disabled when it is not used. This is done by setting the WDTHOLD bit. The C code for this operation is `WDTCTL = WDTPW + WDTHOLD;` or `WDTCTL = WDTPW|WDTHOLD;`. The assembly code for this operation is `mov.w #WDTPW+WDTHOLD,WDTCTL`. In fact, we have been using one of these lines in all our previous C and assembly codes. Now they should make sense. As a matter of fact, we never asked the watchdog timer to operate in our previous codes.

The MSP430 header file has predefined constants for the time intervals of the watchdog timer. They are given in Tables 10.11 and 10.12. In the first table, the clock source for the WDT+ is selected as SMCLK (at 1 MHz). In the second table, the clock source is selected as ACLK (at 32 kHz).

10.5.2 *WDT+ Used as an Interval Timer*

The WDT+ can also be used as an interval timer by setting the WDTTMSEL bit. When the WDT+ is used in the timer mode, a periodic interrupt will be generated instead of the system reset signal. This interrupt is controlled by the WDTIE bit in the IE1 register. This bit must be set in order to request an interrupt. In this mode, the WDT+ interrupt is maskable. Therefore, the global interrupt enable (GIE) bit also must be set. The occurrence of the interrupt can be observed by the watchdog timer interrupt flag (WDTIFG bit in the IFG1 register). This bit is set when the WDTCNT reaches its limit. WDTIF is automatically reset after the ISR is performed.

As in the watchdog timer mode, there are predefined time interval constants for the WDT+ used in the interval timer mode. These are given in Tables 10.13 and 10.14. As in the previous tables, here the clock source for the WDT+ is selected as either SMCLK or ACLK.

Table 10.12

WDT+ constants when ACLK (32 kHz) is used.

Shortcut	Setting	Time (ms)
WDT_ARST_1000	WDTPW+WDTCNTCL+WDTSSEL	1000.0
WDT_ARST_250	WDTPW+WDTCNTCL+WDTSSEL+WDTIS0	250.0
WDT_ARST_16	WDTPW+WDTCNTCL+WDTSSEL+WDTIS1	16.0
WDT_ARST_1_9	WDTPW+WDTCNTCL+WDTSSEL+WDTIS1+WDTIS0	1.9

Table 10.13

WDT+ constants when used in the timer mode with SMCLK (1 MHz).

Shortcut	Setting	Time (ms)
WDT_MDLY_32	WDTPW+WDTTMSEL+WDTCNTCL	32.000
WDT_MDLY_8	WDTPW+WDTTMSEL+WDTCNTCL+WDTIS0	8.000
WDT_MDLY_0_5	WDTPW+WDTTMSEL+WDTCNTCL+ WDTIS1	0.500
WDT_MDLY_0_064	WDTPW+WDTTMSEL+WDTCNTCL+WDTIS10 +WDTIS	0.064

Table 10.14

WDT+ constants when used in the timer mode with ACLK (32 kHz).

Shortcut	Setting	Time (ms)
WDT_ADLY_1000	WDTPW+WDTTMSEL+WDTCNTCL+WDTSSEL	1000.0
WDT_ADLY_250	WDTPW+WDTTMSEL+WDTCNTCL+WDTSSEL+WDTIS0	250.0
WDT_ADLY_16	WDTPW+WDTTMSEL+WDTCNTCL+WDTSSEL+WDTIS1	16.0
WDT_ADLY_1_9	WDTPW+WDTTMSEL+WDTCNTCL+WDTSSEL+WDTIS1 +WDTIS0	1.9

10.5.3 *Coding Practices for the WDT+ Module*

We first provide the C code in which the WDT+ is used as a watchdog in Listing 10.2. In this code, initially the WDT+ and the red LED are off. The red LED turns on and the WDT+ starts to run as we press the push button. We use the VLO for the WDT+. This gives a 2.8 s delay. The program counter goes to `main` as the WDT+ resets the microcontroller. Then, the WDT+ and the red LED are turned off. The system waits for another button press to repeat the procedure. In Listing 10.3, we provide the assembly code doing the same job.

Listing 10.2 Usage of the WDT+ in watchdog mode in C.

```
#include <msp430.h>

#define RedLED BIT0
#define Button BIT3

void main(void)
{
 WDTCTL = WDTPW|WDTHOLD;

 P1DIR = RedLED;
 P1OUT = 0x00;
 P1IE = Button;
 P1IES = Button;
 P1IFG = 0x00;

 _enable_interrupts();

 LPM4;
}

#pragma vector=PORT1_VECTOR
__interrupt void PORT1_ISR(void){
 P1OUT = RedLED;
```

```
 BCSCTL3 |= LFXT1S_2;
 WDTCTL = WDT_ARST_1000;
 P1IFG = 0x00;
}
```

Listing 10.3 Usage of the WDT+ in Watchdog Mode in assembly.

```
 .cdecls C,LIST,"msp430.h"

 .text
 .retain
 .retainrefs

RESET
 mov.w #WDTPW|WDTHOLD,WDTCTL
 mov.w #__STACK_END,SP

 mov.b #01h,P1DIR
 mov.b #00h,P1OUT
 bis.b #08h,P1IE
 bis.b #08h,P1IES
 bic.b #08h,P1IFG

 bis.w #LPM4+GIE,SR

;------------------------------------------------
P1_ISR ;Set P1.0 high, Start WDT in watchdog mode
;------------------------------------------------
 mov.b #01h,P1OUT
 mov.w #LFXT1S_2,BCSCTL3 ;12 Khz VLO as ACLK source
 mov.w #WDT_ARST_1000,WDTCTL
;Use WDT as watchdog to reset the system
;after 1/(12000/32768) = approx. 2.8 sec.
 bic.b #08h,P1IFG
 reti

;------------------------
;Stack Pointer definition
;------------------------
 .global __STACK_END
 .sect .stack

;-------------------
;Interrupt Vectors
;-------------------
 .sect RESET_VECTOR
 .short RESET
 .sect PORT1_VECTOR
 .short P1_ISR
 .end
```

In Listing 10.4, we provide a sample code for the usage of the WDT+ in timer mode. Here, the red and green LEDs toggle every 256 msec by using the SMCLK (divided by eight). The sample code in Listing 10.5 does the same job in assembly language.

Listing 10.4 Usage of the WDT+ in timer mode in C.

```
#include <msp430.h>

#define RedLED BIT0
#define GreenLED BIT6
#define ToggleLeds (P1OUT ^= RedLED|GreenLED)

void main(void)
{
 BCSCTL2 |= DIVS_3;
 WDTCTL = WDT_MDLY_32;
 IE1 |= WDTIE;

 P1DIR = RedLED|GreenLED;
 P1OUT = RedLED;

 _enable_interrupts();

 LPM1;
}

#pragma vector=WDT_VECTOR
__interrupt void WDT(void){
 ToggleLeds;
}
```

Listing 10.5 Usage of the WDT+ in timer mode in assembly.

```
 .cdecls C,LIST,"msp430.h"

 .text
 .retain
 .retainrefs

RESET
 mov.w #__STACK_END,SP

 mov.w #DIVS_3,BCSCTL2 ;(SMCLK frequency)/8
 mov.w #WDT_MDLY_32,WDTCTL
;Use WDT as 32x8 = 256ms time interval
 bis.b #WDTIE,IE1 ;Enable WDT interrupt
 mov.b #41h,P1DIR
 mov.b #01h,P1OUT

 bis.w #LPM1+GIE,SR ;Enable interrupts and LPM

;-----------------------------
```

```
WDT_ISR ;Toggle P1.0 and P1.6
;----------------------------
 xor.b #41h,P1OUT
 reti

;------------------------
;Stack Pointer definition
;------------------------
 .global __STACK_END
 .sect .stack

;-------------------
;Interrupt Vectors
;-------------------
 .sect RESET_VECTOR
 .short RESET
 .sect WDT_VECTOR
 .short WDT_ISR
 .end
```

10.6 WDT+ in Grace

The WDT+ configurations under Grace can be done by clicking the **Watchdog WDT+ block** in the Device Overview window (given in Fig. 5.11). This block should also be enabled first by checking the "Enable WDT+ in my configuration" box. Then, it can be configured by three modes as follows.

10.6.1 *The Basic User Mode*

In the basic user mode of the WDT+ (shown in Fig. 10.6), the WDT+ modes can be selected. These modes are Stop Watchdog Timer, Interval Timer Mode, and Watchdog Timer Mode. When one of the last two modes is selected, a new menu appears in the same window. Here, the clock source can be selected from the Clock Source drop-down list. The frequency divider for this clock source can also be set from the Divider drop-down list. As a reminder, the frequency value for the selected clock source can be changed from the BCM+ module. The WDT+ interrupt can be enabled by checking the WDT+ Interrupt Enable box. Then the WDT+ based ISR prototype can be generated in InterruptVectors_init.c by pressing the Generate Interrupt Handler Code button.

10.6.2 *The Power User Mode*

There is only an extra RST/NMI Pin Configuration menu in the power user mode as shown in Fig. 10.7. The function of this pin can be defined as reset or NMI. This interrupt can be enabled by checking the NMI Pin Interrupt Enable box. The user can set the signal edge to trigger the interrupt. Then the ISR prototype can be generated in InterruptVectors_init.c by pressing the related Generate Interrupt Handler Code button. All other configurations are the same as those provided by the basic user mode.

Figure 10.6

The basic user mode for the WDT+.

Grace (MSP430) ▸ WDT+ - Basic User Mode

Overview Basic User Power User Registers

Interrupt Enables

WDT+ Interrupt Enable

Generate Interrupt Handler Code

View All Interrupt Handlers

Figure 10.7

The power user mode for the WDT+.

Grace (MSP430) ▸ WDT+ - Power User Mode

Overview Basic User Power User Registers

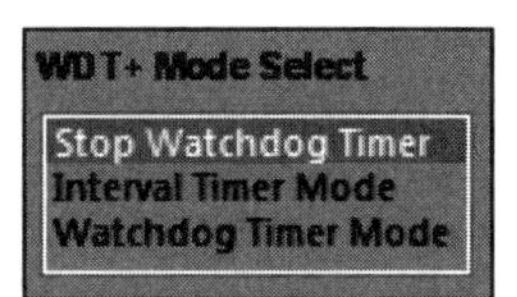

Interrupt Enables

WDT+ Interrupt Enable

Generate Interrupt Handler Code

View All Interrupt Handlers

RST/NMI Pin Configuration

RST/NMI Pin Functionality:

Reset function
NMI function

Interrupt Enables

NMI Pin Interrupt Enable

Generate Interrupt Handler Code

NMI Edge Select:

NMI Rising Edge
NMI Falling Edge

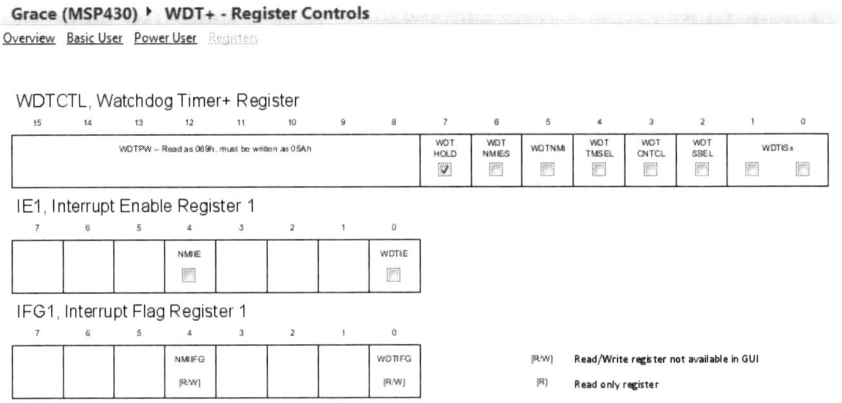

Grace (MSP430) ‣ WDT+ - Register Controls

Overview Basic User Power User Registers

WDTCTL, Watchdog Timer+ Register

15–8	7	6	5	4	3	2	1–0
WDTPW – Read as 069h, must be written as 05Ah	WDT HOLD	WDT NMIES	WDTNMI	WDT TMSEL	WDT CNTCL	WDT SSEL	WDTISx

IE1, Interrupt Enable Register 1

7	6	5	4	3	2	1	0
			NMIIE				WDTIE

IFG1, Interrupt Flag Register 1

7	6	5	4	3	2	1	0
			NMIIFG [R/W]				WDTIFG [R/W]

[R/W] Read/Write register not available in GUI

[R] Read only register

Figure 10.8 The register controls mode for the WDT+.

10.6.3 The Register Controls Mode

Finally, the register controls mode of the WDT+ module is shown in Fig. 10.8. As in the previous section, the WDT+ registers can be directly configured in this mode.

10.6.4 Coding Practices

In this section, we redo the WDT+ time interval interrupt application given in Listing 10.4 using Grace. As a reminder, this application toggles the red and green LEDs (connected to P1.0 and P1.6 on the MSP430 LaunchPad) every 2.8 s. We start by generating a Grace project. Then we configure the pins P1.0 and P1.6 under Grace. Both pins should be set as GPIO output. Moreover, the red LED should be initially set. The green LED should be initially reset. Do not forget to make necessary adjustments on the BCS+ module. The power user mode of the GPIO block should be used for this purpose. We should select the Interval Timer Mode from the WDT+ Mode Select list under the basic user mode. In the same tab, we should select the clock source as "low-speed clock" and the divider as /32768. To generate interrupts, we should check the WDT+ Interrupt Enable box.

In this application, we do not add any code lines to the main.c file. Since we are using the ISR, we generate an ISR prototype using the Generate Interrupt Handler Code button. We fill the prototype WDT+ ISR block under InterruptVectors_init.c as given in Listing 10.6. After compiling the project, we can run our application. The power user mode is not very different for the WDT+. Therefore, we did not give an example on its usage here.

Listing 10.6 The WDT+ timer mode under Grace in basic user mode.

```
#pragma vector=WDT_VECTOR
__interrupt void WDT_ISR_HOOK(void)
{
 P1OUT ^= BIT0|BIT6;
}
```

10.7 Timers

The MSP430 timer module is called the Timer_A. In fact, the MSP430G2553 has two identical Timer_A modules called TA0 and TA1. The first Timer_A module (TA0) is set as the default timer. This timer is also called TA in the header file definitions. Therefore, we will use TA instead of TA0 throughout this book. In Sec. 10.7.4, we also provide a sample code using both TA0 (TA) and TA1.

A block diagram of the TA module is given in Fig. 10.9. As can be seen in this figure, there are two blocks under TA, the timer and capture/compare. The capture/compare block is also divided into three subblocks as capture/compare blocks 0, 1, and 2. These blocks have the same characteristics. Therefore, they are

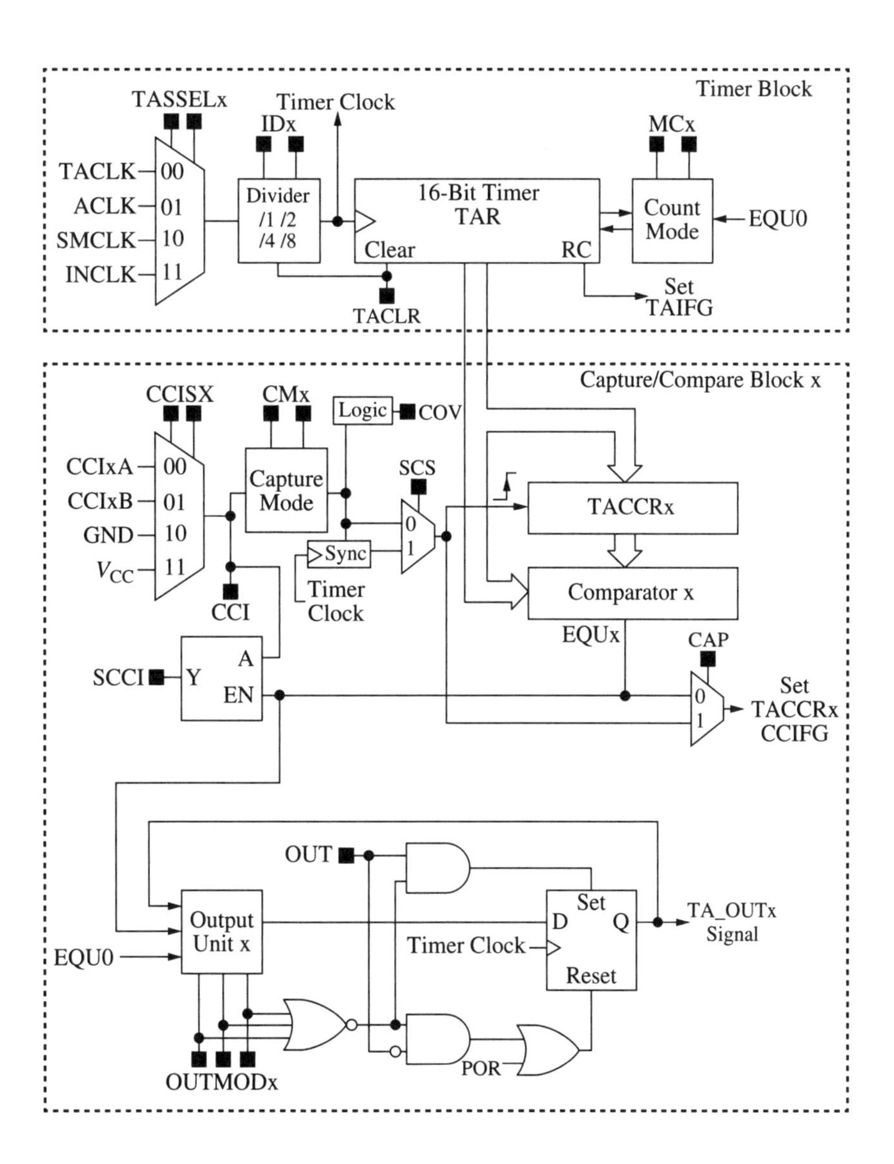

Figure 10.9 Block diagram of the Timer_A module.

Table 10.15

Timer_A control register (TACTL).

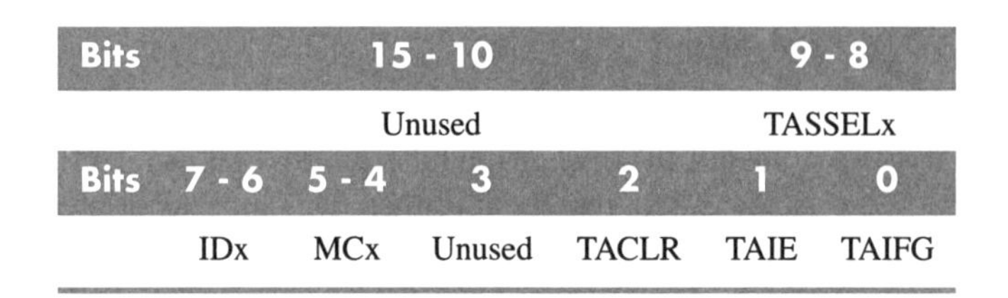

Bits	15 - 10	9 - 8
	Unused	TASSELx

Bits	7 - 6	5 - 4	3	2	1	0
	IDx	MCx	Unused	TACLR	TAIE	TAIFG

represented as capture/compare block x in Fig. 10.9. In the following sections, we will focus on the timer and the capture/compare blocks separately.

10.7.1 *The Timer Block*

The core of the timer block is the 16-bit **TAR** register. Timer count results are kept in this register. The timer block of TA is controlled by the **TACTL** control register. Properties of this register are given in Table 10.15.

In Table 10.15, the **TASSELx** bits are used to select the clock source for TA. Constants for these bits are TASSEL_0, TASSEL_1, TASSEL_2, and TASSEL_3. They correspond to TACLK, ACLK, SMCLK, and INCLK (inverse of the TACLK) as the clock source for TA. **IDx** bits are used for frequency division. Constants for these bits are ID_0, ID_1, ID_2, and ID_3. They correspond to frequency division by 1, 2, 4, and 8 respectively. When the **TACLR** bit is set, the TAR, the clock divider, and the count direction are reset. But resetting the clock divider and count direction does not mean resetting IDx and MCx bits. Resetting the clock divider means current prescaler counter is reset to 0. Resetting the count direction means if TAR in counting down part in up/down mode, it is reset to counting up part. The **TAIE** bit is used to enable the Timer_A interrupt. When an interrupt comes from the timer module, the **TAIFG** bit is set. The **MCx** bits are used for selecting the mode of the timer. Constants for these bits are MC_0, MC_1, MC_2, and MC_3. They correspond to *stop*, *up*, *continuous*, and *up/down* modes. These are listed below.

- **Stop Mode:** The timer stops counting and TAR retains its value to continue later when this mode is selected. Timer_A is initially in this mode to save power.
- **Continuous Mode:** The timer counts up until it reaches FFFFh (65535), then restarts from zero again as shown in Fig. 10.10. The TAIFG bit is set when the TAR value changes from FFFFh to zero. The time period for this mode can be calculated as period = $65536/f_{CLK}$ where f_{CLK} stands for the frequency of the timer clock. The continuous mode is generally used for generating output with different frequencies or independent time intervals. In this mode, four different output frequencies or time intervals can be produced by using three capture/compare and TAR register entries.
- **Up Mode:** The timer counts up until it reaches the value in TACCR0 (to be explained in Sec. 10.7.2) in this mode. Then it restarts from zero again as shown in Fig. 10.11. The TAIFG bit is set when the TAR value changes from TACCR0 to zero. Also, the capture/compare interrupt flag (CCIFG) bit is set when the

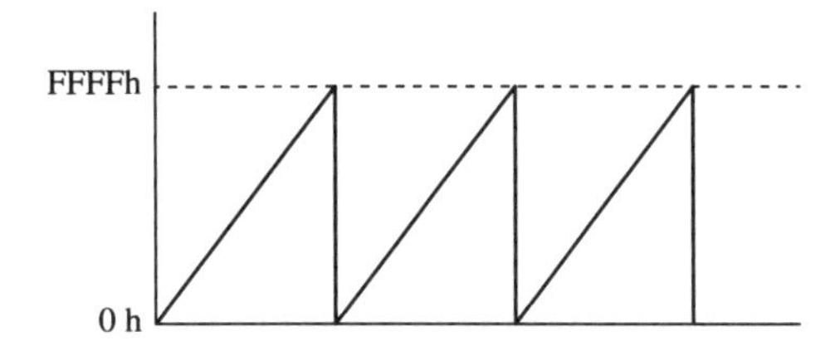

Figure 10.10

Timer_A continuous mode.

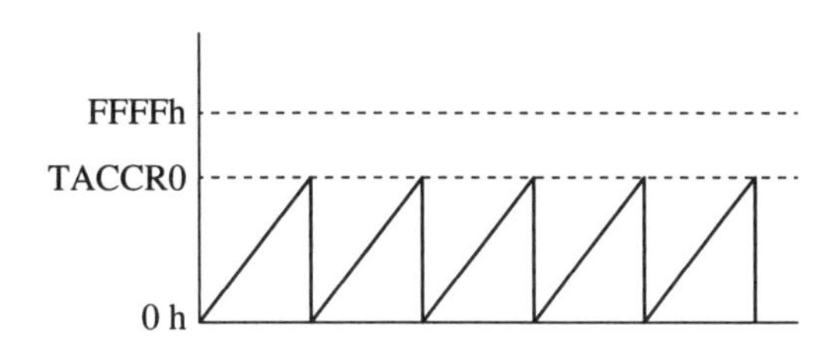

Figure 10.11

Timer_A up mode.

TAR value changes from TACCR0-1 to TACCR0. The timer period for this mode can be calculated as period = (TACRR0+1)/f_{CLK}.

- **Up/Down Mode:** In this mode, first the timer counts up until it reaches the value in the TACCR0. Then counting is inverted, and the timer counts down from TACCR0 to zero as shown in Fig. 10.12. The TAIFG bit is set when the TAR value changes from one to zero in counting down. Also, the CCIFG bit is set when the TAR value changes from TACCR0-1 to TACCR0 in counting up. The timer period for this mode can be calculated as period = (2×TACRR0)/f_{CLK}.

10.7.2 *The Capture/Compare Block*

Timer_A has three capture/compare blocks, 0, 1, and 2. The capture/compare block 0 can also be used by the timer module in counting up or up/down modes. Therefore, the user should be careful when using it. Each capture/compare block is controlled by a separate 16-bit control register **TACCTLx**. The entries of this register are given in Table 10.16.

In Table 10.16, the **CMx** bits are used to select the edge sensitivity in the capture mode. The constants for these bits are CM_0 (no capture), CM_1 (capture on

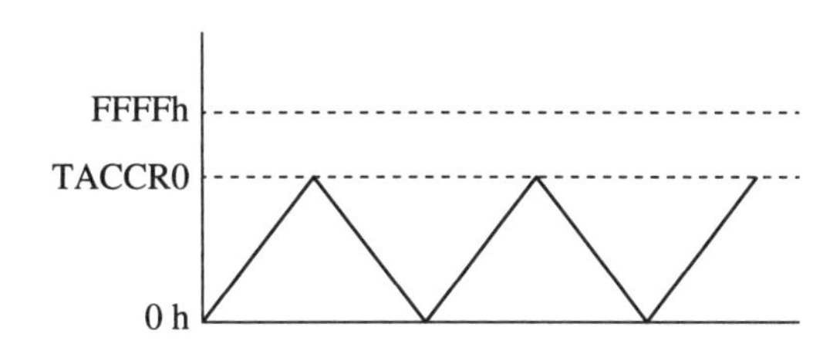

Figure 10.12

Timer_A up/down mode.

Table 10.16 Timer_A capture/compare control register (TACCTLx).

Bits	15-14	13-12	11	10	9	8
	CMx	CCISx	SCS	SCCI		CAP
Bits	**7-5**	**4**	**3**	**2**	**1**	**0**
	OUTMODx	CCIE	CCI	OUT	COV	CCIFG

rising edge), CM_2 (capture on falling edge), and CM_3 (capture on both edges). The **CCISx** bits are used to select the capture/compare input select. They are for external pins and internal signals. (These are listed in Table 10.18.) Constants for the CCISx bits and their values are CCIS_0 (CCIxA), CCIS_1 (CCIxB), CCIS_2 (GND), and CCIS_3 (V_{CC}). The **SCS** bit is used to synchronize the timer clock and the capture signal (to eliminate the race condition). The **SCCI** bit is used to observe the synchronized input. The **CAP** bit is used to select the capture or compare mode. When this bit is reset, the compare mode is selected. When it is set, the capture mode is selected. The CAP bit is initially in the compare mode. The **OUTMODx** bits are used to select the output modes for the compare operation. Constants and their values are OUTMOD_0 (OUT bit value), OUTMOD_1 (set), OUTMOD_2 (toggle/reset), OUTMOD_3 (set/reset), OUTMOD_4 (toggle), OUTMOD_5 (reset), OUTMOD_6 (toggle/set), and OUTMOD_7 (reset/set). These modes will be explained in detail next. The **CCIE** bit is used to enable the capture/compare interrupt. The **CCI** bit is used to observe the capture/compare input. The **OUT** bit (when in OUTMOD_0) directly controls the output. When this bit is reset, the output is low. When it is set, the output is high. The **COV** bit indicates whether a capture overflow has occurred or not. It should be cleared by software to observe a new overflow. The **CCIFG** is the capture/compare interrupt flag.

For each capture/compare block, there is also a separate **TACCRx** register. This register holds the data for the comparison of the timer value in the TAR in compare mode. In capture mode, the TAR value is copied to this register when a capture is performed.

The Capture Mode

The purpose of the capture mode is to link the changes in the input signal with TAR values. We should first set the CAP bit of the TACCTLx register to use this mode. Then the input signal source should be selected by CCISx bits. The capture edge type (rising or falling) of this selected signal is set by CMx bits. When a capture occurs, the value in the TAR register is copied to the related TACCRx register. The CCIFG is set to indicate that the capturing is done. Also, the timer ISR is called if the CCIE bit is set. Then, the time interval between the two time instants can be calculated by these captured TACCRx values.

There are synchronization and overflow issues to be considered in the capture mode. If the input changes its state at the same time as the timer clock, this may cause a race condition when the TAR value is copied to the TACCRx. The

SCS bit should be set and the input should be synchronized with the timer clock in order to prevent this. Also, another capture may occur before the first one is processed. When this happens, the COV bit is set to indicate that an overflow occurred. Therefore, the COV bit must be cleared by software to catch subsequent overflows.

The Compare Mode

The purpose of the compare mode is to generate interrupts at specific time intervals. This can be used to form pulse width modulation (PWM) signals. The interrupt time intervals or the frequency of the PWM can be adjusted by the TACCRx register. When the timer counts up (until the value in the TAR reaches the TACCRx value), the internal signal EQUx (which can be seen at the output of the comparator) is set. Afterwards, the interrupt flag CCIFG is set and the EQUx signal triggers (by the changing of the output signal TA_OUTx) according to the output mode selected. Also, the input signal of the compare block CCI is latched into the SSCI bit. There are eight different output types in the compare mode. They are briefly described below.

- **Out bit value** (OUTMOD_0): OUT bit controls the output signal.
- **Set** (OUTMOD_1): The output is set only once when the TAR reaches the TACCRx value.
- **Toggle/Reset** (OUTMOD_2): The output is toggled when the TAR reaches the TACCRx value. It is reset when the TAR reaches the TACCR0 value.
- **Set/Reset** (OUTMOD_3): The output is set when the TAR reaches the TACCRx value. It is reset when the TAR reaches the TACCR0 value.
- **Toggle** (OUTMOD_4): The output is toggled when the TAR reaches the TACCRx value.
- **Reset** (OUTMOD_5): The output is reset only once when the TAR reaches the TACCRx value.
- **Toggle/Set** (OUTMOD_6): The output is toggled when the TAR reaches the TACCRx value. It is set when the TAR reaches the TACCR0 value.
- **Reset/Set** (OUTMOD_7): The output is reset when the TAR reaches the TACCRx value. It is set when the TAR reaches the TACCR0 value.

The compare mode output types are given in Figs. 10.13, 10.14, and 10.15. In the first figure, the timer is in the up mode. In the second figure, the timer is in the continuous mode. In the third figure, the timer is in the up/down mode. These figures clearly show that the compare mode can be used in PWM generation. In Sec. 11.5 we will use the compare mode to generate PWM signals.

10.7.3 *Timer_A Based Interrupts*

Either the timer or the capture/compare blocks can generate interrupts. An interrupt is generated when the TAR register overflows in the timer block. An interrupt is generated when the timer value in the TACCRx register is captured in the capture mode. An interrupt is generated when the TAR equals the value in the TACCRx register in the compare mode. All these interrupts are maskable. Therefore, the

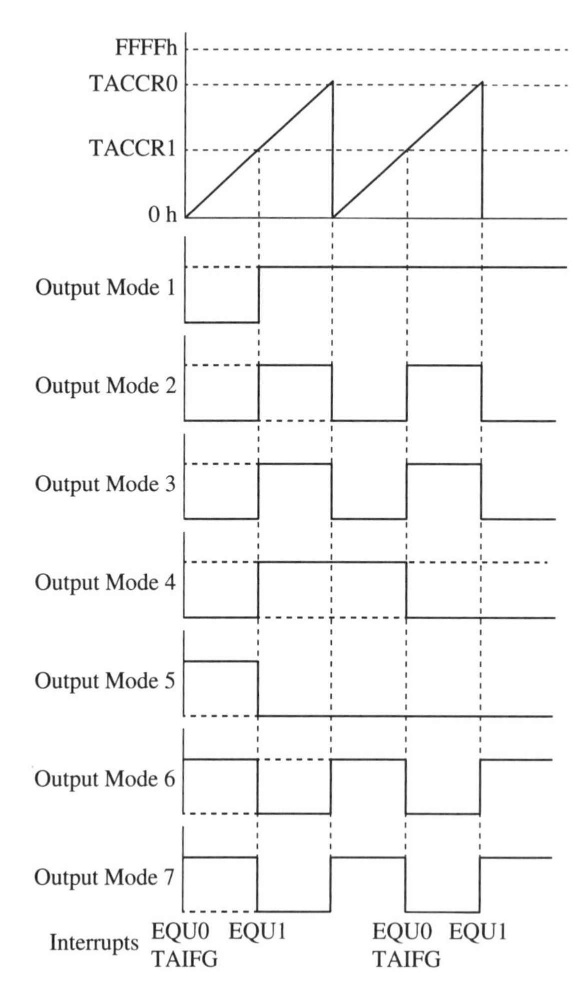

Figure 10.13

Compare mode outputs when the timer is in the up mode.

GIE must be set with corresponding interrupt enable bits. These are **TAIE** for the timer block and **CCIE** for the capture/compare block.

As can be seen in Table 9.2, there are two interrupt vectors for the TA module. These are TIMERx_A0_VECTOR and TIMERx_A1_VECTOR. The TIMERx_A0_VECTOR is associated with the TACCR0 capture/compare register and has the highest priority. The TIMERx_A1_VECTOR, also known as the TAIV interrupt vector, is associated with the TACCR1, TACCR2 capture/compare registers and the timer block. The TAIV interrupt vector register (Table 10.17) is used to control this interrupt vector. When more than one of these sources requests an interrupt, TAIV is loaded with the content of the highest priority interrupt. Other interrupts will be pending until this interrupt is handled.

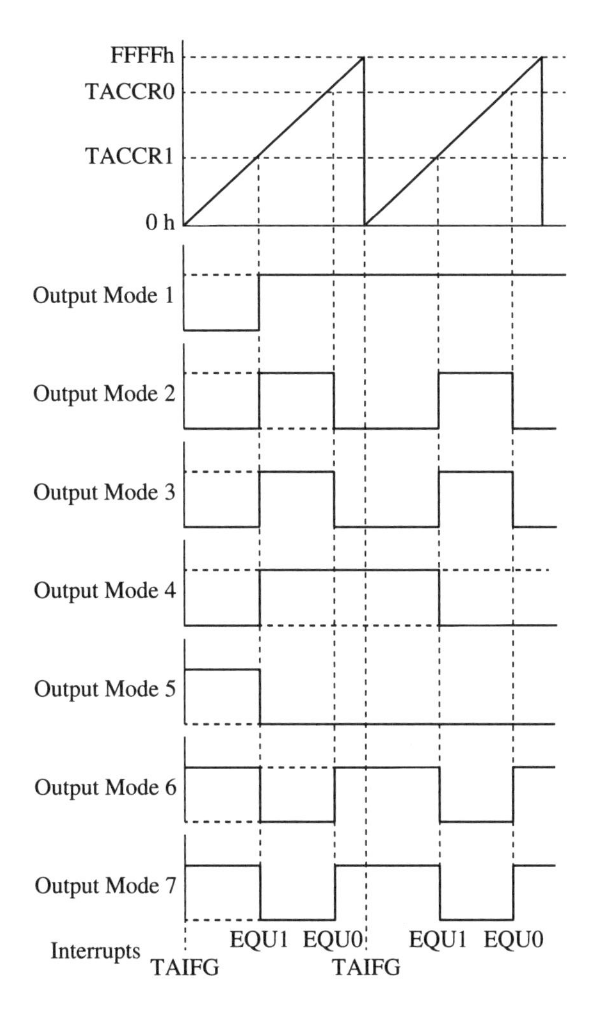

Figure 10.14

Compare mode outputs when the timer is in the continuous mode.

10.7.4 *Coding Practices for the Timer_A Module*

In this section, we provide sample C and assembly codes for the TA module. We first provide a C code on the usage of the capture mode in Listing 10.7. For this program to work, pin P1.3 should be connected to pin P1.1 on the MSP430 LaunchPad since pin P1.3 cannot generate the capture input. In Listing 10.7, the capture mode is used to measure the time difference between successive button presses. The maximum time difference that can be calculated here is approximately 41 s due to the clock and oscillator settings. In the assembly code, given in Listing 10.8, we again measure the time difference between successive button presses. The time difference between the button press and release time can also be calculated by both codes. To do so, the disabled code section should be enabled instead of the current setting.

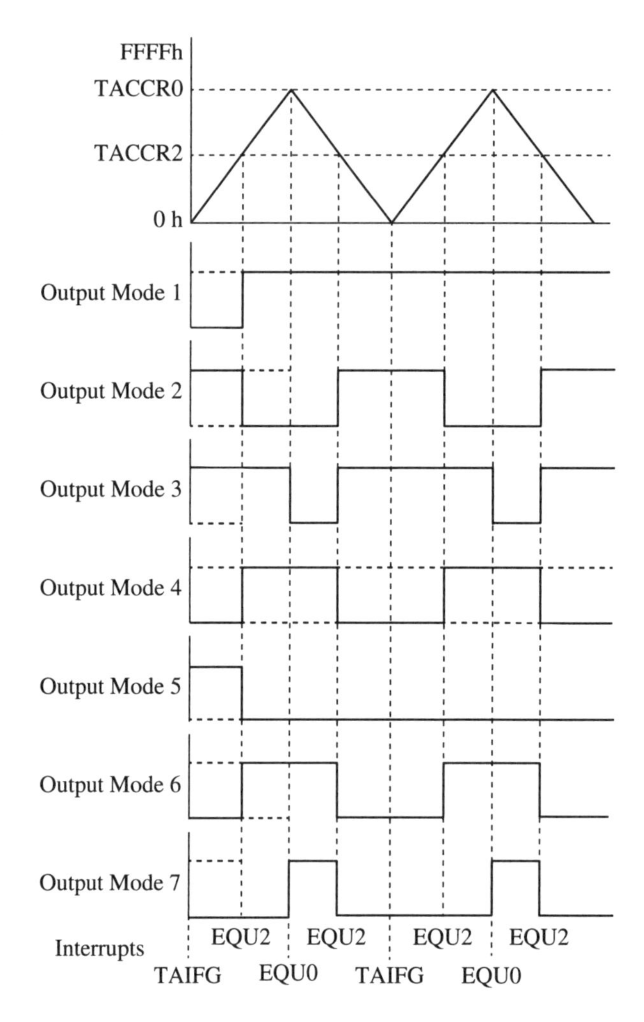

Figure 10.15

Compare mode outputs when the timer is in the up/down mode.

Table 10.17

The TAIV register.

TAIV Content	Interrupt Source	Interrupt Flag	Interrupt Priority
00h	No interrupt pending	-	
02h	Capture/Compare 1	TACCR1 CCIFG	Highest
04h	Capture/Compare 2	TACCR2 CCIFG	
06h	Reserved	-	
08h	Reserved	-	
0Ah	Timer overflow	TAIFG	
0Ch	Reserved	-	
0Eh	Reserved	-	Lowest

Listing 10.7 Usage of the capture mode in C.

```
#include <msp430.h>
// 40 sec. max. //connect P1.1 to P1.3
 int count = 0;
 int result = 0;
 float sec = 0.0;
void main(void)
{
 WDTCTL = WDTPW|WDTHOLD;
 BCSCTL3 |= LFXT1S_2;
 P1SEL = 0x02;
// TACCTL0 = CAP|CM_3|SCS|CCIE|CCIS_0;
// capture at rising and falling edges
 TACCTL0 = CAP|CM_2|SCS|CCIE|CCIS_0;
// capture at falling edge
 TACTL = TASSEL_1|ID_3|MC_2|TACLR;
 _enable_interrupts();
 LPM1;
}
#pragma vector=TIMER0_A0_VECTOR
__interrupt void Timer_A(void){
 count++;
// clear TAR at first press
 if(count == 1)TACTL |= TACLR;
// get the final value at second press
 if(count == 2){
 result = TACCR0;
 sec = (float)result/1500;
 count = 0;
 }
}
```

Listing 10.8 Usage of the capture mode in assembly.

```
 .cdecls C,LIST,"msp430.h"
 .text
 .retain
 .retainrefs
RESET
 mov.w #WDTPW|WDTHOLD,WDTCTL
 mov.w #__STACK_END,SP
 mov.w #LFXT1S_2,BCSCTL3
 mov.b #02h,P1SEL
mov.w #CAP+CM_1+SCS+CCIE+CCIS_0,TACCTL0
```

```
;capture at rising edge
; mov.w #CAP+CM_3+SCS+CCIE+CCIS_0,TACCTL0
;capture at rising and falling edges
 mov.w #TASSEL_1+ID_3+MC_2+TACLR,TACTL
 clr.w R6
 clr.w R7
 bis.w #GIE+LPM1,SR
;----------------------
TA0_ISR;
;----------------------
 inc.w R6
 cmp.w #1d,R6
 jne Loop1
 bis.w #TACLR,TACTL
Loop1:
 cmp.w #2d,R6
 jne Loop2
 mov.w &TACCR0,R7
 clr.w R6
Loop2:
 reti
;------------------------
;Stack Pointer definition
;------------------------
 .global __STACK_END
 .sect .stack
;-------------------
;Interrupt Vectors
;-------------------
 .sect RESET_VECTOR
 .short RESET
 .sect TIMER0_A0_VECTOR
 .short TA0_ISR
 .end
```

Next, we provide C and assembly codes on the usage of TAIV in Listing 10.9. Here the red LED toggles when an overflow occurs. This happens every second (based on the parameter settings). Here, the capture/compare block is not used. The same operation is done in assembly language in Listing 10.10.

Listing 10.9 Usage of the TAIV in C.

```
#include <msp430.h>
#define RedLED BIT0
```

```
#define RedLEDToggle (P1OUT ^= RedLED)

void main(void)
{
 WDTCTL = WDTPW|WDTHOLD;

 P1DIR = RedLED;
 P1OUT = RedLED;

 TACTL = TASSEL_2|ID_3|MC_3|TAIE;
 TACCR0 = 62500;

 _enable_interrupts();
 LPM1;// enter low power mode
}
#pragma vector=TIMER0_A1_VECTOR
__interrupt void Timer_A(void){
 switch(TAIV)
 {
 case 0x02: break;
 case 0x04: break;
 case 0x0A: RedLEDToggle;
 break;
 }
}
```

Listing 10.10 Usage of the TAIV in assembly.

```
  .cdecls C,LIST,"msp430.h"
  .text
  .retain
  .retainrefs

RESET
 mov.w #WDTPW|WDTHOLD,WDTCTL
 mov.w #__STACK_END,SP

 mov.b #01h,P1DIR
 mov.b #01h,P1OUT

 mov.w #TASSEL_2+ID_3+MC_3+TAIE,TACTL
;SMCLK, f/8, up/down mode
;timer overflow interrupt enabled
 mov.w #62500d,TACCR0 ;Add Offset to TACCR0

 bis.w #GIE+LPM1,SR; Enable interrupts and enter LPM
;----------------------
T0_A1_ISR ;Toggle P1.0
;----------------------
 add.w &TAIV,PC ;Use PC for observing TAIV
 reti ; No interrupt pending
 reti ; TACCR0
 reti ; TACCR1
 reti ; Reserved
```

```
 reti ; Reserved
 xor.b #01h,P1OUT ;Toggle P1.0
 reti

;------------------------
;Stack Pointer definition
;------------------------
 .global __STACK_END
 .sect .stack

;-------------------
;Interrupt Vectors
;-------------------
 .sect RESET_VECTOR
 .short RESET
 .sect TIMER0_A1_VECTOR
 .short T0_A1_ISR
 .end
```

We can use the capture/compare blocks instead of using TAIV, as given in Listing 10.9. Now the C code becomes as given in Listing 10.11. Here the ISR toggles the red LED based on the compare mode configuration. The assembly code version of this operation is given in Listing 10.12.

Listing 10.11 Toggling the red LED using the timer interrupt in compare mode in C.

```
#include <msp430.h>

#define LED BIT0

void main(void)
{
 WDTCTL = WDTPW|WDTHOLD;

 P1DIR = LED;
 P1OUT = LED;

 TACCR0 = 49999; // Upper limit of count for TAR
 TACCTL0 = CCIE; // Enable interrupts on compare 0

 TACTL = MC_1|ID_3|TASSEL_2|TACLR;
// Setup and start Timer A

_enable_interrupts();

 LPM1;// enter low power mode
}

#pragma vector = TIMER0_A0_VECTOR
__interrupt void TA0_ISR(void){
 P1OUT ^= LED;
}
```

Listing 10.12 Toggling the red LED using the timer interrupt in compare mode in assembly.

```
 .cdecls C,LIST,"msp430.h"

 .text
 .retain
 .retainrefs

RESET
 mov.w #WDTPW|WDTHOLD,WDTCTL
 mov.w #__STACK_END,SP

 mov.w #LFXT1S_2,&BCSCTL3 ;12 Khz VLO as ACLK source

 mov.b #41h,P1DIR
 mov.b #01h,P1OUT

 mov.w #CCIE,TACCTL0 ;TACCR0 interrupt enabled
 mov.w #999d,TACCR0 ;TACCR0 counts to 1000
 mov.w #TASSEL_1+MC_1,TACTL ;ACLK, upmode

 bis.w #LPM1+GIE,SR ;LPM1, enable interrupts

;----------------------
TA0_ISR ;Toggle P1.0
;----------------------
 xor.b #41h,P1OUT
 reti

;------------------------
;Stack Pointer definition
;------------------------
 .global __STACK_END
 .sect .stack

;-------------------
;Interrupt Vectors
;-------------------
 .sect RESET_VECTOR
 .short RESET
 .sect TIMER0_A0_VECTOR
 .short TA0_ISR
 .end
```

We provide the code in Listing 10.13 to observe the effects of different timer settings. Here, the red and green LEDs toggle. Each disabled setting can be enabled to see its effect. Do not forget to disable the previous active setting when a new one is enabled.

Listing 10.13 Toggling the red and green LEDs with various options in C.

```
#include <msp430.h>

#define RedLED BIT0
```

```
#define GreenLED BIT6

void main(void)
{
 WDTCTL = WDTPW|WDTHOLD;

 P1DIR = RedLED|GreenLED;
 P1OUT = RedLED;

 TACCTL0 = CCIE; // Enable interrupts on compare 0

// TACTL = TASSEL_2|ID_3|MC_2|TACLR;
// Use clock from SMCLK, divide clock by 8,
// up mode, clear Timer A

// TACTL = TASSEL_2|ID_0|MC_2|TACLR;
// The effect of the frequency divider
// TACCR0 = 19999; // Upper limit for count
// TACTL = TASSEL_2|ID_3|MC_1|TACLR;

 TACCR0 = 0xFFFF; // Upper limit for count
 TACTL = TASSEL_2|ID_3|MC_3|TACLR;
// up down mode

 _enable_interrupts();

 LPM1;
}

#pragma vector = TIMER0_A0_VECTOR
__interrupt void TA0_ISR(void){
 P1OUT ^= RedLED|GreenLED;
}
```

We next provide the usage of two timers together in C language in Listing 10.14. Here, the red and green LEDs toggle by interrupts generated by TA0 and TA1 separately. We provide the assembly code doing the same operation in Listing 10.15. Again, the two timers are used together in this code block.

Listing 10.14 Using TA0 and TA1 together in C.

```
#include <msp430.h>

#define RedLED BIT0
#define GreenLED BIT6

void main(void)
{
 WDTCTL = WDTPW|WDTHOLD;

 BCSCTL3 |= LFXT1S_2;

 P1DIR = RedLED|GreenLED;
 P1OUT = RedLED|GreenLED;

 TACCR0 = 62500;
 TA1CCR0 = 6000;
```

```
 TACCTL0 = CCIE;
 TA1CCTL0 = CCIE;
 TACTL = MC_2|ID_3|TASSEL_2|TACLR;
 TA1CTL = MC_3|ID_3|TASSEL_1|TACLR;

 _enable_interrupts();

 LPM1;
}

#pragma vector = TIMER0_A0_VECTOR
__interrupt void TA0_ISR(void){
 P1OUT ^= RedLED;
}

#pragma vector = TIMER1_A0_VECTOR
__interrupt void TA1_ISR(void){
 P1OUT ^= GreenLED;
}
```

Listing 10.15 Using TA0 and TA1 together in assembly.

```
 .cdecls C,LIST,"msp430.h"

 .text
 .retain
 .retainrefs

RESET
 mov.w #WDTPW|WDTHOLD,WDTCTL
 mov.w #__STACK_END,SP

 bis.w #LFXT1S_2,BCSCTL3

 mov.b #41h,P1DIR
 mov.b #41h,P1OUT

 mov.w #CCIE,TACCTL0 ;TACCR0 interrupt enabled
 mov.w #CCIE,TA1CCTL0 ;TA1CCR0 interrupt enabled
 mov.w #62500d,TACCR0 ;TACCR0 counts to 62500
 mov.w #6000d,TA1CCR0 ;TA1CCR0 counts to 6000
 mov.w #TASSEL_2+MC_2+ID_3+TACLR,TACTL ;ACLK, upmode
 mov.w #TASSEL_1+MC_3+ID_3+TACLR,TA1CTL ;ACLK, upmode

 bis.w #LPM1+GIE,SR ;LPM1, Enable interrupts

;----------------------
TA0_ISR
;----------------------
 xor.b #01h,P1OUT
 reti

;----------------------
TA1_ISR
;----------------------
 xor.b #40h,P1OUT
```

```
 reti

;------------------------
;Stack Pointer definition
;------------------------
 .global __STACK_END
 .sect .stack

;-------------------
;Interrupt Vectors
;-------------------
 .sect RESET_VECTOR
 .short RESET
 .sect TIMER0_A0_VECTOR
 .short TA0_ISR
 .sect TIMER1_A0_VECTOR
 .short TA1_ISR
 .end
```

The code given in Listing 10.16 is an example of the joint usage of the port and timer interrupts. Here, the program counts the timer interrupts. If the sum reaches five, the red and green LEDs toggle. The sum is reset if the user presses the button during operation. This operation is done under the port ISR.

Listing 10.16 Jointly using the port and the timer interrupts, the first example in C.

```
#include <msp430.h>

#define RedLED BIT0
#define GreenLED BIT6
#define Button BIT3

 int count = 0;

void main(void)
{
 WDTCTL = WDTPW|WDTHOLD;

 P1DIR = RedLED|GreenLED;
 P1OUT = RedLED;

 P1IE = Button;
 P1IES = Button;
 P1IFG = 0x00;

 TACCTL0 = CCIE; // Enable interrupts on compare 0
 TACCR0 = 0xFFFF; // Upper limit for count
 TACTL = TASSEL_2|ID_3|MC_3|TACLR;
// Use clock from SMCLK, divide clock by 8,
// up down mode, clear Timer A

 _enable_interrupts();

 LPM1;
```

```
}

#pragma vector = TIMER0_A0_VECTOR
__interrupt void TA0_ISR(void){
 count += 1;
 if (count == 5){
 P1OUT ^= RedLED|GreenLED;
 count = 0;
 }
}

#pragma vector=PORT1_VECTOR
__interrupt void PORT1_ISR(void){
 P1OUT = RedLED;
 count = 0;
 TACTL |= TACLR;
 P1IFG = 0x00;
}
```

The code given in Listing 10.17 is another example of the joint usage of the port and timer interrupts. Here the program counts how many times the button is pressed while the red and green LEDs are on separately. In addition, the system disables all the interrupts and goes to LPM4 after a certain time. This means the system is turned off using low-power modes.

Listing 10.17 Jointly using the port and the timer interrupts, the second example in C.

```
#include <msp430.h>

#define RedLED BIT0
#define GreenLED BIT6
#define Button BIT3

 int count = 0;
 int redcount = 0;
 int greencount = 0;
 int state = 0;
 int done = 0;

void main(void)
{
 WDTCTL = WDTPW|WDTHOLD;

 P1DIR = RedLED|GreenLED;
 P1OUT = RedLED;

 P1IE = Button;
 P1IES = Button;
 P1IFG = 0x00;

 TACCTL0 = CCIE; // Enable interrupts on compare 0
 TACCR0 = 0xFFFF; // Upper limit for count
 TACTL = TASSEL_2|ID_3|MC_3|TACLR;
// Use clock from SMCLK, divide clock by 8,
```

```
// continuous up down mode, clear Timer A
 _enable_interrupts();
 while(1)
 {
 if (done > 5){
 P1OUT = 0x00;
 _disable_interrupts();
 LPM4;
 }
 else
 LPM1;
 }
}
#pragma vector = TIMER0_A0_VECTOR
__interrupt void TA0_ISR(void){
 LPM1_EXIT;
 count++;
 if (count == 30){
 P1OUT ^= RedLED|GreenLED;
 state = !state;
 count = 0;
 done++;
 }
}
#pragma vector=PORT1_VECTOR
__interrupt void PORT1_ISR(void){
 if (state == 0)
 redcount++;
 else
 greencount++;
 P1IFG = 0x00;
}
```

Finally, we provide the assembly code in Listing 10.18. Here, the timer and port interrupts are jointly used. The application here is similar to Listing 10.17.

Listing 10.18 Jointly using the port and the timer interrupts in assembly.

```
 .cdecls C,LIST,"msp430.h"
 .text
 .retain
 .retainrefs
RESET
 mov.w #WDTPW|WDTHOLD,WDTCTL
 mov.w #__STACK_END,SP
 mov #0d,R5 ;used as count
 mov #0d,R6 ;used as done
```

```
 mov #0h,R7 ;used as state
 mov #0d,R8 ;used as redcount
 mov #0d,R9 ;used as greencount

 mov.b #41h,P1DIR
 mov.b #08h,P1REN
 mov.b #09h,P1OUT

 bis.b #08h,P1IE
 bis.b #08h,P1IES
 bic.b #08h,P1IFG

 mov.w #0FFFFh,TACCR0
 mov.w #TASSEL_2+MC_3+ID_3,TACTL
 mov.w #CCIE,TACCTL0

 bis.w #GIE,SR ;Enable interrupts
Loop:
 cmp #06d,R6
 jl Subloop
 clr.b P1OUT
 bic.w #GIE,SR
 bis.w #LPM4,SR
Subloop:
 bis.w #LPM1,SR
 jmp Loop
;---------------------------
P1_ISR ;Check the button
;---------------------------
 tst.w R7
 jeq Red
 inc R9
 jmp Ei
Red:
 inc R8
Ei:
 bic.b #08h,P1IFG
 reti

;----------------------
TA0_ISR ;Toggle P1.0
;----------------------
 bic.w #LPM1,0(SP)
 inc R5
 cmp.w #05d,R5
 jl EndISR
 xor.b #41h,P1OUT
 inc R6
 mov #0d,R5
 xor #01d,R7
EndISR:
 reti
;------------------------
```

```
;Stack Pointer definition
;------------------------
 .global __STACK_END
 .sect .stack

;-------------------
;Interrupt Vectors
;-------------------

 .sect RESET_VECTOR
 .short RESET
 .sect PORT1_VECTOR
 .short P1_ISR
 .sect TIMER0_A0_VECTOR
 .short TA0_ISR
 .end
```

10.8 The Pin Layout for the BCM+ and TA Modules

We provide the pin layout of the MSP430G2553 in Fig. 10.16 (again to be compact). The usage of these in BCM+ and TA perspective are tabulated in Table 10.18. Do not forget to set these pins with the appropriate PxSEL bits before using them.

In Table 10.18, TA0CLK is the external clock input used to supply the clock signal for the timer block. ACLK and SMCLK can be fed to output to control other connected devices. If an external crystal is to be connected, the XIN and XOUT should be used. All TAx.x pins can be used as output of compare blocks. Also, some of these pins can be configured as input for capture blocks. Pins 3, 4 (TA0.0, TA0.1), and 8 to 13 (TA1.0, TA1.1, TA1.2) can be used for both purposes. Don't forget that each of these pins can only be connected to one capture or compare block.

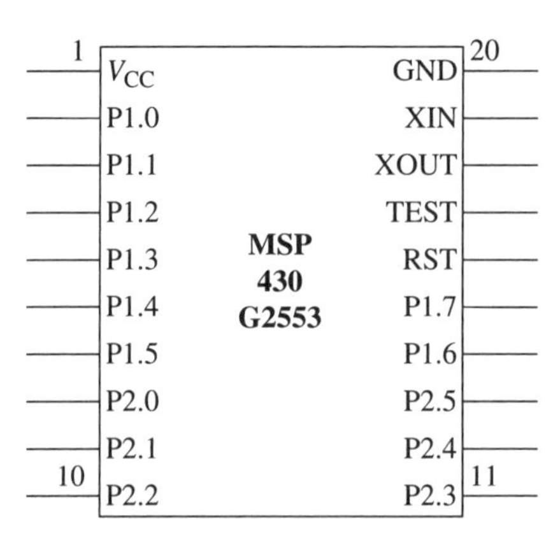

Figure 10.16 Pin layout of the MSP430G2553.

Table 10.18

BCM+ and Timer_A properties for the pins of the MSP430G2553.

Pin	Port Name		Usage Area
1	V_{CC}		Source voltage
2	P1.0/	TA0CLK	Timer0_A, clock signal TACLK input
		ACLK	ACLK signal output
3	P1.1/	TA0.0	Timer0_A, capture: CCI0A input, compare: Out0 output
4	P1.2/	TA0.1	Timer0_A, capture: CCI1A input, compare: Out1 output
5	P1.3		
6	P1.4 /	SMCLK	SMCLK signal output
7	P1.5/	TA0.0	Timer0_A, compare: Out0 output
8	P2.0/	TA1.0	Timer1_A, capture: CCI0A input, compare: Out0 output
9	P2.1/	TA1.1	Timer1_A, capture: CCI1A input, compare: Out1 output
10	P2.2/	TA1.1	Timer1_A, capture: CCI1B input, compare: Out1 output
11	P2.3/	TA1.0	Timer1_A, capture: CCI0A input, compare: Out0 output
12	P2.4/	TA1.2	Timer1_A, capture: CCI2A input, compare: Out2 output
13	P2.5/	TA1.2	Timer1_A, capture: CCI2B input, compare: Out2 output
14	P1.6/	TA0.1	Timer0_A, compare: Out1 output
15	P1.7		
16	RST		Reset
17			
18	P2.7/	XOUT	Output terminal of crystal oscillator
19	P2.6/	XIN	Input terminal of crystal oscillator
		TA0.1	Timer0_A, compare: Out1 output
20	V_{SS}		Ground voltage

10.9 Timer_A in Grace

The Timer_A (TA) module can be configured by the **Timer0_A3** and **Timer1_A3** blocks in the Device Overview window (shown in Fig. 5.11). Configurations for these two timer blocks are identical. Therefore, we will only explain the Timer0_A3 block in this section. As in the previous sections, the Timer_A block should be enabled first. Then it can be configured as follows.

10.9.1 *The Basic User Mode*

The basic user mode of the Timer0_A3 block is given in Fig. 10.17. The Timer Selection list can be used as follows: The Timer OFF option can be used to disable the timer if it is not needed. Initially, this option is chosen. The Interval Mode option can be used to create a time interval. This time interval is obtained by entering a value to the Desired Timer Period box. The generated signal can be fed to P1.1 or P1.5 from the neighboring list. These settings are done under Timer Capture/Compare Block #0.

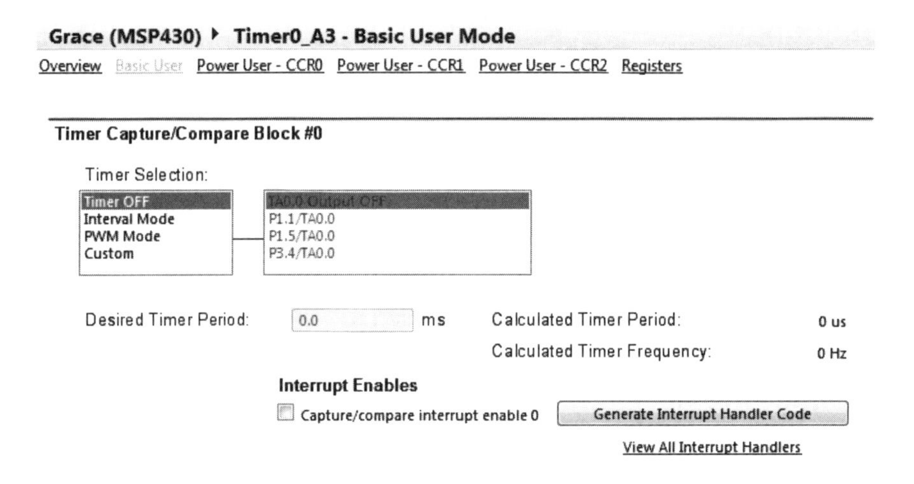

Figure 10.17 Basic user mode for Timer0_A3.

The PWM Mode option in the Timer Selection list is used to generate a PWM signal. When this option is selected, Timer Capture/Compare Block #1 and Timer Capture/Compare Block #2 are also enabled as shown in Fig. 10.18. In this mode, the PWM period is determined by the value entered in the Desired Timer Period box. As the PWM Duty Cycle option is selected from the Mode Selection list under the Timer Capture/Compare Block #1, the Desired PWM Duty Cycle box becomes enabled. The user can enter the desired duty cycle here. The generated PWM signal can be fed to the output pin selected from the neighboring list.

The user can set interrupts for each block (Timer Capture/Compare Block #0, #1, or #2) separately. To do so, first the related Capture/compare interrupt enable box should be checked. Then, the related Generate Interrupt Handler Code button can be pressed to generate the prototype ISR under the InterruptVectors_init.c file.

10.9.2 *The Power User Mode*

In the Power User-CCR0 mode (given in Fig. 10.19), the clock source for the timer block can be set from the Clock Source list. The frequency divider for the selected clock source can be set from the neighboring Divider list. Then, the counting mode can be chosen from the Counting Mode list. The TAR register can be cleared by checking the Clear box. the Timer_A overflow interrupt enable box can be checked to enable the interrupt. The prototype ISR for this interrupt can be generated under InterruptVectors_init.c by the Generate Interrupt Handler Code button.

The desired time interval can be obtained by entering the time value into the Desired Timer Period box under Timer Capture/Compare Block #0. The desired time can also be entered into the Capture Register box in terms of clock ticks. Also Input Capture or Output Compare/Period modes can be selected from the Mode list. If the capture mode is selected, it is configured from the Input Selection and Capture Mode lists. If the compare mode is selected, the Output Pins list and the Output Mode drop-down list should be used for necessary configurations. In this mode, the capture/compare interrupt can be enabled similar to the Timer_A overflow interrupt.

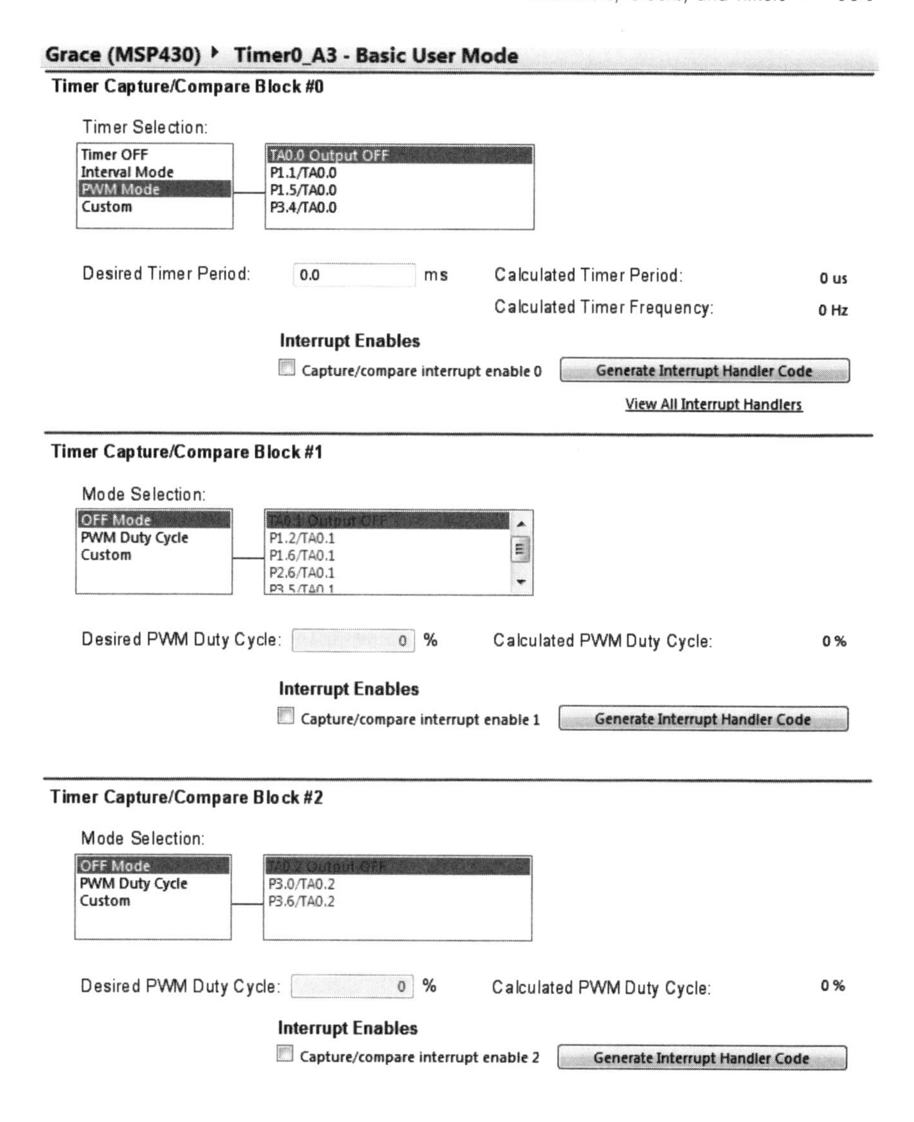

Figure 10.18

The basic user mode for Timer0_A3 specifically for PWM generation.

Configurations for the Power User-CCR1 and Power User-CCR2 views are identical. Therefore, only the Power User-CCR1 is explained in detail here. In the Power User-CCR1 mode (shown in Fig. 10.20), capture and compare mode configurations can be done as in the Power User-CCR0 mode. In this mode, the shape of the generated PWM signal can also be seen. Please note that the configurations in the power user mode interact with each other. Therefore, one setting in one mode affects another setting in the next mode.

10.9.3 *The Register Controls Mode*

In the register controls mode of the Timer0_A3 block (given in Fig. 10.21), all available timer registers can be configured. Also, all capture/compare register values can be configured.

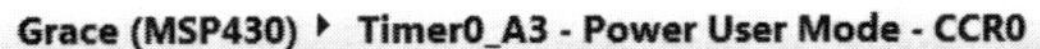

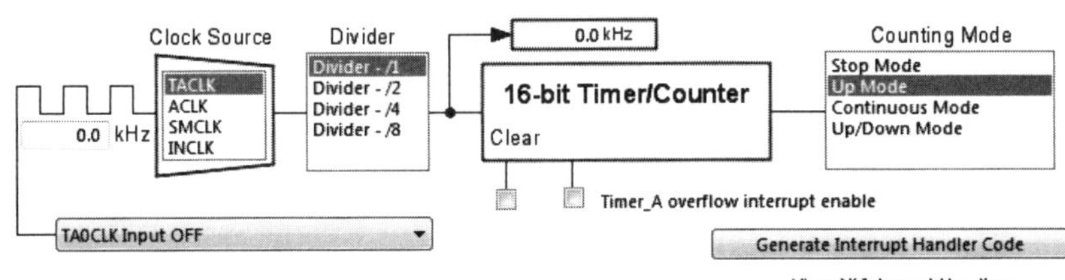

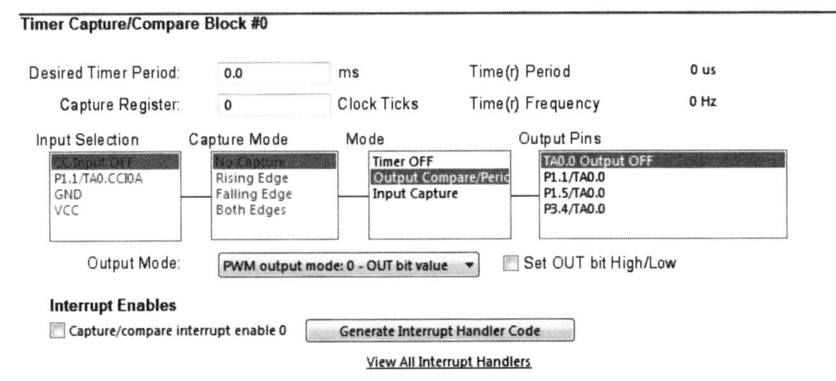

Figure 10.19 The power user-CCR0 mode for Timer0_A3.

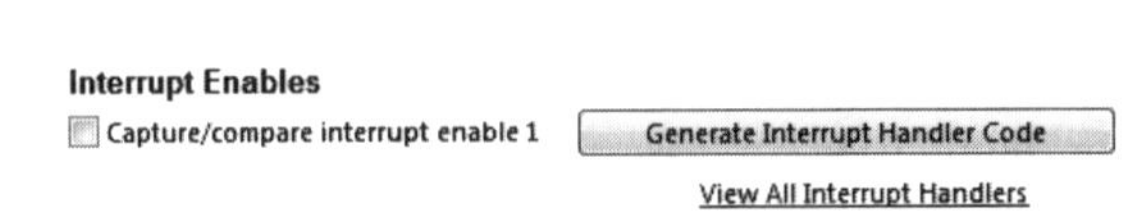

Figure 10.20 The power user CCR1 mode for Timer0_A3.

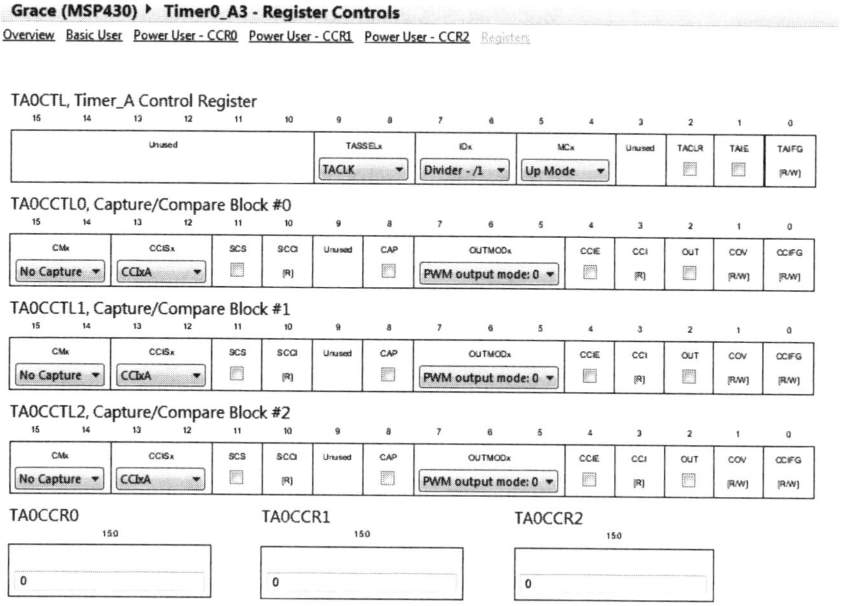

Figure 10.21

The register controls mode for Timer0_A3.

Grace (MSP430) ▸ Timer0_A3 - Register Controls

Overview Basic User Power User - CCR0 Power User - CCR1 Power User - CCR2 Registers

TA0CTL, Timer_A Control Register

TA0CCTL0, Capture/Compare Block #0

TA0CCTL1, Capture/Compare Block #1

TA0CCTL2, Capture/Compare Block #2

TA0CCR0 TA0CCR1 TA0CCR2

10.9.4 *Coding Practices*

In our first example, we use Grace to toggle red and green LEDs every 3 s as in Sec. 10.6.4. We use the basic user mode here. We first select the Interval Mode under the Timer Selection list of the Timer_A block. Then we select the output as TA0.0 Output Off. We enter the desired timer period as 3000 ms. We should enable the capture/compare interrupt by checking the Capture/compare interrupt enable 0 box. Then, we generate the ISR by pressing the Generate Interrupt Handler button. In the InterruptVector_init.c file, we fill the Timer_A ISR as in Listing 10.19. As we run the application, the red and green LEDs toggle on every 3 s. Do not forget to enable the BCM+ for this application.

Listing 10.19 The Timer_A application ISR under Grace, the basic user mode.

```
#pragma vector=TIMER0_A0_VECTOR
__interrupt void TIMER0_A0_ISR_HOOK(void)
{
 P1OUT ^= BIT6|BIT0;
}
```

We repeat the same application now using the power user mode. Here, we set the LED toggle time to be 1 s. We perform this operation as follows: First, we select the ACLK from the Clock Source menu. We set the Divider to Divider-/8. We enable the Timer_A overflow interrupt by checking the Timer_A overflow interrupt enable box. We select the Up Mode as the Counting mode. We enter 1000 ms into the Desired Timer Period box. We select the Output Compare/Period from the

Mode list. We generate the ISR by pressing the Generate Interrupt Handler Code button. In the InterruptVector_init.c file, we fill the Timer_A ISR with Listing 10.20. As we run the application, the red and green LEDs toggle on every second. Do not forget to enable the BCM+ for this application.

Listing 10.20 The Timer_A application ISR under Grace, the power user mode.

```
#pragma vector=TIMER0_A1_VECTOR
__interrupt void TIMER0_A1_ISR_HOOK(void)
{
 switch(TA0IV)
 {
 case 0x02:  break;
 case 0x04:  break;
 case 0x0A: P1OUT ^= BIT0|BIT6;
 break;
 }
}
```

Finally, we can feed the PWM output to adjust the brightness of the green LED using Grace. We do not need any code blocks for this application. The settings for this application are as follows: We select the PWM Mode in the Timer Selection list. We adjust the desired timer period to 1 ms. In the Timer Capture/Compare Block # 1, we select the PWM Duty Cycle under the Mode Selection list. We select P1.6/TA0.1 as output. As we enter the Desired PWM Duty Cycle (between 1% and 100%), we can run our application. The brightness of the green LED is adjusted by duty cycle here.

10.10 Chronometer Application

The aim in this application is to learn how to set and use the timer module and low-power modes of the MSP430 microcontroller. As a real-world application, we design a chronometer using a liquid crystal display (LCD) and push buttons. In this section, we provide the equipment list, layout of the circuit, procedure, and system design specifications.

10.10.1 *Equipment List*

Following is the equipment list to be used in this application.

- One 12-V dc adaptor
- One LM7805 voltage regulator
- One 16×2 character LCD (with a Samsung processor)
- One 10-kΩ potentiometer
- Two push buttons
- One 330-ηF capacitor
- One 10-μF electrolytic capacitor
- Two 100-ηF capacitors

Table 10.19

The pin description of the LCD.

Pin	Name	Function
1	V_{ss}	Power supply (GND)
2	V_{dd}	Power supply (+5 V)
3	Vo	Contrast adjust
4	RS	Register select
5	R/W	Data read/write
6	E	Enable signal
7	DB0	Data bus line0
8	DB1	Data bus line1
9	DB2	Data bus line2
10	DB3	Data bus line3
11	DB4	Data bus line4
12	DB5	Data bus line5
13	DB6	Data bus line6
14	DB7	Data bus line7
15	A	Power supply for the LED (+)
16	K	Power supply for the LED (-)

16×2 Character LCD: In this application, an LCD with two lines (each having 16 characters) will be used to show the time. Although another LCD brand can be used, we picked the one with a Samsung processor due to its availability. When using another LCD, the reader should obtain its pin description and the header file. The pin description for our LCD is given in Table 10.19.

As can be seen in Table 10.19, pins 1 and 2 are used to supply power to the LCD. Pin 3 is used to adjust the contrast of the LCD. This is done by changing the voltage at this pin between 0 and 5 V with a suitable potentiometer. Pin 4 is used to identify the data type. When this pin is low, the data transferred to the LCD is recognized as an instruction. When this pin is high, the data transferred to the LCD is recognized as a character (to be displayed). Pin 5 is used to set the state of the LCD. When this pin is low, the LCD is in write state. When it is high, the LCD is in read state. Pin 6 is used to start the data transfer. When a high-to-low transition occurs at this pin, the data is sent to the LCD. When a low-to-high transition occurs, the data is read from the LCD. Pins 7 to 14 are used to transfer 8-bit data to the LCD. When the LCD is used in a 4-bit mode, only the last four pins (from 11 to 14) are used. Pins 15 and 16 are used to supply power to the LED backlight of the LCD.

We collected the functions to control the LCD properly in a header file given in Listing 10.21. First the definitions are done in this file. The `LCD_Change()` function is used to generate the necessary high-to-low transition for pin 6. The functions `SendCommand` and `SendCharacter` are used to provide low and high signals for pin 4. The `LCD_Data` function represents port 2. But only four pins of this port are used for data transfer. The `delay_ms()` function is used to generate the required delay times for the system. The data is sent to the LCD (first upper 4 bits, then lower 4 bits) with the `lcd_send()` function. The functions `lcd_writestr()` and `lcd_writechr()` write string and character variables to the LCD. Integer values cannot be sent directly to the LCD. They should be converted to the corresponding character values first. The function `itoa(i,buffer,base)` does the job. This function converts the integer `i` to the corresponding character value, saved in the string `buffer`. The variable `base` represents the base number of the integer `i`. The `lcd_goto(x,y)` function is used to set the starting point on the LCD. In this function, `x` represents the column number, `y` represents the row number. The `lcd_init()` function initializes the LCD.

Listing 10.21 The header file for the LCD.

```
#define LCD_Data P2OUT
#define LCD_Change() ((P1OUT |= BIT7),(P1OUT &= ~BIT7))
#define SendCommand (P1OUT &= ~BIT6)
#define SendCharacter (P1OUT |= BIT6)

void delay_ms(int a){
 while(a != 0)
 {
 _delay_cycles(1000);
 a--;
 }
}

void lcd_send(unsigned char data){
 LCD_Data=((data & 0xF0) >> 4);
 LCD_Change();
 LCD_Data=((data & 0x0F));
 LCD_Change();
}

void lcd_writestr(const char *str){
 SendCharacter;
 while(*str)
 lcd_send(*str++);
}

void lcd_writechr(char chr){
 SendCharacter;
 lcd_send(chr);
}

void lcd_goto(char x,char y){
 SendCommand;
 if(x == 1)
 lcd_send(0x80+((y-1)));
 else
```

```
 lcd_send(0xC0+((y-1)));
}

void lcd_init(void){
 SendCommand;
 delay_ms(40);
 lcd_send(0x30);
 delay_ms(1);
 lcd_send(0x28);
 delay_ms(1);
 lcd_send(0x28);
 delay_ms(1);
 lcd_send(0x0C);
 delay_ms(1);
 lcd_send(0x01);
 delay_ms(2);
 lcd_send(0x06);
}

char* itoa(int value, char* result, int base){
 if (base<2 || base>36){*result = '\0';return result;}
 char* ptr = result, *ptr1 = result, tmp_char;
 int tmp_value;

 do{
 tmp_value = value;
 value /= base;
 *ptr++ = "zyxwvutsrqponmlkjihgfedcba9876543210123\
 456789abcdefghijklmnopqrstuvwxyz"\
 [35+(tmp_value_value*base)];
 } while (value);

 if (tmp_value < 0) *ptr++ = '-';
 *ptr-- = '\0';
 while(ptr1<ptr){
 tmp_char = *ptr;
 *ptr-- = *ptr1;
*ptr1++ = tmp_char;
 }
 return result;
}
```

10.10.2 *Layout*

The layout of this application is given in Fig. 10.22. For more information about the voltage supply block, please see Fig. 9.3.

10.10.3 *System Design Specifications*

The chronometer will be controlled by two push buttons. The first button will be used to start and stop the chronometer. When it is pressed once, the chronometer will start counting. The time will be displayed on the LCD. When the first button is pressed again, the chronometer will stop. The second button will reset the chronometer. If the chronometer is running, pressing this button will also stop it.

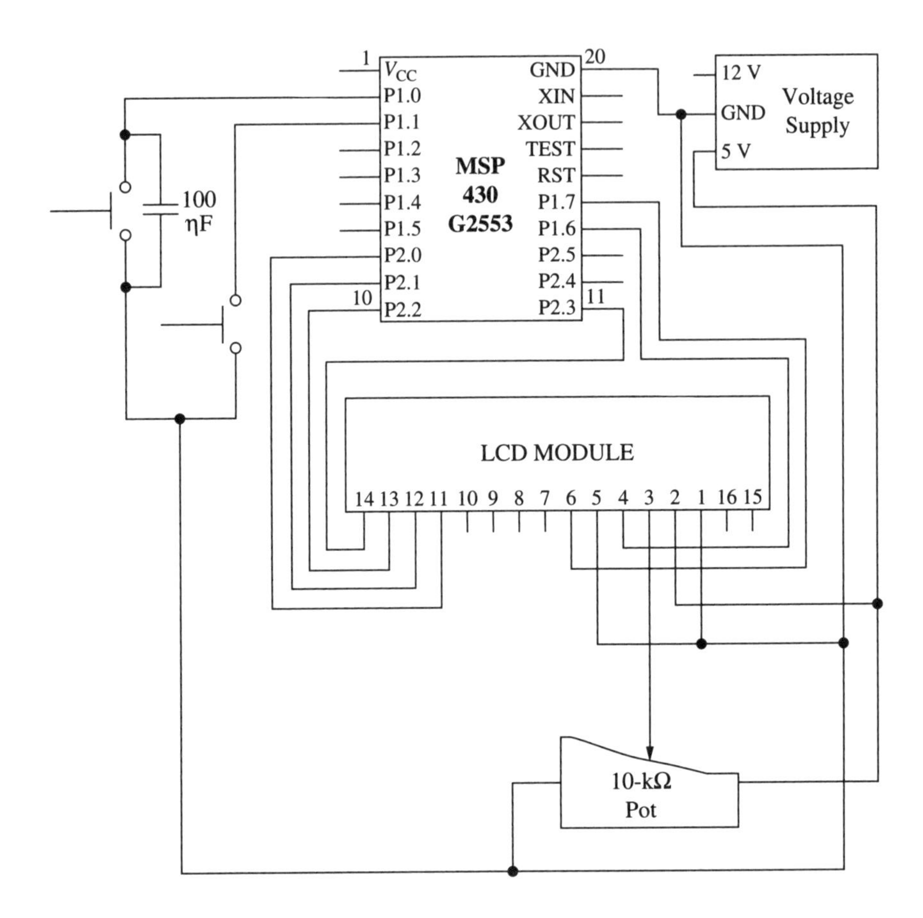

Figure 10.22 Layout of the chronometer application.

In designing the chronometer, we need to use the proper clock frequency for the timer. Since this is a prototype system, we will use the RC-based oscillators in this application. As mentioned in previous sections, the internal RC-based oscillators of the MSP430 are not precise. Therefore, the designed chronometer will not be precise. An external oscillator can always be connected to increase the precision. Besides, the TAR value should also be taken into account.

10.10.4 *The C Code for the System*

In the first part of the code, constants for interrupts are defined. This is done to make the code more readable. The code block for this part is given in Listing 10.22.

Listing 10.22 Chronometer, the C code part I.

```
#define STARTSTOP ((P1IFG & 0x01) == 0x01)
#define RESET ((P1IFG & 0x02) == 0x02)
```

In the second part of the code, given in Listing 10.23, global variables to be used in the code are defined. In this code block, the `Start` variable keeps the stop or start state of the chronometer. Initially, the chronometer is stopped.

Therefore, the Start variable has the value 0. The variables secondh and minuteh are used to keep the tens digit of the second and minute. Similarly, the variables secondl and minutel are used to keep the ones digit of the second and minute. The arrays lcdsecondh, lcdminuteh, lcdsecondl, and lcdminutel are used to keep the character values of the second and minute values. The lcd array is used to keep the complete time value (to be sent to the LCD).

Listing 10.23 Chronometer, the C code part II.

```
int start = 0;
int secondh = 0;
int secondl = 0;
int minuteh = 0;
int minutel = 0;
char lcdsecondh[1];
char lcdsecondl[1];
char lcdminuteh[1];
char lcdminutel[1];
char lcd[5];
```

In the third part of the code, the hardware setup for digital input and output (I/O) is done. This code block is given in Listing 10.24 in terms of the PinConfig() function. In the first line of the function PinConfig(), port P2 is set as digital I/O. In the second and third lines, pin directions are determined for ports P1 and P2. As can be seen in Fig. 10.22, Register Select and Enable pins of the LCD are connected to pins P1.6 and P1.7 of the MSP430G2553. Two push buttons are also connected to pins P1.0 and P1.1 of the microcontroller. In software, the corresponding code line is P1DIR=0xFC. Here also, all other pins of port P1 are set as output. As can be seen in Fig. 10.22, Data pins of the LCD are connected to pins P2.0, P2.1, P2.2, and P2.3 of the microcontroller. In software, the corresponding code line is P2DIR=0xFF. Again, unused pins of port P2 are set as output. In the fourth line, pull-up/down resistors for button-connected pins of port P1 are enabled. In the fifth and sixth lines, output registers are set as P1OUT=0x03 and P2OUT=0x00. Unnecessary power consumption is also prevented for unused output pins by this procedure. On the other hand, high bits of the P1OUT register are used for choosing pull-up resistors for input pins. In the next three lines, interrupt settings for port P1 are done. Interrupts are obtained from two push buttons connected to pins P1.0 and P1.1. Therefore, P1IE=0x03. These interrupts are triggered by a high-to-low transition. Therefore, P1IES=0x03. Also, all interrupt flags are cleared initially by P1IFG=0x00.

Listing 10.24 Chronometer, the C code part III.

```
void PinConfig(void){
 P2SEL = 0x00;
 P1DIR = 0xFC;
 P2DIR = 0xFF;
```

```
 P1REN = 0x03;
 P1OUT = 0x03;
 P2OUT = 0x00;
 P1IE = 0x03;
 P1IES = 0x03;
 P1IFG = 0x00;
}
```

In the fourth part of the code, the hardware setup for the timer block is done. This code block is given in Listing 10.25 in terms of the `TimerConfig()` function. In the first line of the function `TimerConfig()`, the watchdog timer is disabled. In the second line, the VLO is chosen to source the ACLK at 12 kHz. In the third line, the timer interrupt is enabled. In the fourth line, the ACLK is chosen as the clock source with `TASSEL_1`. Its frequency is divided by 8 with `ID_3`. This 1.5 kHz clock is used in the timer block. The TAR register is cleared with `TACLR`. Also, the timer is stopped with `MC_0` since the system should not start until the Start button is pressed. In the fifth line, the time interval is set as 1 s by assigning 1499 to `TACCR0`. Remember, the period is $(\text{TACCR0}+1)/f_{\text{CLK}}$.

Listing 10.25 Chronometer, the C code part IV.

```
void TimerConfig(void){
 WDTCTL = WDTPW|WDTHOLD;

 BCSCTL3 |= LFXT1S_2;

 TACCTL0 = CCIE;
 TACTL = MC_0|ID_3|TASSEL_1|TACLR;
 TACCR0 = 1499;
}
```

ISR settings for port- and timer-based interrupts are given in Listing 10.26. As a port interrupt comes from the Start button when `start==0`, the system exits from LPM4. The system goes to the ISR `int_timer` every second by the timer-based interrupt. Then the variable `secondl` is increased by one within this ISR. `secondl` is cleared when it equals 10. Then, `secondh` is increased by one. There can be two different interrupts coming from the two buttons for port P1. The necessary actions for these two button presses are separately defined in the ISR `int_button`. After the ISR completes its task, the related interrupt flag is cleared to get a new interrupt.

Listing 10.26 Chronometer, the C code part V.

```
#pragma vector=TIMER0_A0_VECTOR
 __interrupt void int_timer(void){
 secondl++;
 if(secondl == 10){
 secondl = 0;
 secondh++;
```

```
 if(secondh == 6){
 secondh = 0;
 minutel++;
 if(minutel == 10){
 minutel = 0;
 minuteh++;
 }}}
}

#pragma vector=PORT1_VECTOR
 __interrupt void int_button(void){
 LPM4_EXIT;
 if(STARTSTOP){
 start ^= 1;
 P1IFG &= ~0x01;
 }
 if(RESET){
 secondh = 0;
 secondl = 0;
 minuteh = 0;
 minutel = 0;
 start = 0;
 TACTL |= TACLR;
 P1IFG &= ~0x02;
 }
}
```

Finally, the C code for the system (with all its components) is given in Listing 10.27. The code block doing the operation is put in an infinite loop. Every time the system returns from the ISR `int_timer` or `int_button`, it goes to this loop. Then the system calls the `Write_to_LCD()` function. This function converts the second and minute digits to corresponding character values, places them into the `lcd` character array, then writes to the LCD. Also, if the `start` variable equals one, the timer starts by writing `MC_1` to the `TACTL` register. Otherwise, the system goes to LPM4 and the timer stops. The LCD is initialized by the `lcd_init()` function in the main code. The GIE bit is also set to enable maskable interrupts.

Listing 10.27 Chronometer, the C code.

```
#include <msp430.h>
#include "lcd.h"

#define STARTSTOP ((P1IFG & 0x01) == 0x01)
#define RESET ((P1IFG & 0x02) == 0x02)

 int start = 0;
 int secondh = 0;
 int secondl = 0;
 int minuteh = 0;
 int minutel = 0;
 char lcdsecondh[1];
```

```
 char lcdsecondl[1];
 char lcdminuteh[1];
 char lcdminutel[1];
 char lcd[5];

void PinConfig(void);
void TimerConfig(void);
void Write_to_LCD(void);

void main(void)
{
 PinConfig();
 TimerConfig();
 lcd_init();
 _enable_interrupts();

 while(1)
 {
 if(start == 1){
 Write_to_LCD();
 TA0CTL |= MC_1;
 }
 if(start == 0){
 Write_to_LCD();
 LPM4;
}}
}

void PinConfig(void){
 P2SEL = 0x00;
 P1DIR = 0xFC;
 P2DIR = 0xFF;
 P1REN = 0x03;
 P1OUT = 0x03;
 P2OUT = 0x00;
 P1IE = 0x03;
 P1IES = 0x03;
 P1IFG = 0x00;
}

void TimerConfig(void){
 WDTCTL = WDTPW|WDTHOLD;

 BCSCTL3 |= LFXT1S_2;

 TACCTL0 = CCIE;
 TACTL = MC_0|ID_3|TASSEL_1|TACLR;
 TACCR0 = 1499;
}

void Write_to_LCD(void){
 itoa(secondh,lcdsecondh,10);
 itoa(secondl,lcdsecondl,10);
 itoa(minuteh,lcdminuteh,10);
 itoa(minutel,lcdminutel,10);
```

```
    lcd[0] = lcdminuteh[0];
    lcd[1] = lcdminutel[0];
    lcd[2] = ':';
    lcd[3] = lcdsecondh[0];
    lcd[4] = lcdsecondl[0];
    lcd_goto(1,1);
    lcd_writestr(lcd);
}

#pragma vector=TIMER0_A0_VECTOR
 __interrupt void int_timer(void){
 secondl++;
 if(secondl == 10){
 secondl = 0;
 secondh++;
 if(secondh == 6){
 secondh = 0;
 minutel++;
 if(minutel == 10){
 minutel = 0;
 minuteh++;
 }}}
}

#pragma vector=PORT1_VECTOR
 __interrupt void int_button(void){
 LPM4_EXIT;
 if(STARTSTOP){
 start ^= 1;
 P1IFG &= ~0x01;
 }
 if(RESET){
 secondh = 0;
 secondl = 0;
 minuteh = 0;
 minutel = 0;
 start = 0;
 TACTL |= TACLR;
 P1IFG &= ~0x02;
 }
}
```

10.11 Summary

This chapter was on time-based operations. We started with the oscillators available in the MSP430. Then we considered the BCM+ and available clocks. Since the user can select more than one clock for different operations, the clock properties should be known. We also considered the BCM+ under Grace. Next, we focused on low-power modes. Using low-power modes is extremely important for battery-based systems. Through them energy can be saved by disabling the CPU or peripherals when they are not needed. Then we considered the watchdog timer

(WDT+) module of the MSP430. This module can be used as a watchdog or as a timer. We provided sample C and assembly codes on both operations. We also explored the WDT+ operation under Grace. We looked at the TA module of the MSP430 afterwards. There are two identical timer modules in the MSP430G2553. Each module has a separate timer and three capture/compare blocks. We explored the operation of each block in detail. We also considered the usage of these blocks under Grace. Finally, we provided the design of a simple chronometer application by using concepts considered in this chapter.

10.12 Problems

10.1 Write a program in C such that the MSP430 is in low-power mode most of the time. Initially the green LED (connected to P1.6 on the MSP430 LaunchPad) is on and the red LED (connected to P1.0 on the MSP430 LaunchPad) is off. The CPU wakes up in periodic time intervals of

a. 10 s
b. 1 min
c. 1 h (if possible)
d. 1 day (if possible)

As the CPU wakes up, the LEDs toggle. Then the CPU will go to the low-power mode again.

10.2 Repeat Prob. 10.1 in assembly language.

10.3 Add a push button (connected to P1.3 on the MSP430 LaunchPad) to Prob. 10.1. When this button is pressed, the timer will reset itself and the LEDs will go to their initial states.

10.4 Repeat Prob. 10.3 in assembly language.

10.5 Repeat Prob. 10.3 using Grace.

10.6 Write a subroutine to display the characters on the LCD in assembly language.

10.7 Write international characters like ü, ı, and ç to the LCD in C and assembly languages.

11 Mixed Signal Systems

Chapter Outline

This chapter deals with analog-to-digital and digital-to-analog conversion. In microcontrollers, there are specific modules to convert analog signals to digital form. These perform sampling in time and quantization in amplitude. They are generally called analog-to-digital converters (ADCs). The ADC module under the MSP430G2553 is called ADC10. There is also a Comparator_A+ module which can be taken as a 1-bit ADC under the MSP430. For digital-to-analog conversion, there are also specific modules in microcontrollers. These perform interpolation between digital samples. They are generally called digital-to-analog converters (DACs). Unfortunately, the MSP430G2553 does not have a DAC module. Therefore, we will use the PWM operation to obtain the approximate analog representation of the corresponding digital signal. Next, we will start with the general explanation of analog and digital signals.

11.1 Analog and Digital Signals

A value changing with time or another dependent variable can be taken as a signal. There are two signal types, analog and digital. By definition, an analog signal can have its amplitude represented with infinite precision. It can also be defined for any time value. A digital signal can represent samples of the analog signal in time. Moreover, its amplitude values are also quantized. This means that the amplitude is represented by only certain values. More information on these signal types (in terms of theory) can be found in [8, 4].

The digital signal is the sampled and quantized form of the analog signal. Therefore, digital signal representation contains less information than its analog counterpart. This may seem a disadvantage for digital signal representation. In practice, this is not the case. Analog signals are hard to store and process. They are also prone to noise. Besides, the system for processing an analog signal is usually static. On the other hand, a digital signal is very robust to noise. The system to process a digital signal can be a code block. Hence, all the system parameters (or the system itself) can easily be changed by replacing a code block. That is why most recent systems are in digital form. In this book, we take the microcontroller as the digital system. Next, we will consider the most primitive module to convert an analog signal to digital form.

11.2 The Comparator

The comparator has two inputs called positive and negative. One of these inputs can be used for the reference voltage (either external or internal). The other is used for the input voltage. The comparator compares these two values. Let's assume that the input voltage is fed to the positive input and the reference voltage is fed to the negative input. If the input voltage is higher than the reference, the comparator output will be one. Otherwise, it will be zero. In other saying, the comparator output is just 1 bit. This operation can be taken as a 1-bit ADC.

11.2.1 *The Comparator_A+ Module*

The comparator module under the MSP430 is called Comparator_A+. A block diagram of Comparator_A+ is given in Fig. 11.1. As in other modules, Comparator_A+ is controlled by specific registers. They are explored next.

11.2.2 *Comparator_A+ Registers*

The Comparator_A+ module has three control registers. These are the control register 1, **CACTL1**, control register 2, **CACTL2**, and port disable register, **CAPD**. Their entries are shown in Tables 11.1, 11.2, and 11.3 in detail.

In Table 11.1, setting the **CAEX** bit exchanges the comparator inputs and inverts the comparator output. The **CARSEL** bit selects the terminal for the reference

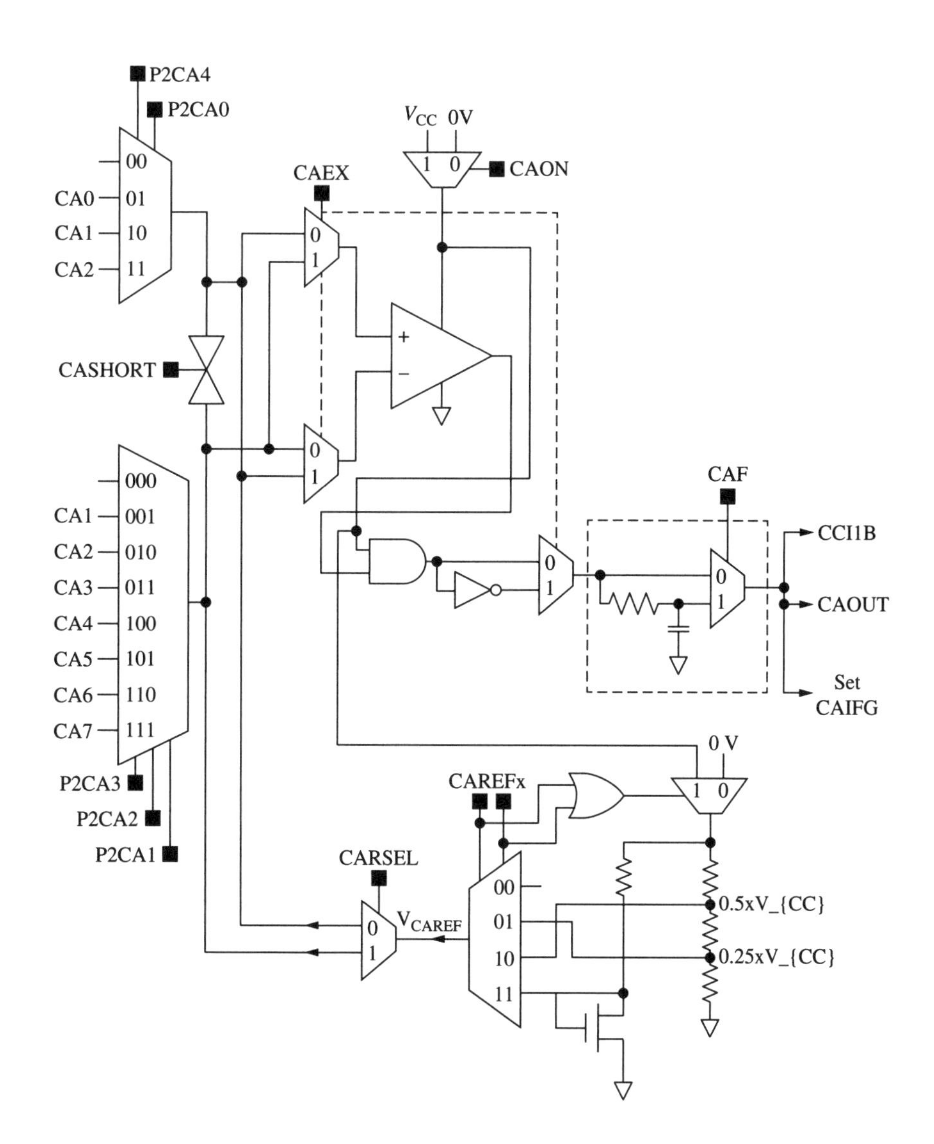

Figure 11.1 Block diagram of the Comparator_A+ module.

Table 11.1 The Comparator_A+ control register 1 (CACTL1).

Bits	7	6	5 - 4	3	2	1	0
	CAEX	CARSEL	CAREF	CAON	CAIES	CAIE	CAIFG

Table 11.2 The Comparator_A+ control register 2 (CACTL2).

Bits	7	6	5	4	3	2	1	0
	CASHORT	P2CA4	P2CA3	P2CA2	P2CA1	P2CA0	CAF	CAOUT

voltage V_{CAREF}. If CARSEL is zero, V_{CAREF} is applied to the positive terminal. Otherwise, it is applied to the negative terminal. For this scenario, CAEX is assumed to be zero. The **CAREF** bits select the source for the reference voltage V_{CAREF}. Based on the binary values from 00 to 11, the reference voltage is applied as follows: internal reference off (an external reference can be applied), $0.25 \times V_{CC}$, $0.5 \times V_{CC}$, and diode reference. Predefined constants for these are CAREF_0, CAREF_1, CAREF_2, and CAREF_3 respectively. The **CAON** bit turns on the comparator. The **CAIES** bit sets the interrupt edge select (low to high or reverse). The **CAIE** bit is used to enable the comparator interrupt. The **CAIFG** bit represents the interrupt flag.

In Table 11.2, setting the **CASHORT** bit short-circuits the positive and negative inputs of the comparator. Bits **P2CA4** and **P2CA0** are used to select the positive input of the comparator among pins CA0, CA1, and CA2. Similarly, bits **P2CA3**, **P2CA2**, and **P2CA1** are used to select the negative input of the comparator among pins CA1 to CA7. Please see Table 11.4 for the pin layout for the Comparator_A+ module. The **CAF** bit sets the comparator's output filter. The **CAOUT** bit keeps the comparator's output.

In Table 11.3, each bit disables the input buffer for the pins associated with the Comparator_A+ module. This reduces the current consumption for certain input voltage levels. For more detail on this issue, please see [17].

11.2.3 *The Pin Layout for the Comparator_A+ Module*

As in the previous chapters, we provide the pin layout of the MSP430G2553 in Fig. 11.2. This figure will serve for both the Comparator_A+ and ADC10 modules (to be explained in the following section). The pin usage table for the Comparator_A+ module is given in Table 11.4. As can be seen in this table, pins

Table 11.3 The Comparator_A+ port disable register (CAPD).

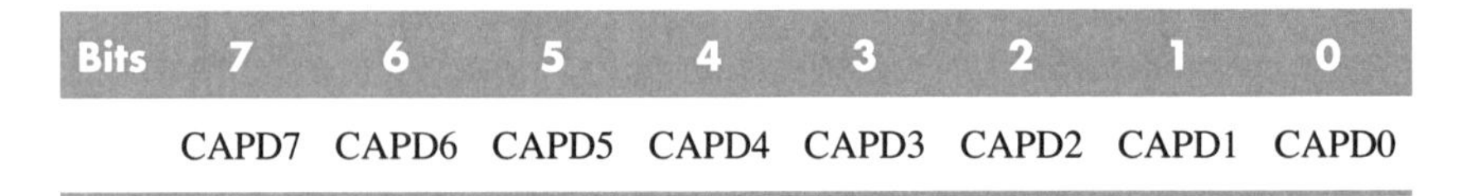

Bits	7	6	5	4	3	2	1	0
	CAPD7	CAPD6	CAPD5	CAPD4	CAPD3	CAPD2	CAPD1	CAPD0

Table 11.4

Pin usage table for the Comparator_A+ module.

Pin	Port Name	Usage Area
1	V_{CC}	Source voltage
2	P1.0/ CA0	Comparator_A+, CA0 input
3	P1.1/ CA1	Comparator_A+, CA1 input
4	P1.2/ CA2	Comparator_A+, CA2 input
5	P1.3/ CA3	Comparator_A+, CA3 input
	CAOUT	Comparator_A+, output
6	P1.4/ CA4	Comparator_A+, CA4 input
7	P1.5/ CA5	Comparator_A+, CA5 input
8	P2.0	
9	P2.1	
10	P2.2	
11	P2.3	
12	P2.4	
13	P2.5	
14	P1.6/ CA6	Comparator_A+, CA6 input
15	CA7	Comparator_A+, CA7 input
	P1.7/ CAOUT	Comparator_A+, output
16	RST	Reset
17		
18	P2.7	
19	P2.6	
20	V_{SS}	Ground voltage

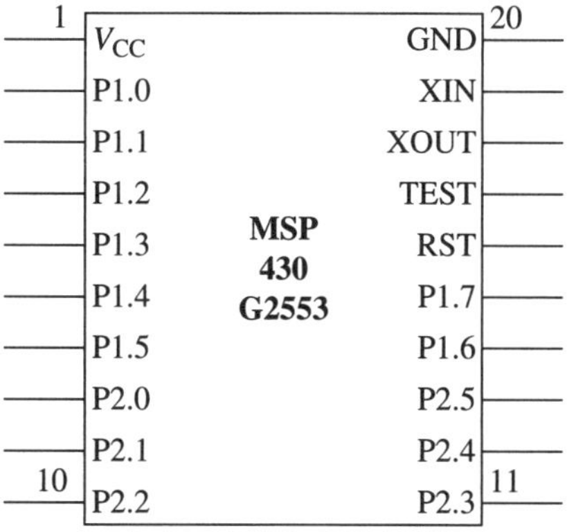

Pin	Left	MSP 430 G2553	Right	Pin
1	V_{CC}		GND	20
	P1.0		XIN	
	P1.1		XOUT	
	P1.2		TEST	
	P1.3		RST	
	P1.4		P1.7	
	P1.5		P1.6	
	P2.0		P2.5	
	P2.1		P2.4	
10	P2.2		P2.3	11

Figure 11.2

Pin layout of the MSP430G2553.

2 to 7, 13, and 14 can be used for comparator input. Only pins 5 and 15 can be used for comparator output.

11.2.4 Coding Practices for the Comparator_A+ Module

Below, we provide sample C and assembly codes for the usage of the Comparator_A+ module. The same operation is done in both codes. The input from CA1 is compared with the internal reference voltage, which is 0.25 × V_{CC} here. If the input voltage is greater than this reference voltage, the green LED (connected to the P1.6 on the MSP430 LaunchPad) will turn on. Otherwise, the red LED (connected to the P1.0 on the MSP430 LaunchPad) will turn on. This application can be taken as a simple battery check system.

In Listing 11.1, we first set the comparator parameters. Then we check the input values and turn on the appropriate LED in an infinite loop. To do so, we have to check the register CACTL2 since we cannot reach the CAOUT bit alone. In Listing 11.2, we perform the same operation in assembly language.

Listing 11.1 Usage of the Comparator_A+ module in C language.

```
#include <msp430.h>

#define REDLED BIT0
#define GREENLED BIT6

void main (void)
{
 WDTCTL = WDTPW|WDTHOLD;

 P1DIR = BIT0|BIT6;

 CACTL1 = CARSEL+CAREF_1+CAON;
//  0.25 Vcc = -comp, on
 CACTL2 = P2CA4; //  P1.1/CA1 = +comp

 while(1)
 {
 if(CAOUT & CACTL2)
 P1OUT = GREENLED;
 else P1OUT = REDLED;
 }
}
```

Listing 11.2 Usage of the Comparator_A+ module in assembly language.

```
 .cdecls C,LIST,"msp430.h"

 .text
 .retain
 .retainrefs

RESET
```

```
 mov.w #WDTPW|WDTHOLD,WDTCTL
 mov.w #__STACK_END,SP

 bis.b #41h,P1DIR

 mov.b #CARSEL+CAREF_1+CAON,CACTL1
 mov.b #P2CA4,CACTL2

Mainloop:
 bit.b #CAOUT,CACTL2
 jz REDLED

GREENLED:
 bis.b #40h,P1OUT
 bic.b #01h,P1OUT
 jmp Mainloop

REDLED:
 bis.b #01h,P1OUT
 bic.b #40h,P1OUT
 jmp Mainloop

;------------------------
;Stack Pointer definition
;------------------------
 .global __STACK_END
 .sect .stack

;-------------------
;Interrupt Vectors
;-------------------
 .sect RESET_VECTOR
 .short RESET
 .end
```

11.3 Comparator_A+ in Grace

The Comparator_A+ module can be used under Grace by clicking the **Comp_A+ 8 Channels** block shown in Fig. 5.11. Do not forget to check the box "Enable Comparator_A+ in my configuration" first to configure the Comparator_A+ module under Grace.

11.3.1 *The Basic User Mode*

The Comparator_A+ module can be configured basically by setting inputs, output, and the reference voltage in the basic user mode (given in Fig. 11.3). There are two drop-down lists for positive and negative inputs to the comparator. There is a third drop-down list for the reference voltage under Voltage Reference. There is also a list to select which input will use this reference voltage. Here, the two options are + Channel and - Channel. The output of the comparator module can be directed either to Timer_A CCI1B or to either of the two options from the drop-down list. The Comparator_A+ based interrupts can be adjusted in this mode also.

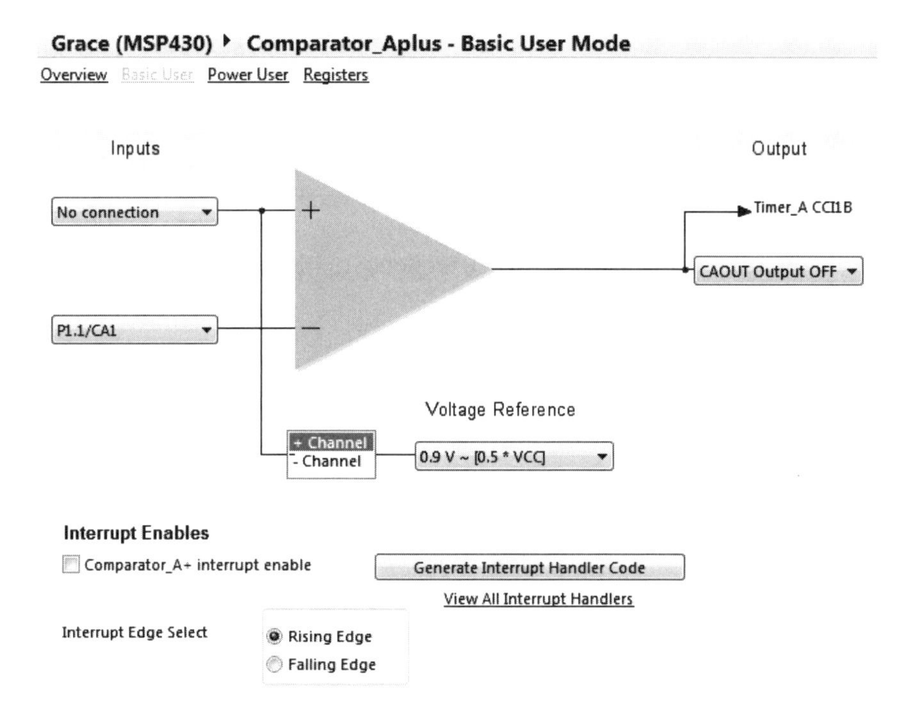

Figure 11.3 The basic user mode for the Comp_A+ block under Grace.

First, the user should check the Comparator_A+ interrupt enable box. The user can select the interrupt edge select type (whether it will occur on the rising or the falling edge) from the two check boxes. Then the prototype interrupt service routine (ISR) can be added to the InterruptVectors_init.c file by pressing the Generate Interrupt Handler Code button.

11.3.2 *The Power User Mode*

The power user mode for the Comparator_A+ module is given in Fig. 11.4. There are three additional check boxes in the power user mode. Two of them are related to the inputs. The Short inputs check box can be used to short-circuit inputs. The Flip Inputs, Inverse Output check box can be used to flip inputs and inverse output. The third check box, Enable Filter is related to the output. The user can enable the comparator's output filter by checking this box.

11.3.3 *The Register Controls Mode*

Finally, all of the preceding Comparator_A+ module settings can be done in the register controls mode given in Fig. 11.5. In this mode, Comparator_A+ registers CACTL1, CACTL2, and CAPD can be adjusted by appropriate check boxes. There is also a drop-down list to adjust the reference voltage value in CACTL1.

11.3.4 *Coding Practices*

In this section, we redo the comparator application given in Listing 11.1 using Grace. As a reminder, this application is a simple battery checker. We start by generating a Grace project. Then, we configure pins P1.0 and P1.6 under Grace.

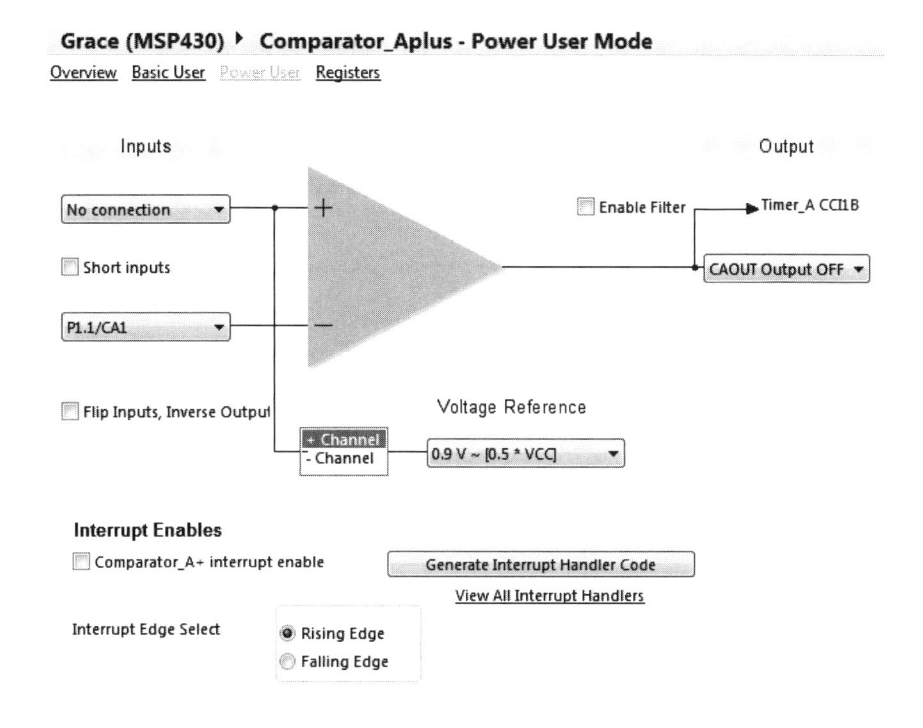

Figure 11.4

The power user mode for the Comp_A+ block under Grace.

Grace (MSP430) ▸ Comparator_Aplus - Register Controls

Overview Basic User Power User Registers

CACTL1, Comparator_A+ Control Register 1

7	6	5	4	3	2	1	0
CAEX	CARSEL	CAREFx		CAON	CAIES	CAIE	CAIFG
☐	☐	0.50 * VCC		☑	☐	☐	[R/W]

CACTL2, Comparator_A+ Control Register 2

7	6	5	4	3	2	1	0
CA SHORT	P2CA4	P2CA3	P2CA2	P2CA1	P2CA0	CAF	CAOUT
☐	☐	☐	☐	☑	☐	☐	[R]

CAPD, Comparator_A+, Port Disable Register

7	6	5	4	3	2	1	0
CAPD7	CAPD6	CAPD5	CAPD4	CAPD3	CAPD2	CAPD1	CAPD0
☐	☐	☐	☐	☐	☐	☑	☐

Figure 11.5

The register controls mode for the Comp_A+ block under Grace.

These should be set as GPIO output. These settings can be done by any of the three GPIO views. The internal reference voltage is connected to the positive input of the Comparator_A+ module. The voltage input is connected to the negative input of the Comparator_A+ module. In this application, we reconfigure the main.c file as given in Listing 11.3. After compiling the project, we can run our application.

Listing 11.3 Usage of the Comparator_A+ module under Grace, basic user mode.

```
/*
 * ========  Standard MSP430 includes ========
 */
#include <msp430.h>

/*
 * ========  Grace related includes ========
 */
#include <ti/mcu/msp430/Grace.h>

/*
 *  ========  main ========
 */
int main(void)
{
 Grace_init();
//  Activate Grace-generated configuration

 while(1)
 {
 if (CAOUT & CACTL2)
 P1OUT=BIT0;
 else P1OUT=BIT6;
 }

 return (0);
}
```

11.4 Analog-to-Digital Conversion

There are several analog-to-digital conversion (ADC) methods. Each has advantages and disadvantages [2]. The ADC10 module in the MSP430G2553 uses the successive approximation register (SAR) conversion method. Therefore, we will only deal with it in this section. Then we will focus on the properties of the ADC10 module.

11.4.1 *Successive Approximation Register Converter*

As the name implies, the SAR converter works iteratively in obtaining the digital form of the analog signal. In iteration, the MSB of the digital form is obtained first. Then, step-by-step, remaining lower-order bit values are obtained until the LSB is reached.

The SAR circuitry is shown in Fig. 11.6. In this figure, V_{IN} stands for the analog voltage value to be converted to digital form. The working principle of the

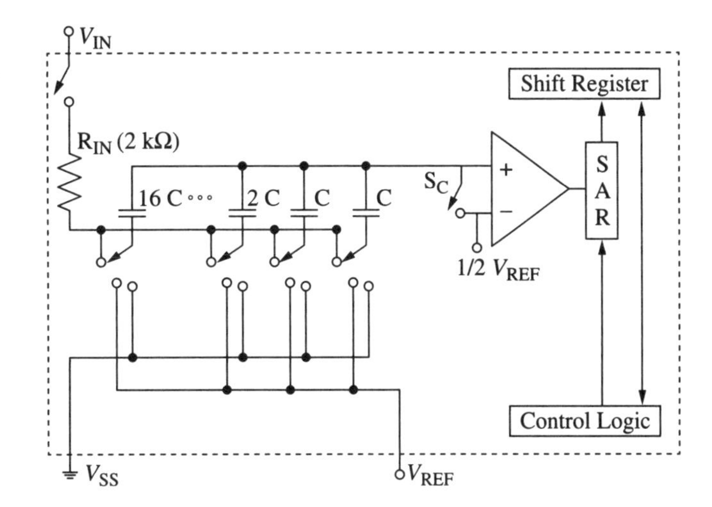

Figure 11.6

The circuit diagram of the SAR converter.

given SAR circuitry is as follows: Initially, all the capacitors will be discharged to the offset voltage of the comparator. As the analog signal is fed to the input, it will be kept at that value by a sample and hold circuit. Then this voltage is applied to all capacitors. Since each capacitor has a different capacitance (in powers of two), they will be charged accordingly. These values are compared with the reference voltage. Based on the comparison, the bit value (either zero or one) is generated and saved in the shift register. Then the reference voltage (V_{REF}) is updated and the conversion continues until the desired accuracy is obtained.

To explain how the SAR conversion works, we simulate it with the code given in Listing 11.4. Here, the constant `bitsize` represents the bit size of the digital form to be obtained. The variable `Vin` stands for the analog voltage to be converted to digital form. The variable `Vref` stands for the reference voltage. The array `bits` holds the digital form obtained. Finally, the variable `quantized` holds the quantized approximate form of the analog input voltage. We can observe the `bits` and `quantized` variables from the Expressions window in CCS.

Listing 11.4 The simulation program for the SAR conversion.

```
#include <msp430.h>

#define bitsize 10

 float Vref = 3.6;
 float Vin = 3.3;
 float thresh;
 float quantized = 0;
 int count;
 int bitval;
 int bits[bitsize];

void main (void)
{
```

```
WDTCTL = WDTPW|WDTHOLD;

Vref /= 2;
thresh = Vref;

for(count=0; count<bitsize; count++){
Vref /= 2;
if (Vin >= thresh)
{bitval = 1;
thresh += Vref;}
else
{bitval = 0;
thresh -= Vref;}

bits[count] = bitval;
quantized += 2*Vref*bitval;
}

while(1);
}
```

Let's consider an example of the use of this simulation program. Assume that we take the reference voltage (V_{REF}) as 3.6 (V). We set the bit size to 10 (bits). Assume the input voltage (V_{IN}) to be 1.9 (V). We will get the digital representation 1000011100b or 21Ch as we run the simulation program. As can be observed from the CCS Expressions window, the input voltage is approximated by 1.898438 V. This simulation program also gives an insight into the working principles of the SAR conversion.

11.4.2 *The ADC10 Module*

The layout of the ADC10 module is shown in Fig. 11.7. In this figure, the input channels A12 to A15 are connected to channel A11 internally. The input channel A10 is connected to the internal temperature sensor. In this figure, *TA_OUT0*, *TA_OUT1*, and *TA_OUT2* represent the timer block output (TA_OUTx) given in Fig. 10.9. Therefore, ADC10 can be directly triggered by any channel of the timer module.

11.4.3 *ADC10 Registers*

As can be seen in Fig. 11.7, the ADC10 module has several parameters. These are controlled by the two registers called **ADC10CTL0** and **ADC10CTL1**. We will first focus on the ADC10CTL0 register given in Table 11.5. As can be seen in this table, the user can select the voltage reference values, sample and hold time, sampling rate, reference voltage output, internal reference buffer, multiple sample and conversion, reference generators, turning on the SAR core, interrupt capabilities, and the conversion operation through the ADC10CTL0 register.

In Table 11.5, the most significant 3 bits of the ADC10CTL0 register, **SREFx**, are used for voltage reference values. In connection with Fig. 11.7, these correspond to *SREF2* (MSB), *SREF1*, and *SREF0* (LSB) respectively. These values

Figure 11.7

Block diagram of the ADC10 module.

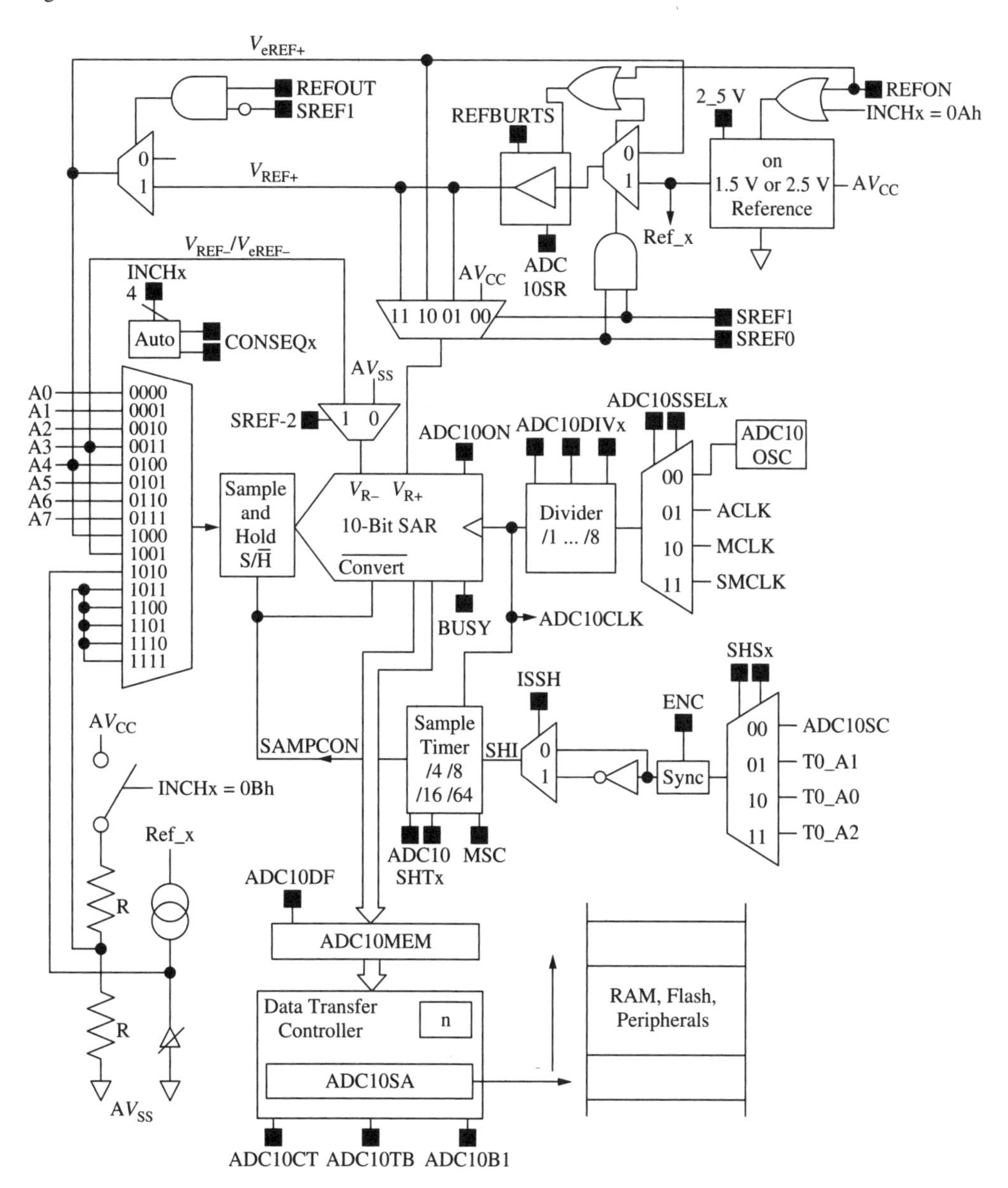

Table 11.5

ADC10 control register 0 (ADC10CTL0).

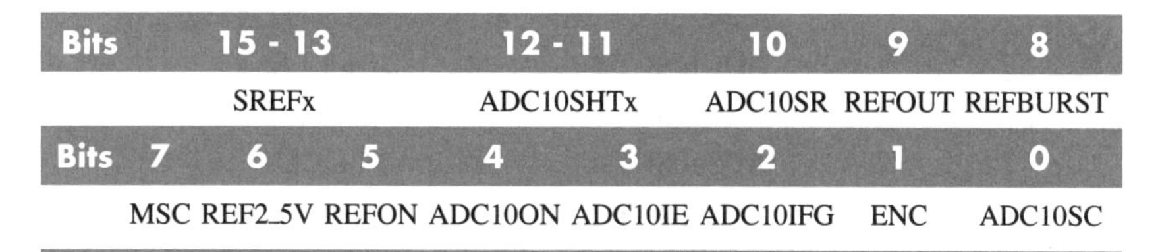

Bits	15 - 13	12 - 11	10	9	8
	SREFx	ADC10SHTx	ADC10SR	REFOUT	REFBURST

Bits 7	6	5	4	3	2	1	0
MSC	REF2_5V	REFON	ADC10ON	ADC10IE	ADC10IFG	ENC	ADC10SC

Table 11.6

SREFx values and corresponding constants.

Constant	V_{R+}	V_{R-}
SREF_0	V_{CC}	V_{SS}
SREF_1	V_{REF+}	V_{SS}
SREF_2	Ve_{REF+}	V_{SS}
SREF_3	Buffered Ve_{REF+}	V_{SS}
SREF_4	V_{CC}	V_{REF-}/Ve_{REF-}
SREF_5	V_{REF+}	V_{REF-}/Ve_{REF-}
SREF_6	Ve_{REF+}	V_{REF-}/Ve_{REF-}
SREF_7	Buffered Ve_{REF+}	V_{REF-}/Ve_{REF-}

are shown in detail in Table 11.6. Here, we provide the constants (from SREF_0 to SREF_7) defined in the MSP430 header file (given in the Appendix) instead of giving the individual values for these bits. In this table, V_{REF} represents the built-in reference voltage and Ve_{REF} represents the external reference voltage.

In Table 11.5, bits 12 and 11 (**ADC10SHTx**) are reserved for sample and hold times. The constants and the corresponding sample and hold times related to these bits are tabulated in Table 11.7. As can be seen in this table, the sample and hold time is related to the clock used in the ADC10 module.

The remaining bits for the ADC10CTL0 register have the following properties. The **ADC10SR** bit adjusts the sampling rate. When this bit is reset, the sampling rate can be up to 200 ksps. When it is set, the sampling rate may go up to 50 ksps. The **REFOUT** bit enables the reference voltage output. The **REFBURST** bit controls the internal reference buffer. When this bit is reset, the reference buffer will be fed to output continuously, independent of the status of the ADC10 module. When this bit is set, the reference voltage is fed to output only when the ADC10 module is active. The **MSC** bit allows multiple sample and conversion operations for valid sampling modes (to be explained next). The **REFON** bit enables the reference generator. The **REF2_5V** bit selects the reference voltage as either 1.5 or 2.5 V when it is reset and set, respectively. The REFON bit must also be enabled for this purpose. The **ADC10IE** bit enables the interrupts related to the ADC10 module. The **ADC10IFG** bit stands for the interrupt flag. The **ENC** bit enables the conversion. Finally, setting the **ADC10SC** bit starts the analog-to-digital conversion.

Table 11.7

ADC10SHTx values and corresponding constants.

Constant	Sample and Hold Time
ADC10SHT_0	4 × ADC10CLK
ADC10SHT_1	8 × ADC10CLK
ADC10SHT_2	16 × ADC10CLK
ADC10SHT_3	64 × ADC10CLK

Table 11.8

ADC10 control register 1 (ADC10CTL1).

Bits	15 - 12	11 - 10	9	8
	INCHx	SHSx	ADCDF	ISSH
Bits	**7 - 5**	**4 - 3**	**2 - 1**	**0**
	ADC10DIVx	ADC10SSELx	CONSEQx	ADC10BUSY

The second control register for the ADC10 module is ADC10CTL1, given in Table 11.8. As can be seen in this table, the user can select the input channel, sample and hold source, data format, whether or not to invert the sample and hold signal, ADC10 clock divider, ADC10 clock source, and conversion sequence mode through the ADC10CTL1 register.

In Table 11.8, the most significant 4 bits in the ADC10CTL1 control register, **INCHx**, are reserved for the input pins to the ADC core. Pins A0 to A7 (explained in Table 11.10) can be selected as inputs for the ADC. The constants corresponding to these inputs are INCH_0, INCH_1,..., INCH_7 respectively. The corresponding pin should also be enabled in the **ADC10AE0** register for analog input. The input should be selected as INCH_10 to use the internal temperature sensor of the MSP430. The **SHSx** bits select the sample and hold source. The constants for these bits are defined as SHS_0, SHS_1, SHS_2, and SHS_3. They correspond to ADC10SC, Timer_A output units 1, 0, and 2 respectively. The **ADC10DF** bit sets whether the data format for the ADC10 will be in binary or two's complement form. In the first form, the result will be right-justified binary in the 0000h–03FFh range. Zero corresponds to the bottom of the input range. In the second form, the lowest 6 bits are always clear, and bit 15 gives the sign. Zero corresponds to the middle of the input range, and lower inputs give negative values. The **ISSH** bit enables or disables the sample input signal inversion.

The **ADC10DIVx** bits set the clock divider values. The clock in the ADC10 module can be divided into 1, 2,..., 8 based on the constants ADC10DIV_0, ADC10 DIV_1,..., ADC10DIV_7. The **ADC10SSELx** bits are used to select the clock source for the ADC10 module. This clock can be taken from the module's internal oscillator ADC10OSC, ACLK, MCLK, or SMCLK. These can be selected by constants ADC10SSEL_0, ADC10SSEL_1, ADC10SSEL_2, and ADC10SSEL_3 respectively. The default clock is ADC10OSC. It runs nominally at 5 MHz. It is automatically enabled when needed and disabled when conversions have finished. This makes it the most convenient source for most applications. The **CONSEQx** bits are used to select the conversion mode. These are:

- **Single channel, single conversion**: Single conversion for the channel selected by INCHx bits. This mode is represented by the constant CONSEQ_0.
- **Sequence of channels**: One conversion in multiple channels, beginning with the channel selected by INCHx bits and decrementing to channel A0. The operation stops after the conversion of channel A0. This mode is represented by the constant CONSEQ_1.
- **Repeat single channel**: A single channel selected by INCHx bits is converted repeatedly until stopped. This mode is represented by the constant CONSEQ_2.

- **Repeat sequence of channels**: Repeated conversions for multiple channels, beginning with the channel selected by INCHx bits and decrementing to channel A0. The sequence ends after conversion of channel A0. The next trigger signal restarts the sequence. This mode is represented by the constant CONSEQ_3.

The final bit in the ADC10CTL1 register is **ADC10BUSY**. It will be set while the conversion is in progress.

As the conversion is done, the result will be written to the **ADC10MEM** register. Based on the previously mentioned reference selections, this will be in the form of

$$N = nint\left(\frac{1024 \times (V_{\mathrm{IN}} - V_{\mathrm{R-}})}{V_{\mathrm{R+}} - V_{\mathrm{R-}}}\right) \tag{11.1}$$

where N is the output of the conversion operation. $nint(\cdot)$ stands for the nearest integer function. V_{R+} and V_{R-} are the upper and lower reference voltages given in Table 11.6. V_{IN} is the input voltage applied to the ADC. Here, the constant 1024 corresponds to the maximum level, 2^{10}, that can be obtained from the ADC10 module.

In its basic form, three steps are needed to perform a single conversion with the ADC10 module. First, the ADC10 should be configured through its registers ADC10CTL0 and ADC10CTL1. Meanwhile, the ADC10ON bit should be set to enable the ADC10 module. ADC10 control registers can only be adjusted when the ENC bit is reset. Second, the ENC bit should be set to enable conversion. Third, the conversion should be started either by setting the ADC10SC bit or by an edge from the Timer_A (TA) module. The last two steps must be repeated for each conversion, which requires clearing and setting the ENC bit again. This two-step sequence is relaxed for conversions triggered by software. In this case, the first conversion can be triggered by setting the ENC and ADC10SC bits together in a single instruction. Subsequent conversions can be triggered by setting the ADC10SC bit alone without toggling the ENC bit. The interrupt flag ADC10IFG is set when the result is written to the ADC10MEM except when the data transfer controller is used. This will be explained next.

11.4.4 Multiple Conversions Using the Data Transfer Controller

In some applications, more than one conversion may be needed. Instead of performing these conversions within a loop, the data transfer controller (DTC) inside the ADC10 module can be used. The DTC automatically transfers the conversion results from the ADC10MEM to specified memory locations.

DTC can be controlled through two registers, **ADC10DTC0** and **ADC10 DTC1**. The entries of the ADC10DTC0 register are shown in Table 11.9.

In Table 11.9, the most significant 4 bits of the ADC10DTC0 register are reserved. The **ADC10TB** bit is used to select the transfer mode. When this bit is reset, one-block transfer mode will be active. When it is set, two-block transfer mode will be active. When the **ADC10CT** bit is reset, data transfer stops when one block (in one-block mode) or two blocks (in two-block mode) have completed. When this bit is set, data is transferred continuously. The DTC operation is stopped only if the ADC10CT bit is reset or ADC10SA (holding the data transfer start address) is written to. The **ADC10B1** bit indicates which block is filled with the

Table 11.9

ADC10 DTC control register 0 (ADC10DTC0).

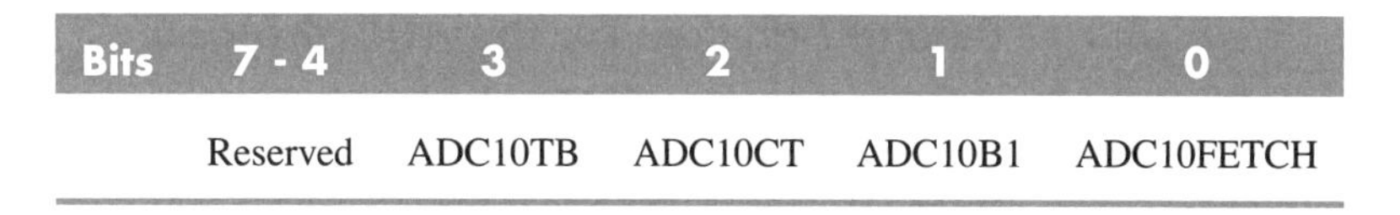

Bits	7 - 4	3	2	1	0
	Reserved	ADC10TB	ADC10CT	ADC10B1	ADC10FETCH

ADC10 conversion results (in the two-block mode). When this block is reset, it indicates that block 2 is filled. When it is set, it indicates that block 1 is filled. For this bit to be valid, ADC10IFG and ADC10TB should be set. The **ADC10FETCH** bit should normally be reset.

The ADC10DTC1 register is used to define the number of transfers per block. The user should also declare the data transfer start address through the **ADC10SA** register. We will show how to make this declaration in C and assembly languages in Sec. 11.4.6.

11.4.5 *The Pin Layout for the ADC10 Module*

The pin usage for the ADC10 module is given in Table 11.10. As can be seen in this table, pins 2 to 7 and 14, 15 can be used for ADC input. These are labeled as A0–A7. Reference voltages for conversion can be fed through pins 5 and 6.

Table 11.10

Pin usage table for the ADC10 module.

Pin	Port Name	Usage Area
1	V_{CC}	Source voltage
2	P1.0/ A0	ADC10 analog input A0
3	P1.1/ A1	ADC10 analog input A1
4	P1.2/ A2	ADC10 analog input A2
5	P1.3/	ADC10 analog input A3
	A3 V_{REF-}/Ve_{REF-}	ADC10 negative reference voltage
6	P1.4/	ADC10 analog input A4
	A4 V_{REF+}/Ve_{REF+}	ADC10 positive reference voltage
7	P1.5/A5	ADC10 analog input A5
8	P2.0	
9	P2.1	
10	P2.2	
11	P2.3	
12	P2.4	
13	P2.5	
14	P1.6/A6	ADC10 analog input A6
15	P1.7/A7	ADC10 analog input A7
16	RST	Reset
17		
18	P2.7	
19	P2.6	
20	V_{SS}	Ground voltage

11.4.6 *Coding Practices for the ADC10 Module*

In this section, we will provide several examples using the ADC10 module. Our examples will be in C and assembly languages. We will also deal with the usage of the DTC for multiple conversions.

Our first ADC example in the C language is given in Listing 11.5. Here, the ADC10 module is basically used as a comparator. The input voltage level at pin A1 (pin P1.1) is checked within an infinite loop. If the value there is above a certain level, the red LED is turned on.

Listing 11.5 The first ADC example in C language.

```
#include <msp430.h>

#define LED BIT0

void main(void)
{
 WDTCTL = WDTPW|WDTHOLD;

 P1DIR = LED;
 P1OUT = 0x00;

 ADC10CTL0 = SREF_0|ADC10SHT_2|ADC10ON;
// Vcc and Vss references,sample for 16 cycles
// ADC on

 ADC10CTL1 = INCH_1|SHS_0|ADC10DIV_0|ADC10SSEL_0\
 |CONSEQ_0;
// Input channel1 (A1), trigger using ADC10SC bit,
// use internal ADC clock, single channel
// and single conversion

 ADC10AE0 = BIT1; //Enable conversion
 ADC10CTL0 |= ENC;

 while(1)
 {
 ADC10CTL0 |= ADC10SC; //Trigger new conversion
 while (ADC10CTL1 & BUSY);
// Wait if ADC10 core is active

 if(ADC10MEM >= 0x0200)
 P1OUT = LED;
 else P1OUT = 0x00;
 }
}
```

In the second ADC example, given in Listing 11.6, the voltage level at pin A1 is checked within an infinite loop. This level is converted to the corresponding floating-point representation. Then it is saved in the variable `voltage`. This variable can be observed through the Watch window.

Listing 11.6 The second ADC example in C language.

```
#include <msp430.h>

void main(void)
{
 WDTCTL = WDTPW|WDTHOLD;

 volatile float voltage;

 P1DIR = BIT4;

 ADC10CTL0 = SREF_0|ADC10SHT_2|ADC10ON;
 ADC10CTL1 = INCH_1|SHS_0|ADC10DIV_0|ADC10SSEL_0\
 |CONSEQ_0;

 ADC10AE0 = BIT1;
 ADC10CTL0 |= ENC;

 while(1){
 ADC10CTL0 |= ADC10SC;
 while (ADC10CTL1 & BUSY);
 //  Wait if ADC10 core is active
 voltage = ((ADC10MEM*3.55)/0x03FF);
 }
}
```

In the third ADC example, given in Listing 11.7, the internal temperature sensor of the ADC10 module is used. The temperature is measured 20 times using the DTC module. The average temperature value is calculated. Again, the average temperature is converted to the true (scaled) value. Then it is saved in the `avgtemp` variable. This can also be observed in the Watch window.

Listing 11.7 The third ADC example in C language.

```
#include <msp430.h>

#define nsamp 20

 float avgtemp = 0;
 void main( void)
{
 WDTCTL = WDTPW|WDTHOLD;

  int count;
 unsigned int temparr[nsamp];

 ADC10CTL1 = CONSEQ_2 + INCH_10 + ADC10DIV_7;
 ADC10CTL0 = SREF_1 + ADC10SHT_3 + REFON + ADC10ON \
 + ADC10IE + MSC;
// Repeat single channel, Temp Sensor, ADC10CLK/8

 ADC10DTC1 = nsamp; // number of conversions

 while (ADC10CTL1 & BUSY);
// Wait if ADC10 core is active
```

```
 ADC10SA = (unsigned int) temparr;
// Data buffer start address

 ADC10CTL0 |= ENC + ADC10SC;
// Sampling and conversion start

 _enable_interrupts();

 LPM0; // LPM0, ADC10_ISR will force exit

 for(count=0; count<nsamp; count++)
 avgtemp += temparr[count];

 avgtemp = avgtemp/nsamp;
 avgtemp = ((avgtemp-673)*423)/1024;

 while(1);
}

#pragma vector=ADC10_VECTOR
__interrupt void ADC10_ISR(void){
 LPM0_EXIT;
}
```

The first assembly code for the ADC10 module is given in Listing 11.8. Here, the ADC10 module is used as a comparator. If the input voltage is greater than a predefined value, the red LED on the MSP430 LaunchPad will turn on.

Listing 11.8 The first ADC example in assembly.

```
 .cdecls C,LIST,"msp430.h"

 .text
 .retain
 .retainrefs

RESET
 mov.w #WDTPW|WDTHOLD,WDTCTL
 mov.w #__STACK_END,SP

 mov.w #ADC10SHT_2+ADC10ON+ADC10IE,ADC10CTL0
;16x, enable int.
 mov.w #INCH_1,ADC10CTL1
 bis.b #02h,ADC10AE0
;P1.1 ADC10 option select
 bis.b #01h,P1DIR

Mainloop:
 bis.w #ENC+ADC10SC,ADC10CTL0
;Start sampling/conversion

 bis.w #LPM0+GIE,SR
;LPM0, ADC10_ISR will force exit

 bic.b #01h,P1OUT
 cmp.w #01FFh,ADC10MEM
```

```
;ADC10MEM = A1 > 0.5AVcc?
 jlo Mainloop ;Again
 bis.b #01h,P1OUT
 jmp Mainloop

;---------------------------
ADC10_ISR ;Exit LPM0 on reti
;---------------------------
 bic.w #LPM0,0(SP) ;Exit LPM0 on reti
 reti

;-----------------------
;Stack Pointer definition
;-----------------------
 .global __STACK_END
 .sect .stack

;------------------
;Interrupt Vectors
;------------------
 .sect RESET_VECTOR
 .short RESET
 .sect  ADC10_VECTOR
 .short ADC10_ISR
 .end
```

In the second ADC example, given in Listing 11.9, the DTC module and the internal temperature sensor of the ADC10 module are used. The temperature is measured 16 times. If the average temperature value is greater than 27°C, then the red LED on the MSP430 LaunchPad is turned on.

Listing 11.9 The second ADC example in assembly.

```
 .cdecls C,LIST,"msp430.h"
 .text
 .retain
 .retainrefs
RESET
 mov.w #WDTPW|WDTHOLD,WDTCTL
 mov.w #__STACK_END,SP
 mov.w #CONSEQ_2+INCH_10+ADC10DIV_7,ADC10CTL1
 mov.w #SREF_1+MSC+ADC10SHT_3+REFON,ADC10CTL0
 bis.w #ADC10ON+ADC10IE,ADC10CTL0
 mov.b #10h,ADC10DTC1 ;16 conversions
 bis.b #41h,P1DIR
 bic.b #41h,P1OUT
Mainloop:
 bic.w #ENC,ADC10CTL0
 mov.w #0200h,ADC10SA ;Data buffer start
 bis.w #ENC+ADC10SC,ADC10CTL0
```

```
;Sampling and conversion start

 bis.w #LPM0+GIE,SR
;LPM0, ADC10_ISR will force exit

 call #Average

 cmp.w #02E2h,R6 ;Temp  > 27C?
 jlo Less
 bis.b #01h,P1OUT
 jmp Mainloop
Less:
 bis.b #40h,P1OUT
 jmp Mainloop

Average:
 mov.w #0200h,R5 ;set as pointer
 mov.w #0000h,R6 ;set as sum

Total:
 add.w @R5,R6
 incd.w R5
 cmp.w #0220h,R5
 jlo Total
 rra.w R6
 rra.w R6
 rra.w R6
 rra.w R6
 ret

;------------------------
ADC10_ISR
;------------------------
 bic.b #41h,P1OUT
 bic.w #LPM0,0(SP)
 reti

;------------------------
;Stack Pointer definition
;------------------------
 .global __STACK_END
 .sect .stack

;-------------------
;Interrupt Vectors
;-------------------
 .sect RESET_VECTOR
 .short RESET
 .sect ADC10_VECTOR
 .short ADC10_ISR
 .end
```

11.5 Digital-to-Analog Conversion

To convert a digital signal to analog form, a digital-to-analog conversion (DAC) is needed. Unfortunately, the MSP430G2553 does not have such a module. Therefore, we will use pulse width modulation (PWM) for this purpose.

11.5.1 *Pulse Width Modulation*

The output signal in pulse width modulation (PWM) is a high-frequency digital pulse sequence. The width of the pulses changes depending on the setting. As this high-frequency signal is smoothed by a low-pass filter (such as a simple RC circuit), we will get an average voltage which is approximately a dc signal. This average voltage (V_{avg}) obtained from the system will be

$$V_{avg} = \frac{t_{on}}{t_{period}} \times V_{CC} = D \times V_{CC} \tag{11.2}$$

where t_{on} is the duration the pulse will be on and t_{period} is the period of the pulse. The ratio t_{on}/t_{period} is called the duty cycle (D) of the PWM signal. By changing the duty cycle (changing t_{on} and keeping t_{period} constant), an approximate dc signal can be generated.

The Timer_A capture/compare mode can be used to generate PWM signals in the MSP430. This mode is explained in detail in Sec. 10.7.2. Based on the definitions there, the duty cycle of the PWM becomes

$$D = \frac{TACCRx}{TACCR0 + 1} \tag{11.3}$$

where TACCRx stands for the xth Timer_A capture/compare register. To note here, the timer will be in the *up mode* for this equation to be valid.

The arrangement for PWM while using the Timer_A capture/compare block (in output mode 7) is as follows: The output is turned on when the TAR value reaches zero. It is turned off when the TAR value reaches TACCRx. This means that increasing the value in TACCRx increases the duty cycle. The period of PWM is the same as that of timer. Therefore, its frequency is

$$f_{PWM} = \frac{f_{CLK}}{TACCR0 + 1} \tag{11.4}$$

where f_{CLK} stands for the frequency of the timer clock.

Finally, the generated PWM signal can be taken out from ports P1 and P2. Please see Table 10.18 for specific pins. Do not forget to set the corresponding bit in the PxSEL register for analog output.

11.5.2 *Coding Practices for PWM*

In this section, we provide sample C and assembly codes for PWM generation. We benefit from the capture/compare block of the TA module in generating the PWM. In Listing 11.10, the period and the duty cycle of the generated PWM can be adjusted by two variables. The output is fed to the green LED of the MSP430

LaunchPad. Therefore, the PWM signal can be observed by the dimness of the LED.

Listing 11.10 The PWM generation example in C language.

```
#include <msp430.h>

void main(void)
{
 WDTCTL = WDTPW|WDTHOLD;

 int period = 0x0FFF; // period of the PWM
 float D = .8; // duty cycle, max value 1

 P1DIR |= BIT6;
 P1SEL |= BIT6;

 TACCR0 = period-1; //  PWM Period
 TACCR1 = period*D; //  CCR1 PWM duty cycle
 TACCTL1 = OUTMOD_7;  //  CCR1 reset/set
 TACTL = TASSEL_2|MC_1; //  SMCLK, up mode

 LPM1;
}
```

In the assembly code, given in Listing 11.11, we follow the same strategy as in Listing 11.10. The period and the duty cycle of the PWM signal can be adjusted in this example also. The output is fed to the green LED of the MSP430 LaunchPad. Therefore, the PWM signal can be observed by the dimness of the LED.

Listing 11.11 The PWM generation example in assembly.

```
 .cdecls C,LIST,"msp430.h"

 .text
 .retain
 .retainrefs

RESET
 mov.w #WDTPW|WDTHOLD,WDTCTL
 mov.w #__STACK_END,SP

 bis.b #40h,P1DIR
 bis.b #40h,P1SEL

 mov.w #0FFFh-1,TACCR0 ;PWM Period
 mov.w #OUTMOD_7,CCTL1

 mov.w #00FFh,TACCR1 ;PWM Duty Cycle
 mov.w #TASSEL_2+MC_1,TACTL

 bis.w #LPM1,SR ;CPU off

;------------------------
;Stack Pointer definition
;------------------------
```

```
 .global __STACK_END
 .sect .stack

;-------------------
;Interrupt Vectors
;-------------------
 .sect RESET_VECTOR
 .short RESET
 .end
```

11.6 ADC10 in Grace

The ADC10 module can be used under Grace by clicking the ADC10 10-bit block shown in Fig. 5.11. To configure the ADC10 module under Grace, do not forget to check the "Enable ADC10 in my configuration" box first.

11.6.1 *The Basic User Mode*

The basic user mode for ADC10 is shown in Fig. 11.8. In this mode, the ADC10 module can be configured basically by setting the ADC channel, signal bandwidth, impedance, and sampling rate. The sampling time is calculated by the impedance value. Grace directs the user to the MSP430 User's Guide [17] for this issue. The user can select the ADC channel from the ADC Channel drop-down list. Finally, the drop-down list Sampling Rate can be used to select the timer to be used in the sampling operation. If the user wants to sample the signal in an irregular manner, then he or she should choose the Manually Sample option from the drop-down list. The other three options Timer_A3 Channel 0, Timer_A3 Channel 1, and Timer_A3 Channel 2 in fact correspond to TA_OUT0, TA_OUT1, and TA_OUT2 in Fig. 11.7. The ADC10-based interrupts can be adjusted in this mode also. First, the user should check the ADC10 interrupt enable box. Then the prototype ISR can be added to the InterruptVectors_init.c file by pressing the Generate Interrupt Handler Code button.

11.6.2 *The Power User Mode*

The power user mode for the ADC10 block is shown in Fig. 11.9. In this mode, the GPIO pins to be used can be selected from the Enable External GPIO Pin check boxes. Configurations for these pins can be done by the check boxes under the Enable ADC Channel Config menu. The user can also select the sampling operation type (whether it will be from a single channel or a sequence of channels) in this menu. Reference voltages are organized in Negative and Positive Reference Voltage lists in the power user mode. The system ground or the external negative reference voltage values can be selected from the Negative Reference Voltage list. The user can select the positive reference voltage from the Positive Reference Voltage list. The buffer setting for the external voltage can also be done by check boxes there. The conversion type (single, repeated) can be selected from the Conversion Type list. The user can also select the sample and hold time from the Sample & Hold Time list. There are four options for the ADC10CLK here. The user can invert the sample and hold the signal with its check box in this menu. As in the

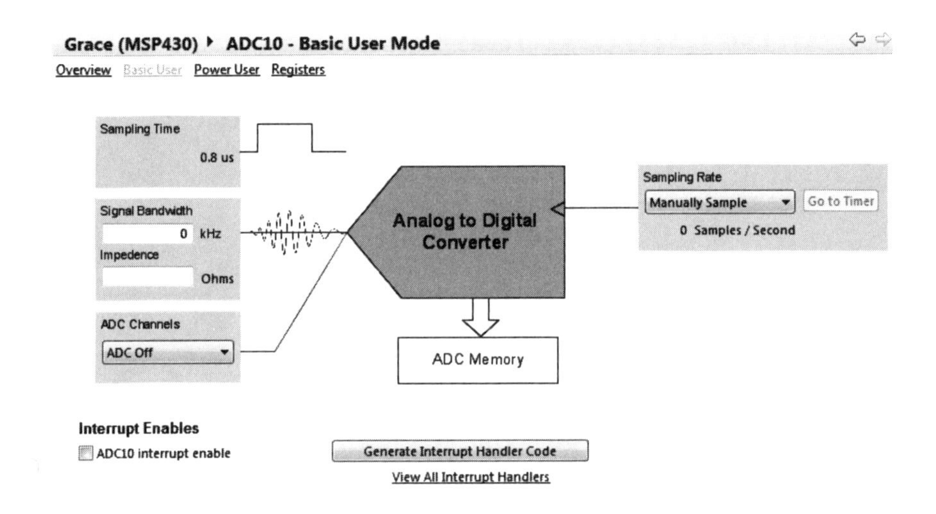

Figure 11.8 The basic user mode for the ADC10 block under Grace.

basic user mode, the user can also enter the impedance value into the related box to set the sample time. Grace selects the suitable sample and hold time from the list based on the entered impedance value. The ADC clock source can be selected from the ADC Clock Source list. The frequency divider for the selected clock can be selected from the Clock Divider drop-down list by the clock source list. The user can also use the ADC Trigger Source & Sampling Rate list to select an appropriate source for this operation. The user can enable the two's complement operation by its check box. The interrupt operations in the power user mode are the same as in the basic user mode.

The DTC block in the ADC10 module can be enabled in the power user mode. Its properties can be set within the Automatic Data Transfer Controller block. Here the user should enter the Starting Memory Address and Memory Block Size values in the appropriate boxes. To note here, the starting memory address can be entered as a global variable defined in the main.c program of the Grace project. Also, one-block, two-block, and one-time or continuous data transfer modes can be selected by checking related boxes under this block.

11.6.3 *The Register Controls Mode*

Finally, all the above ADC10 module settings can be done in the register controls mode shown in Fig. 11.10. In this mode, the ADC10 registers ADC10CTL0, ADC10CTL1, and ADC10AE0 can be adjusted by appropriate check boxes. Also, the DTC registers ADC10DTC0, ADC10DTC1, and ADC10SA can be set in this mode.

11.6.4 *Coding Practices*

In this section, we provide two ADC examples using Grace. In the first application, we use the basic user mode with the internal temperature sensor of the MSP430. Here, either the red or the green LED is turned on, depending on the temperature

Figure 11.9

The power user mode for the ADC10 block under Grace.

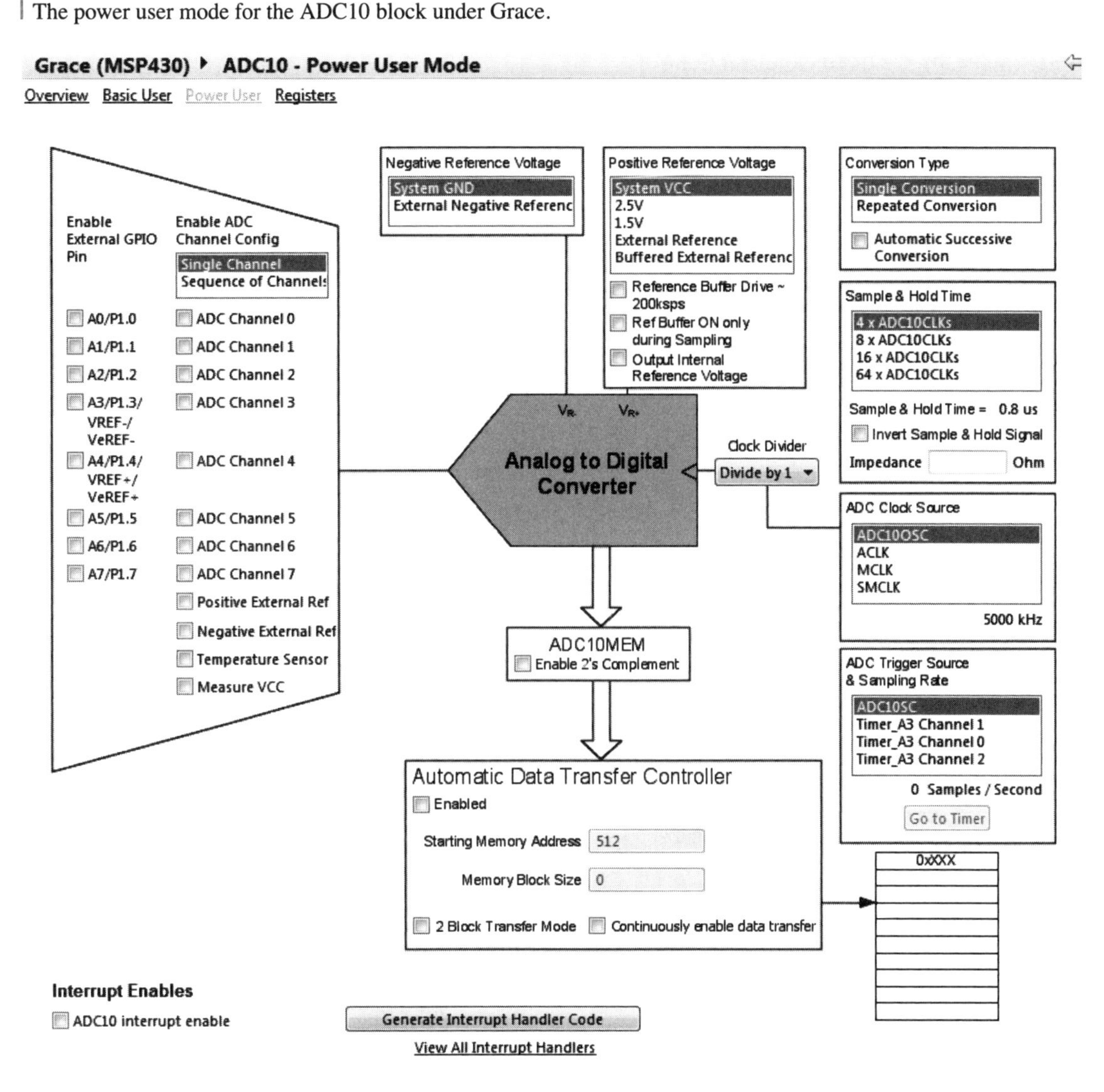

value measured. The settings for this operation are as follows: Red and green LEDs are set as output from the GPIO. The temperature sensor is selected from the ADC Channels drop-down list. Sampling rate is selected as Manually Sample from the associated drop-down list. The main.c file of the Grace project will be as in Listing 11.12 for this application.

In this application, we also use the ADC10 interrupts. Therefore, the ADC interrupts should be enabled. The ADC ISR under InterruptVectors_init.c will be as given in Listing 11.13. As we compile and run the project, the MSP430 will check the temperature value in an infinite loop. Depending on the measured value, either the red or green LED will turn on.

Figure 11.10

The register controls mode for the ADC10 block under Grace.

Grace (MSP430) ▸ ADC10 - Register Controls

Overview Basic User Power User Registers

ADC10CTL0, ADC10 Control Register 0

15–13	12–11	10	9	8	7	6	5	4	3	2	1	0
SREFx	ADC10SHTx	ADC10SR	REFOUT	REF BURST	MSC	REF2_5V	REFON	ADC10 ON	ADC10IE	ADC10 IFG	ENC	ADC10SC
VR+ = VCC and VR- =	4 x ADC10Cl									[R/W]	[R/W]	[R/W]

ADC10CTL1, ADC10 Control Register 1

15–12	11–10	9	8	7–5	4–3	2–1	0
INCHx	SHSx	ADC10DF	ISSH	ADC10DIVx	ADC10SSELx	CONSEQx	ADC10 BUSY
ADC Channel 0	ADC10SC			Divide by 1	ADC10OSC	Single chan	[R]

ADC10AE0, Analog (Input) Enable Control Register 0

7:0

0

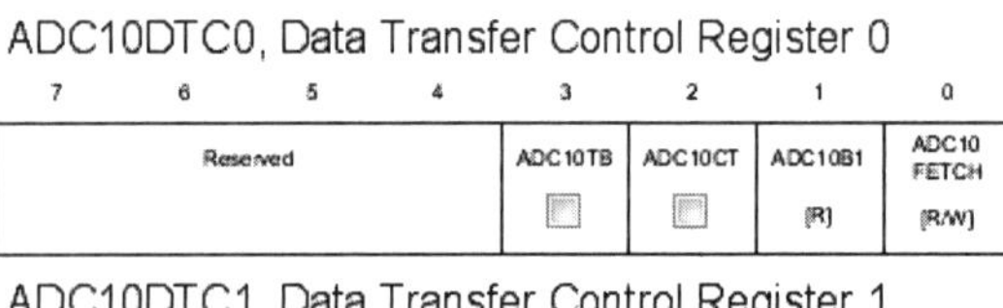
ADC10DTC0, Data Transfer Control Register 0

7–4	3	2	1	0
Reserved	ADC10TB	ADC10CT	ADC10B1	ADC10 FETCH
			[R]	[R/W]

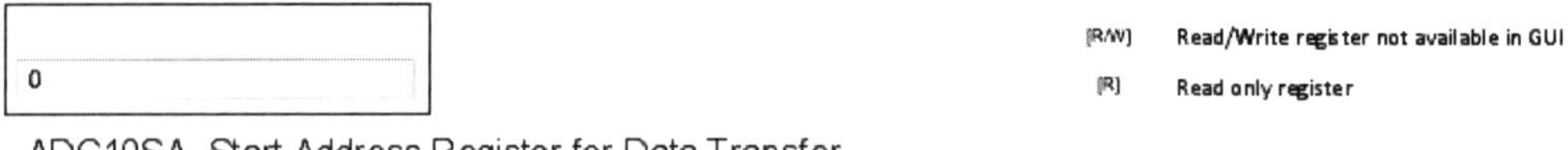
ADC10DTC1, Data Transfer Control Register 1

7:0

0

[R/W] Read/Write register not available in GUI

[R] Read only register

ADC10SA, Start Address Register for Data Transfer

15:1

512

Modifiable only when ENC = 0

Listing 11.12 The main.c file of the first ADC example under Grace, basic user mode.

```
/*
 * ========  Standard MSP430 includes ========
 */
#include <msp430.h>

/*
 * ========  Grace related includes ========
 */
```

```
#include <ti/mcu/msp430/Grace.h>

/*
 *  ========  main ========
 */
int main(void)
{
 Grace_init();
//  Activate Grace-generated configuration
 ADC10CTL0 |= ADC10SC;
//  ADC Start Conversion - Software trigger
 while ((ADC10CTL0 & ADC10IFG) == 0);
//  Loop until ADC10IFG is set indicating ADC
//  conversion is complete

 LPM4;

 return (0);
}
```

Listing 11.13 The ISR file of the first ADC example under Grace, basic user mode.

```
#pragma vector=ADC10_VECTOR
__interrupt void ADC10_ISR_HOOK(void)
{
 if (ADC10MEM > 0x02D5)
 P1OUT = BIT0;
 else
 P1OUT = BIT6;
}
```

In the second example, we redo the DTC-based temperature sensing application given in Listing 11.7 using Grace. Unlike from Listing 11.7, we use the two-block transfer mode here. In the power user mode, we make the following adjustments: We select the Temperature Sensor from the Enable ADC Channel Config list. Then we set the negative and positive reference voltage values as System GND and 1.5 V respectively. We select the conversion type as repeated with the "automatic successive conversion" box checked. The sample and hold time is 64 × ADC10CLK. The ADC10 clock source is selected as ADC10OSC with the clock division value of four.

We enable the DTC by its check box. Then, we enter "temparr" (defined in the main.c file of the Grace project) to the Starting Memory Address box in the DTC block. We enable the two-block transfer mode by checking its box. Since we are using a two-block transfer mode, we enter eight (half of the total number of samples to be taken) into the Memory Block Size box. The main.c file of the Grace project will be as in Listing 11.14 for this application.

Listing 11.14 The main.c file of the second ADC example under Grace, power mode.

```
/*
 * ========  Standard MSP430 includes ========
 */
#include <msp430.h>

/*
 * ========  Grace related includes ========
 */
#include <ti/mcu/msp430/Grace.h>

/*
 *  ======== main ========
 */

 unsigned int temparr[16];
 float avgtemp=0;

int main(void)
{
int count;

 Grace_init();
//  Activate Grace-generated configuration

//  ADC Start Conversion - Software trigger
 ADC10CTL0 |= ADC10SC;

 LPM0;

 for(count=0;count<16;count++)
 avgtemp+=temparr[count];

 avgtemp=avgtemp/16;
 avgtemp=((avgtemp-673)*423)/1024;

 LPM4;

 return(0);
}
```

We use the ADC interrupts in this application also. Therefore, we enable the ADC interrupts by checking a box. The ADC ISR under InterruptVectors_init.c will be as in Listing 11.15. As we compile and run the project, the MSP430 will take 16 temperature samples. The user can observe their (scaled) average value from the `avgtemp` float variable.

Listing 11.15 The ISR file of the second ADC example under Grace, power mode.

```
#pragma vector=ADC10_VECTOR
__interrupt void ADC10_ISR_HOOK(void)
{
 LPM0_EXIT;
}
```

11.7 Non-Touch Paper Towel Dispenser Application

The purpose of this application is to learn how to use the ADC and PWM on the MSP430 microcontroller. As a real-world application, we design a non-touch paper towel dispenser. In this section, we provide the equipment list, layout of the circuit, procedure, and system design specifications.

11.7.1 *Equipment List*

Following is the equipment list to be used in this application.

- One 12-V dc adaptor
- One LM7805 voltage regulator
- One 330-ηF capacitor
- One 10-μF electrolytic capacitor
- One 100-ηF capacitor
- One light-dependent resistor (LDR)
- One LED
- One 12-V dc motor
- One L293D motor driver integrated circuit (IC)
- One 220-Ω resistor
- One 10-kΩ resistor

L293D Motor Driver: In this application, a dc motor will be used. We will use the L293D dual H-bridge motor driver IC to control it. This IC can be used to drive two dc motors simultaneously, both in forward and reverse directions. Pin names and their descriptions for the L293D IC are given in Table 11.11.

In this application, the PWM signal generated by the MSP430G2553 will be fed to the *INH* pin. The *IN1* and *IN2* pins will be used to specify the direction of the rotation. This is done by setting one of these pins and resetting the other.

11.7.2 *Layout*

The layout of this application is shown in Fig. 11.11. For more information on the voltage supply block, please see Fig. 9.3.

11.7.3 *System Design Specifications*

In the first part of the application, we will design a non-touch towel dispenser using an LDR and an LED. When the user crosses his or her hand by the LDR, this will indicate that the paper towel is needed. This should generate a timer interrupt. The LED will turn on for 4 s to indicate that the paper towel is fed. During this time, no other paper towel request is accepted. When the waiting time is over, the LED will turn off. The system will wait for a new paper towel request.

In the second part of the application, we will repeat the first part using a dc motor instead of the LED. To do so, we should set the PWM frequency to 5 kHz.

Table 11.11 Pin names and descriptions for the L293D IC.

Pin No	Name	Function
1	INH1	Enable pin for Motor 1; active high
2	IN1	Input 1 for Motor 1
3	OUT1	Output 1 for Motor 1
4	GND	Ground (0 V)
5	GND	Ground (0 V)
6	OUT2	Output 2 for Motor 1
7	IN2	Input 2 for Motor 1
8	VC	Supply voltage for motors; 9–12 V (up to 36 V)
9	INH2	Enable pin for Motor 2; active high
10	IN3	Input 1 for Motor 2
11	OUT3	Output 1 for Motor 2
12	GND	Ground (0 V)
13	GND	Ground (0 V)
14	OUT4	Output 2 for Motor 2
15	IN4	Input 2 for Motor 2
16	V_CC	Supply voltage; 5 V (up to 36 V)

The duty cycle of the PWM signal should be 50%. The dc motor will rotate for 4 s to simulate the feeding of the paper towel. Again, no other paper towel request is accepted during this time. After the waiting time is over, the motor will stop.

11.7.4 *The C Code for the System*

In the first part of the code, given in Listing 11.16, constants and global variables are defined. This is done to make the code more readable. In this code block, `LedOn` and `LedOff` constants are used to turn on and turn off the LED. The `Count` variable is used for a 4-s delay. The `Control` variable is used to reject any interrupts during operation.

Listing 11.16 Non-touch paper towel dispenser, the C code part I.

```
#define LedOn (P1OUT |= 0x08)
#define LedOff (P1OUT &= ~0x08)

int Count = 0;
int Control = 0;
```

In the second part of the code, given in Listing 11.17, the hardware configurations for the digital input and output (I/O), timer, and ADC modules are done. In this code block, configurations for each hardware module are done in a separate function. In the `PinConfig()` function, pin directions are assigned as `P1DIR=0xFE`

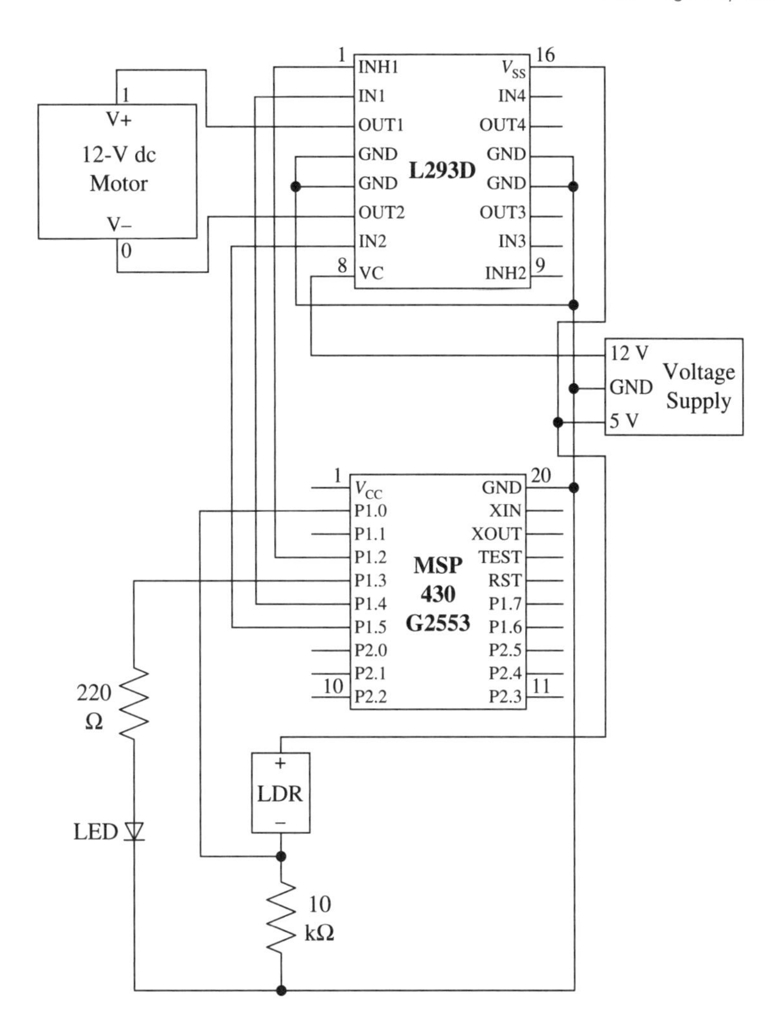

Figure 11.11

Layout of the non-touch paper towel dispenser application.

in the first line since the LED is connected to pin P1.3 and the LDR is connected to pin P1.0. All other unused pins are set as output. In the second line, all output pins are reset. In the `TimerConfig()` function, the watchdog timer is disabled in the first line. In the second line, the VLO is chosen to source the ACLK at 12 kHz. In the third line, the timer interrupt is enabled. In the fourth line, the timer is stopped with `MC_0` because the timer should not start until an input comes from the LDR. In the fifth line, the time interval is set as 1 s by writing 1499 to the TACCR0 register. Remember, period = (TACRR0+1)/f_{CLK}. In the `ADCConfig()` function, the `ADC10CTL0` register is configured in the first line. The `ADC10ON` bit is set to enable the ADC10 module. Reference voltages for the ADC10 are taken from V_{CC} and V_{SS} which are analog power supplies for the microcontroller. `ADC10SHT_3` is used to choose 64 clock cycles to take a sample. In the second line, the `ADC10CTL1` register is configured. First, A0 is chosen as the input channel with `INCH_0`. Then, `ADC10SSEL_0` is used to choose the internal ADC oscillator (with about 5 MHz frequency) as the clock source. ADC10DIV_0 is used for no frequency division. The trigger for a new conversion is set as the ADC10SC bit with SHS_0. Single-channel, single-conversion mode is selected with CONSEQ_0. In the third line, A0 is enabled as the analog input with ADC10AE0 = BIT0. In the fifth line, conversion is enabled with ADC10CTL0 |= ENC.

Listing 11.17 Non-touch paper towel dispenser, the C code part II.

```
void PinConfig(void){
 P1DIR = 0xFE;
 P1OUT = 0x00;
}

void TimerConfig(void){
 WDTCTL = WDTPW|WDTHOLD;

 BCSCTL3 |= LFXT1S_2;

 TACCTL0 = CCIE;
 TACTL = MC_0;
 TACCR0 = 1499;
}

void ADCConfig(void){
 ADC10CTL0 = SREF_0|ADC10SHT_3|ADC10ON;
 ADC10CTL1 = INCH_0|SHS_0|ADC10DIV_0|ADC10SSEL_0\
 |CONSEQ_0;
 ADC10AE0 = BIT0;
 ADC10CTL0 |= ENC;
}
```

In the third part of the code, given in Listing 11.18, the ISR settings for the timer are done as follows: The system generates a timer interrupt at every second. Through the ISR, the counter is increased. If the counter equals four, the LED is turned off, Count and Control variables are cleared, and the timer is stopped with MC_0.

Listing 11.18 Non-touch paper towel dispenser, the C code part III.

```
#pragma vector=TIMER0_A0_VECTOR
__interrupt void isr_name (void){
 Count++;
 if(Count == 4){
 LedOff;
 TACTL = MC_0;
 Control = 0;
 Count = 0;
 }
}
```

Finally, the C code for the system (with all its components) for the first part of the application is given in Listing 11.19. The code block performing the operation is placed in an infinite loop. In this loop, a new conversion is triggered with the code line `ADC10CTL0 |= ADC10SC`. Then the system waits until this conversion is complete with the code line `while((ADC10CTL1 & ADC10BUSY) == ADC10BUSY)`. After the conversion is done, the obtained value is written to the ADC10MEM register. This value changes between 03FFh (at full light) and 01FFh (at no light). The ADC10MEM value is compared with the reference value 0300h. If it is smaller than this reference value, the LED is turned on first. Then the code line `TACTL = MC_1 | ID_3 | TASSEL_1 | TACLR` is used to start the timer. ACLK is set as the clock source, the clock frequency is divided by eight, and the TAR register is reset. Finally, the `Control` variable is set. It is reset again after the 4-s time delay. This disables any new interrupt request for the timer during this period. Also before this while loop, the global interrupt enable (GIE) bit is set to enable maskable interrupts.

Listing 11.19 Non-touch paper towel dispenser, the C code for the first part of the application.

```
#include <msp430.h>

#define LedOn (P1OUT |= 0x08)
#define LedOff (P1OUT &= ~0x08)

 int Count = 0;
 int Control = 0;

void PinConfig(void);
void TimerConfig(void);
void ADCConfig(void);

void main(void)
{
 PinConfig();
 TimerConfig();
 ADCConfig();
 _enable_interrupts();

 while(1){
 ADC10CTL0 |= ADC10SC;

 while((ADC10CTL1&ADC10BUSY)==ADC10BUSY);

 if(Control == 0){
 if(ADC10MEM < 0x0300){
 LedOn;
 TACTL = MC_1|ID_3|TASSEL_1|TACLR;
 Control = 1;
 }}}
```

```
}

void PinConfig(void){
 P1DIR = 0xFE;
 P1OUT = 0x00;
}

void TimerConfig(void){
 WDTCTL = WDTPW|WDTHOLD;

 BCSCTL3 |= LFXT1S_2;

 TACCTL0 = CCIE;
 TACTL = MC_0;
 TACCR0 = 1499;
}

void ADCConfig(void){
 ADC10CTL0 = SREF_0|ADC10SHT_3|ADC10ON;
 ADC10CTL1 = INCH_0|SHS_0|ADC10DIV_0|ADC10SSEL_0|CONSEQ_0;
 ADC10AE0 = BIT0;
 ADC10CTL0 |= ENC;
}

#pragma vector=TIMER0_A0_VECTOR
__interrupt void isr_name (void){
 Count++;
 if(Count == 4){
 LedOff;
 TACTL = MC_0;
 Control = 0;
 Count = 0;
 }
}
```

The C code for the system (with all its components) is modified for the second part of the application. It is given in Listing 11.20. In this code block, first constant definitions for the LED are changed for the motor to `MotorStart` and `MotorStop`. The motor starts to turn when pin P1.2 is set as PWM output (by `P1SEL |= 0x04`). The motor stops when pin P1.2 is set as digital I/O (by `P1SEL &=~0x04`). In the `PinConfig()` function, pins P1.4 and P1.5 are used as inputs to the motor driver. Pin P1.3 is unused this time. `P1OUT = 0x10` is used to set one of the motor driver inputs to turn it in one direction. In the `TimerConfig()` function, the timer block is reconfigured to generate a 5-kHz PWM signal with 50% duty cycle. A 250-kHz clock signal is used for the timer block. A 5-kHz PWM with 50% duty cycle is obtained by `TACCR0=49` and `TACCR1=25`. Also, the reset/set mode is selected with `TACCTL1 = OUTMOD_7` for the PWM. In the timer ISR, the `Count` variable must be equal to 20,000 to obtain 4-s delay since the time interval is 0.2 ms this time. If the `Count` variable equals 20,000, the motor is stopped instead of turning off the LED. In the infinite while loop, if the ADC10MEM value is smaller than the reference value, the motor starts instead of turning on the LED. Also, when the timer starts, the SMCLK is used instead of the ACLK and it is divided by four this time.

Listing 11.20 Non-touch paper towel dispenser, the C code for the second part of the application.

```
#include <msp430.h>

#define MotorStart (P1SEL |= 0x04)
#define MotorStop (P1SEL &= ~0x04)

 int Count = 0;
 int Control = 0;

void PinConfig(void);
void TimerConfig(void);
void ADCConfig(void);

void main(void)
{
 PinConfig();
 TimerConfig();
 ADCConfig();

 _enable_interrupts();

 while(1){
 ADC10CTL0 |= ADC10SC;

 while((ADC10CTL1&ADC10BUSY)==ADC10BUSY);

 if(Control == 0){
 if(ADC10MEM < 0x0300){
 MotorStart;
 TACTL = MC_1|ID_2|TASSEL_2|TACLR;
 Control = 1;
 }}}
}

void PinConfig(void){
 P1DIR = 0xFE;
 P1OUT = 0x10;
}

void TimerConfig(void){
 WDTCTL = WDTPW|WDTHOLD;

 TACCTL0 = CCIE;
 TACCTL1 = OUTMOD_7;
 TACTL = MC_0;
 TACCR0 = 49;
 TACCR1 = 25;
}

void ADCConfig(void){
```

```
  ADC10CTL0 = ADC10ON|SREF_0|ADC10SHT_3;
  ADC10CTL1 = INCH_0|ADC10SSEL_0|ADC10DIV_0|SHS_0|CONSEQ_0;
  ADC10AE0 = BIT0;
  ADC10CTL0 |= ENC;
}

#pragma vector=TIMER0_A0_VECTOR
__interrupt void isr_name (void){
  Count++;
  if(Count == 20000){
  MotorStop;
  TACTL = MC_0;
  Control = 0;
  Count = 0;
  }
}
```

11.8 Summary

The MSP430 can process analog signals as well as digital signals. In this chapter, we considered ADC and DAC operations. We first focused on the Comparator_A+ module. It provides a binary output by comparing its input values. We provided sample codes and Grace usage examples for this module. Then we focused on the ADC10 module. This module provides a 10-bit digital representation of the analog signal fed to it. In analog-to-digital conversion, the ADC10 uses the SAR method. We explored the operation principles of this method through a simulation program. As in the Comparator_A+ module, we provided sample C and assembly codes. We also considered the ADC10 module under Grace. Unfortunately, the MSP430G2553 does not have a DAC module. Therefore, we used PWM to obtain analog signals from digital representations. Although the obtained analog signal is an approximation, for most applications it is sufficient. We used the timer module under Grace to generate PWM signals. We should use an external DAC module to obtain a precise analog signal. We provide such an example in Chap. 14. Finally, we considered the non-touch paper towel dispenser system as a real-world application. It contains both ADC and PWM operations. We designed the system step-by-step both in hardware and software.

11.9 Problems

11.1 Use Listing 11.4 to calculate the 10-bit SAR conversion of the analog voltage levels 1.2, 2.85, and 3.243 V. The reference voltage will be 3.6 V.

11.2 Design a battery charge controller using the MSP430 with the following specifications:

a. The battery will be connected between pin P1.1 and the ground of the MSP430 LaunchPad.
b. The Comparator_A+ module will be used in operation.
c. The control operation will be performed only when the push button (connected to P1.3 on the MSP430 LaunchPad) is pressed.

d. The system will be in an appropriate low-power mode during idle times.

e. If the voltage level of the battery is above a threshold (let's say $0.25 \times V_{CC}$ V), the green LED (connected to P1.6 on the MSP430 LaunchPad) will turn on. Otherwise the red LED (connected to P1.0 on the MSP430 LaunchPad) will turn on.

11.3 Repeat Prob. 11.2 under Grace.

11.4 Repeat Prob. 11.2 using the ADC10 module. Here, set the threshold as $0.32 \times V_{CC}$ V.

11.5 Repeat Prob. 11.4 under Grace.

11.6 Repeat Probs. 11.2 and 11.4 in assembly language.

11.7 Repeat Prob. 11.4 using thc DTC module. Here, take 16 samples and calculate their average using this module. Use this value in operation.

11.8 Repeat Prob. 11.7 in assembly language.

11.9 Repeat Prob. 11.7 under Grace.

12 Digital Communication

Chapter Outline

Data transfer between two (or more) microcontrollers becomes a necessity for complex projects. Moreover, some peripheral devices (such as sensors and digital-to-analog converter [DAC] modules) communicate with the microcontroller through data transfer channels. Therefore, digital communication has become an essential part of a modern microcontroller. In the MSP430, the module responsible for digital communication is called the universal serial communication interface (USCI). This module supports universal asynchronous receiver/transmitter (UART), serial peripheral interface (SPI), and inter integrated circuit (I^2C) communication modes. In this chapter, we will concentrate on the USCI module and the communication modes it provides. Here we will only concentrate on the communication between two devices. For details on communication between more than two devices, please see [17]. We begin with a brief description of the USCI module.

12.1 Universal Serial Communication Interface

There are two USCI modules called USCI_A0 and USCI_B0 in the MSP430. USCI_A0 can support UART and SPI communication modes. Similarly, USCI_B0 can support SPI and I^2C communication modes. In this section, we will describe the general properties of the USCI_A0 and USCI_B0 modules.

12.1.1 USCI Registers

The USCI module has several special function control and status registers for the UART, SPI, and I^2C communication modes. Some of these registers are specific to the communication mode. Some of them share the same name for different communication modes. All USCI registers are listed in Tables 12.1 and 12.2. In these tables, the usage area of each register is also provided.

We will explain the control and status registers given in Tables 12.1 and 12.2 in detail for each communication mode in the following sections. However, receive and transmit buffer registers for the USCI_A0 and USCI_B0 modules deserve

Table 12.1 USCI_A0 control and status registers.

Register Name	Short Form	Used in
USCI_A0 control register 0	UCA0CTL0	UART, SPI
USCI_A0 control register 1	UCA0CTL1	UART, SPI
USCI_A0 baud rate control register 0	UCA0BR0	UART, SPI
USCI_A0 baud rate control register 1	UCA0BR1	UART, SPI
USCI_A0 modulation control register	UCA0MCTL	UART, SPI
USCI_A0 status register	UCA0STAT	UART, SPI
USCI_A0 receive buffer register	UCA0RXBUF	UART, SPI
USCI_A0 transmit buffer register	UCA0TXBUF	UART, SPI
USCI_A0 auto baud rate control register	UCA0ABCTL	UART
USCI_A0 IrDA transmit control register	UCA0IRTCTL	UART
USCI_A0 IrDA receive control register	UCA0IRRCTL	UART

Table 12.2

USCLB0 control and status registers.

Register	Short Form	Used in
USCLB0 control register 0	UCB0CTL0	SPI, I^2C
USCLB0 control register 1	UCB0CTL1	SPI, I^2C
USCLB0 bit rate control register 0	UCB0BR0	SPI, I^2C
USCLB0 bit rate control register 1	UCB0BR1	SPI, I^2C
USCLB0 status register	UCB0STAT	SPI, I^2C
USCLB0 receive buffer register	UCB0RXBUF	SPI, I^2C
USCLB0 transmit buffer register	UCB0TXBUF	SPI, I^2C
USCLB0 I^2C interrupt enable register	UCB0I2CIE	I^2C
USCLB0 I^2C own address register	UCB0I2OA	I^2C
USCLB0 I^2C slave address register	UCB0I2SA	I^2C

specific consideration here. The data to be transmitted should be written to the transmit buffer register for any communication mode. These are **UCA0TXBUF** and **UCB0TXBUF** for the USCLA0 and USCLB0 modules respectively. Similarly, the data received will be read from the receive buffer register. These are **UCA0RXBUF** and **UCB0RXBUF** for the USCLA0 and USCLB0 modules respectively.

There are also two special-function interrupt registers used by the UART, SPI, and I^2C communication modes. These are special function register (SFR) interrupt enable register (IE2) and SFR interrupt flag register (IFG2). These are described in Tables 12.3 and 12.4. The IE2 register is responsible for enabling interrupts. As given in Table 12.3, the **UCA0TXIE** and **UCB0TXIE** bits enable the transmit interrupts for the related USCI module. Similarly, the **UCA0RXIE** and **UCB0RXIE** bits enable the receive interrupts for the related USCI module. Bits **UCA0TXIFG**, **UCB0TXIFG**, **UCA0RXIFG**, and **UCB0RXIFG**, given in Table 12.4, are set when an interrupt occurs from a transmission or reception operation in the related USCI module.

12.1.2 *USCI Clocks*

The USCI module has three clocks, BRCLK, BITCLK, and BITCLK16. The BRCLK represents the selected clock for the USCI module. The UART mode can use UC0CLK, ACLK, and SMCLK as BRCLK. UC0CLK is the external clock

Table 12.3

Interrupt enable register 2 (IE2).

Bits	7 - 2		1	0
Unused	UCB0TXIE	UCB0RXIE	UCA0TXIE	UCA0RXIE

Table 12.4

Interrupt flag register 2 (IFG2).

Bits	7 - 2		1	0
Unused	UCB0TXIFG	UCB0RXIFG	UCA0TXIFG	UCA0RXIFG

for the UART mode. It can be fed through pin P1.4 when this pin is not used by the SPI mode. The SPI mode can use ACLK and SMCLK as BRCLK. Finally, the I^2C mode can use UC1CLK, ACLK, and SMCLK as BRCLK. UC1CLK is the external clock for the I^2C mode. It can be fed through pin P1.5 when this pin is not used by the SPI mode. BITCLK is generated from the BRCLK. It is mainly used in controlling the bit transmission and reception rates. Finally, the BITCLK16 is used as the sampling clock in oversampling mode. In the following sections, we will explain all these clocks in specific communication modes.

There are two registers to divide the clock for the USCLA0 and USCLB0 modules. These are called baud rate control register 0 (**UCA0BR0**) and baud rate control register 1 (**UCA0BR1**) in the USCLA0 module. UCA0BR0 and UCA0BR1 registers form the 16-bit division coefficient for the clock. This is called UCBRx. In this coefficient, UCA0BR0 forms the low byte and UCA0BR1 forms the high byte. In the USCLB0 module, the registers used in clock division are called bit rate control register 0 (**UCB0BR0**) and bit rate control register 1 (**UCB0BR1**). They can be used in the same manner as in the USCLA0 registers to form the UCBRx.

12.1.3 *Common Properties*

The UART, SPI, and I^2C communication modes are initialized by the same steps. Initially, the USCI module must be reset to configure all related USCI registers. The USCI module must be set after this operation. Finally, if the interrupts are used in the USCI module, they should be enabled. We will explore these steps for each communication mode separately in the following sections.

Another common issue for the UART, SPI, and I^2C communication modes is the SMCLK usage with low-power modes. When the USCI module is clocked by SMCLK, it is activated automatically even if it is deactivated by a low-power mode. As a result, all other modules using the SMCLK also restart. This may cause error. Therefore, the SMCLK should be used carefully with the USCI module. Also for SPI and I^2C slave modes, no internal clock is needed since the master device provides the clock. Therefore, the microcontroller can be held in LPM4. It wakes up by a receive or transmit interrupt.

12.1.4 *Pin Layout for USCI*

We provide the pin layout of the MSP430G2553 in Fig. 12.1 (again to be compact). The usage of these in the USCI perspective are listed in Table 12.5. Do not forget to set these pins by appropriate PxSEL bits before using them.

12.2 Universal Asynchronous Receiver/Transmitter

UART is the asynchronous communication mode used between two or more devices. Being asynchronous, there is no need for a common clock in the UART. Hence, connected devices can work independently. In fact, UART is the only asynchronous communication mode in the MSP430. UART is simple to use compared to the synchronous communication modes to be considered in the following sections. In this section, we will only focus on the UART mode for communication between two microcontrollers (or one microcontroller and a host computer). Also,

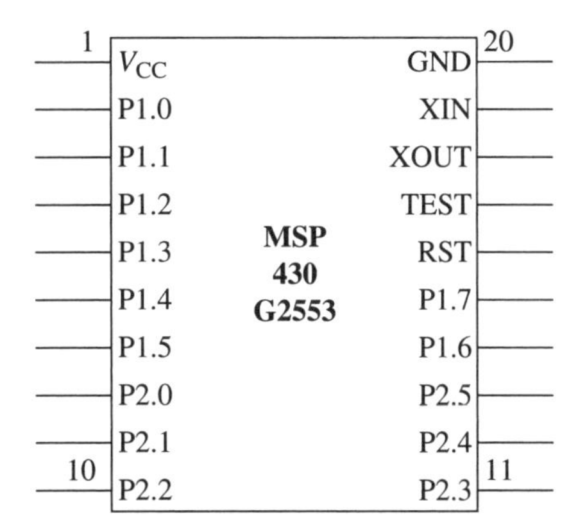

Figure 12.1

Pin layout of the MSP430G2553.

Table 12.5

Pin usage table for the USCI module.

Pin	Port Name	Usage Area
1	V_{CC}	Source voltage
2	P1.0	
3	P1.1/ UC0RX	USCI_A0 receive data input in UART mode
	UCA0SOMI	USCI_A0 slave data out/master in SPI mode
4	P1.2/ UC0TX	USCI_A0 transmit data output in UART mode
	UCA0SIMO	USCI_A0 slave data in/master out in SPI mode
5	P1.3	
6	P1.4/ UCB0STE	USCI_B0 slave transmit enable in SPI mode
	UCA0CLK	USCI_A0 clock input/output
7	P1.5/UCB0CLK	USCI_B0 clock input/output
	UCA0STE	USCI_A0 slave transmit enable in SPI mode
8	P2.0	
9	P2.1	
10	P2.2	
11	P2.3	
12	P2.4	
13	P2.5	
14	P1.6/ UCB0SOMI	USCI_B0 slave out/master in SPI mode
	UCB0SCL	USCI_B0 SCL I^2C clock in I^2C mode
15	P1.7/ UCB0SIMO	USCI_B0 slave in/master out in SPI mode
	UCB0SDA	USCI_B0 SDA I^2C data in I^2C mode
16	RST	Reset
17		
18	P2.7	
19	P2.6	
20	V_{SS}	Ground voltage

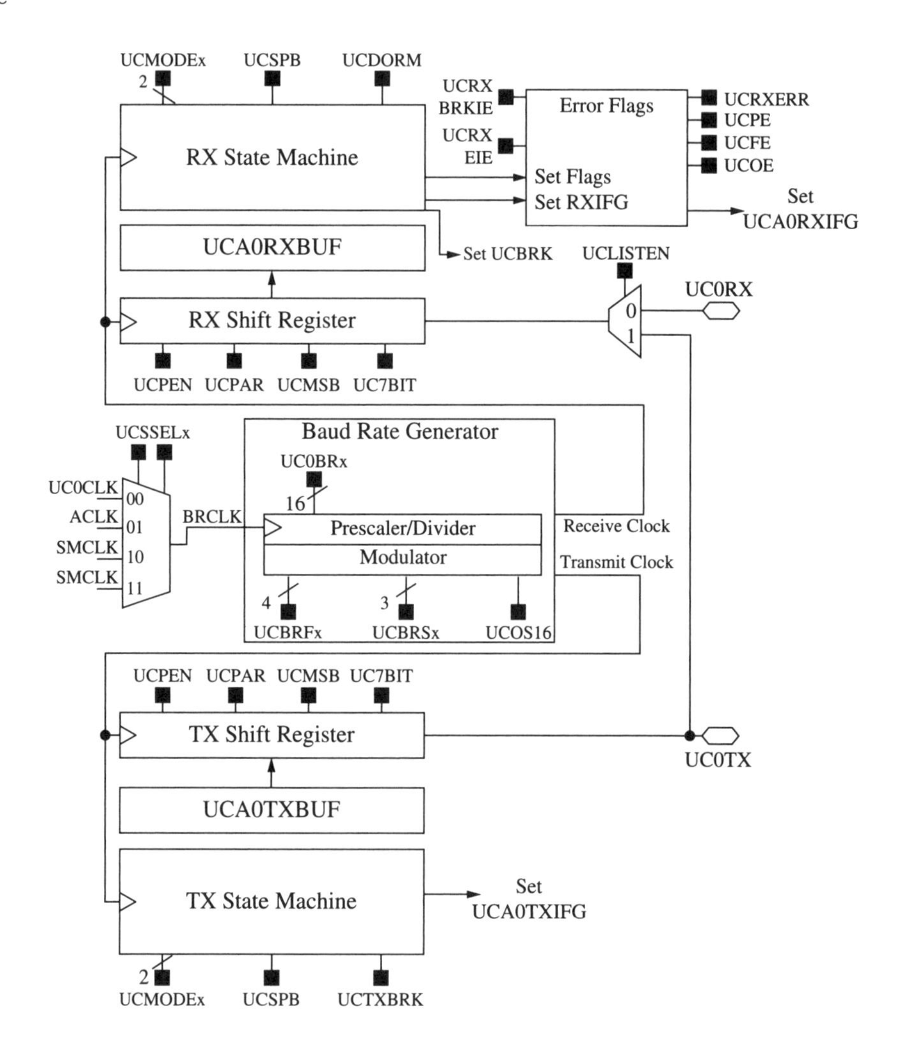

Figure 12.2 Block diagram of the UART mode.

we will not consider the enhanced UART with automatic baud rate detection (local interconnect network, LIN) and infrared data association (IrDA). More information on them can be found in [17].

A block diagram of the UART is given in Fig. 12.2. As can be seen in this figure, the MSP430 UART mode has two pins to communicate with other devices. These are the receive (**UC0RX**) and transmit (**UC0TX**) pins. In this block diagram, the transmit and receive shift registers are not accessible to the user. Instead, the transmit and receive buffers will be used for communication.

The UART is mainly configured by two control registers. These are USCLA0 Control Register 0 (**UCA0CTL0**) and USCLA0 Control Register 1 (**UCA0CTL1**). Their entries are given in Tables 12.6 and 12.7.

In Table 12.6, **UCPEN** and **UCPAR** bits are used for parity bit settings [5]. The **UCPEN** bit is used to enable the parity bit for the system. If this bit is reset, the parity bit is disabled. If it is set, the parity bit is enabled. After the UCPEN

Table 12.6

USCI_A0 control register 0 (UCA0CTL0).

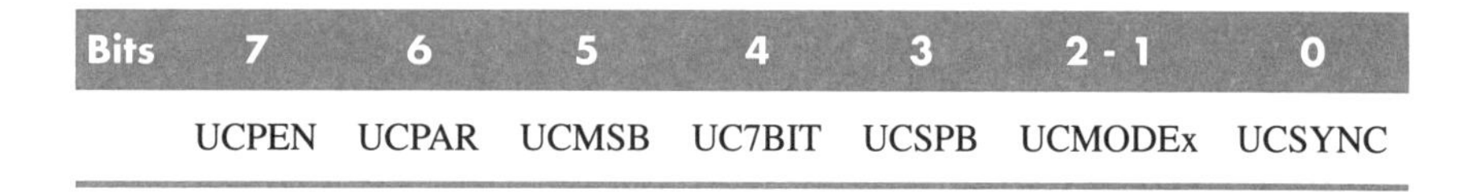

Bits	7	6	5	4	3	2 - 1	0
	UCPEN	UCPAR	UCMSB	UC7BIT	UCSPB	UCMODEx	UCSYNC

Table 12.7

USCI_A0 control register 1 (UCA0CTL1).

Bits	7 - 6	5	4	3	2	1	0
	UCSSELx	UCRXEIE	UCBRKIE	UCDORM	UCTXADDR	UCTXBRK	UCSWRST

bit is set, the **UCPAR** bit is used to decide on the parity type. When this bit is reset, odd parity is used. When it is set, even parity is used. The **UCMSB** bit is used to choose the start bit for the data transfer. When this bit is reset, the transmission starts from the LSB. When it is set, the transmission starts from the MSB. The former configuration is generally selected in the UART mode. The **UC7BIT** bit is used to select the data length. When this bit is reset, the data length is set to eight bits. When it is set, the data length is set to 7 bits. The **UCSPB** bit is used to decide on the number of stop bits. When this bit is reset, one stop bit is used. When it is set, two stop bits are used. **UCMODEx** bits are used to select the asynchronous mode. Constants for these bits are UCMODE_0 (UART mode), UCMODE_1 (idle-line multiprocessor mode), UCMODE_2 (address-bit multiprocessor mode), and UCMODE_3 (automatic baud rate detection mode). The default setting is UCMODE_0 for the UART communication between two devices. UCMODE_1 and UCMODE_2 can be used for the UART communication between more than two devices. The **UCSYNC** bit is used to choose the asynchronous or synchronous communication mode. When this bit is reset, the asynchronous mode (UART) is selected. When it is set, the synchronous mode (SPI) is selected. Therefore, the UCSYNC bit should be reset for the UART mode.

In Table 12.7, **UCSSELx** bits are used to select the UART clock source. Constants for these bits are UCSSEL_0 (for UC0CLK), UCSSEL_1 (for ACLK), UCSSEL_2, and UCSSEL_3 (for SMCLK). The **UCRXEIE** bit is used to enable the interrupt for receiving erroneous characters (detected by parity bit tests). When this bit is reset, received erroneous characters are rejected and the UCA0RXIFG bit (explained in Sec. 12.2.3) is not set. When the UCRXEIE bit is set, received erroneous characters set the UCA0RXIFG bit. The **UCBRKIE** bit is used to enable the interrupt for receiving the break condition. When this bit is reset, the received break character does not set the UCA0RXIFG bit. When it is set, the received break character sets the UCA0RXIFG bit. For the break operation, please see [17]. The **UCDORM** bit is used to decide on which characters will set the UCA0RXIFG bit. When this bit is reset, all received characters will set the UCA0RXIFG bit. When it is set, no character sets the UCA0RXIFG bit in the normal UART mode. The **UCTXADDR** bit is used in communication of more than two devices. Therefore, it is not explained here. The **UCTXBRK** bit is used to inform that the next frame will be transmitted as a break condition. When the UCTXBRK bit is reset, the

Table 12.8

USCI_A0 status register (UCA0STAT).

Bits	7	6	5	4	3	2	1	0
	UCLISTEN	UCFE	UCOE	UCPE	UCBRK	UCRXERR	UCADDR UCIDLE	UCBUSY

next frame is not acknowledged as a break. When it is set, the next frame is acknowledged as a break or break/synch. The **UCSWRST** bit is used to reset the USCI module. When this bit is set, the USCI module is reset. When it is reset, the USCI module will be ready for operation.

The UART mode also has a status register called as **UCA0STAT**. It is specifically used to observe the changes in the system. The entries of this register are given in Table 12.8. In this table, the **UCLISTEN** bit is used to generate an internal loop between the transmitter and receiver on the same device. When this bit is set, the loopback is enabled. When it is reset, the loopback is disabled. This property can be used to troubleshoot the communication codes on a single device. The **UCFE** bit is used to observe the framing error (caused by the low stop bit). When the received character has a low stop bit, UCFE is set. The **UCOE** bit is used to observe the overrun error. When a new character is sent into the receive buffer register before the previous one is read, this bit is set to indicate that there is an overrun in the system. This bit is cleared automatically when the receive buffer register is read. Therefore, the user should not try to clear it by software. The **UCPE** bit is used to observe the parity error. This bit is set when the received character has zeros or ones different from the number stated in the parity bit. The **UCBRK** bit is used to observe the break condition. This bit is set when a break condition occurs. The **UCRXERR** bit is used to observe any error in the received character. This bit is set when one or more than one of the UCPE, UCOE, or UCFE bits are set. **UCADDR** and **UCIDLE** bits are used in the communication of more than two devices. Therefore, they are not explained here. The **UCBUSY** bit shows that whether the USCI module is busy or not. This bit is set when the transmit or receive operation is performed. It is reset when the system is inactive.

12.2.1 *Baud Rate Generation*

Baud rate represents the number of received or sent symbols per second. Desired baud rates can be generated by using the baud rate generator block in the UART mode. This block receives the selected clock (**BRCLK**) as input. The clock frequency can be divided by the 16-bit division coefficient UCBRx (explained in Sec. 12.1.2). The baud rate generator block also has a USCI_A0 Modulation Control Register (**UCA0MCTL**) to set the modulation property. The entries of this register are given in Table 12.9. Depending on the settings and the input clock frequency, the MSP430 UART baud rate generator block can be used in low-or high-frequency modes. We will talk about these in the following paragraphs.

In Table 12.9, **UCBRFx** bits are used to select the modulation pattern for BITCLK16. For more detail on these patterns, please see [17]. This is the first modulation step for the oversampling (high-frequency) mode. This step is not

Table 12.9

USCI_A0 modulation control register (UCA0MCTL).

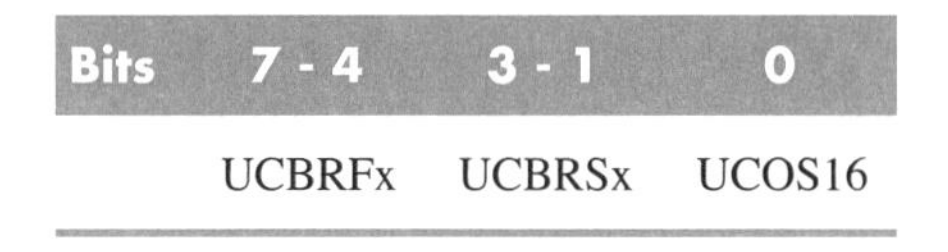

Bits	7 - 4	3 - 1	0
	UCBRFx	UCBRSx	UCOS16

applicable in the low-frequency mode. **UCBRSx** bits are used to select the modulation pattern for BITCLK which has the closest frequency for the desired baud rate. This is the only modulation step for the low-frequency mode. Also, this is the second step for the oversampling mode. The **UCOS16** bit is used to activate the oversampling mode. When this bit is reset, oversampling mode is disabled and the baud rate is generated by using low-frequency clock sources. In this mode, high-frequency clock sources can also be used. However, this is generally not recommended since it decreases the time interval for majority votes (to be explained in the following section). When the UCOS16 bit is set, the oversampling mode is enabled. Here, the baud rate is generated by using only high-frequency clock sources.

In the UART mode, baud rate calculation formulas are given in [17]. However, typical baud rates can be generated by setting the UCOS16, UCBRx, UCBRFx, and UCBRSx values. Based on the status of the UCOS16 bit, the baud rates that can be generated are given in Tables 12.10 and 12.11. In these tables, possible transmission and reception errors (labeled as "TX Error" and "RX Error") are also provided for each baud rate generation scenario.

12.2.2 *UART Transmit/Receive Operations*

Before focusing on the transmit and receive operations, we should mention that these operations are done on a character basis in the UART mode. Also, the character is not sent alone. In Table 12.12, we provide the character format for the UART mode. In this table, D0· · ·D6 stand for the seven data (character) bits. *D7* stands for the eighth data bit. In Table 12.12, italic characters indicate that the mentioned bits are optional to use. Also, the LSB first transmission is typically used in the UART mode. As a reminder, this is achieved by resetting the UCMSB bit in the UCA0CTL0.

Transmit and receive operations are simple in the UART mode. If there is no data written to the UCA0TXBUF, the baud rate generator does not provide any clock to the UART. Hence, it stays in the idle state. The transmit operation starts when data is written to the UCA0TXBUF. Then the baud rate generator starts working. The data within the UCA0TXBUF is moved to the transmit shift register. Meanwhile, the UCA0TXIFG bit in IFG2 is set to indicate that UCA0TXBUF is ready to accept new data. The data in the transmit shift register is sent to the receiver in a serial manner. Then UART returns to the idle state.

The receive operation starts when the falling edge of the start bit is detected. Until then, the baud rate generator does not provide any clock to the UART. Therefore, the UART stays in the idle state as in the transmission operation. The baud rate generator starts working after the falling edge of the start bit is detected. Then the receiver checks the validity of the start bit. If the start bit is not valid,

Table 12.10

Typical baud rates that can be generated when the UCOS16 Bit is reset.

Baud Rate	BRCLK Frequency (Hz.)	UCBRx	UCBRSx	UCBRFx	Maximum TX Error (%)		Maximum RX Error (%)	
1,200	32,768	27	2	0	−2.80	1.40	−5.90	2.00
2,400	32,768	13	6	0	−4.80	6.00	−9.70	8.30
4,800	32,768	6	7	0	−12.10	5.70	−13.40	19.00
9,600	32,768	3	3	0	−21.10	15.20	−44.30	21.30
9,600	1,000,000	104	1	0	−0.50	0.60	−0.90	1.20
9,600	1,048,576	109	2	0	−0.20	0.70	−1.00	0.80
9,600	4,000,000	416	6	0	−0.20	0.20	−0.20	0.40
9,600	8,000,000	833	2	0	−0.10	0.00	−0.20	0.10
9,600	12,000,000	1250	0	0	0.00	0.00	−0.05	0.05
9,600	16,000,000	1666	6	0	−0.05	0.05	−0.05	0.10
19,200	1,000,000	52	0	0	−1.80	0.00	−2.60	0.90
19,200	1,048,576	54	5	0	−1.10	1.00	−1.50	2.50
19,200	4,000,000	208	3	0	−0.20	0.50	−0.30	0.80
19,200	8,000,000	416	6	0	−0.20	0.20	−0.20	0.40
19,200	12,000,000	625	0	0	0.00	0.00	−0.20	0.00
19,200	16,000,000	833	2	0	−0.10	0.05	−0.20	0.10
38,400	1,000,000	26	0	0	−1.80	0.00	−3.60	1.80
38,400	1,048,576	27	2	0	−2.80	1.40	−5.90	2.00
38,400	4,000,000	104	1	0	−0.50	0.60	−0.90	1.20
38,400	8,000,000	208	3	0	−0.20	0.50	−0.30	0.80
38,400	12,000,000	312	4	0	−0.20	0.00	−0.20	0.20
38,400	16,000,000	416	6	0	−0.20	0.20	−0.20	0.40
56,000	1,000,000	17	7	0	−4.80	0.80	−8.00	3.20
56,000	1,048,576	18	6	0	−3.90	1.10	−4.60	5.70
56,000	4,000,000	71	4	0	−0.60	1.00	−1.70	1.30
56,000	8,000,000	142	7	0	−0.60	0.10	−0.70	0.80
56,000	12,000,000	214	2	0	−0.30	0.20	−0.40	0.50
56,000	16,000,000	285	6	0	−0.30	0.10	−0.50	0.20
115,200	1,000,000	8	6	0	−7.80	6.40	−9.70	16.10
115,200	1,048,576	9	1	0	−1.10	10.70	−11.50	11.30
115,200	4,000,000	34	6	0	−2.10	0.60	−2.50	3.10
115,200	8,000,000	69	4	0	−0.60	0.80	−1.80	1.10
115,200	12,000,000	104	1	0	−0.50	0.60	−0.90	1.20
115,200	16,000,000	138	7	0	−0.70	0.00	−0.80	0.60
128,000	1,000,000	7	7	0	−10.40	6.40	−18.00	11.60
128,000	1,048,576	8	1	0	−8.90	7.50	−13.80	14.80
128,000	4,000,000	31	2	0	−0.80	1.60	−3.60	2.00
128,000	8,000,000	62	4	0	−0.80	0.00	−1.20	1.20
128,000	12,000,000	93	6	0	−0.80	0.00	−1.50	0.40
128,000	16,000,000	125	0	0	0.00	0.00	−0.80	0.00
256,000	1,000,000	3	7	0	−29.60	0.00	−43.60	5.20
256,000	1,048,576	4	1	0	−2.30	25.40	−13.40	38.80
256,000	4,000,000	15	5	0	−4.00	3.20	−8.40	5.20
256,000	8,000,000	31	2	0	−0.80	1.60	−3.60	2.00
256,000	12,000,000	46	7	0	−1.90	0.00	−2.00	2.00
256,000	16,000,000	62	4	0	−0.80	0.00	−1.20	1.20

Table 12.11

Typical baud rates that can be generated when the UCOS16 bit is set.

Baud Rate	BRCLK Frequency (Hz.)	UCBRx	UCBRSx	UCBRFx	Maximum TX Error (%)	Maximum RX Error (%)
9,600	1,000,000	6	0	8	−1.8 0.0	−2.2 0.4
9,600	1,048,576	6	0	13	−2.3 0.0	−2.2 0.8
9,600	4,000,000	26	0	1	0.0 0.9	0.0 1.1
9,600	8,000,000	52	0	1	−0.4 0.0	−0.4 0.1
9,600	12,000,000	78	0	2	0.0 0.0	−0.1 0.1
9,600	16,000,000	104	0	3	0.0 0.2	0.0 0.3
19,200	1,000,000	3	0	4	−1.8 0.0	−2.6 0.9
19,200	1,048,576	3	1	6	−4.6 3.2	−5.0 4.7
19,200	4,000,000	13	0	0	−1.8 0.0	−1.9 0.2
19,200	8,000,000	26	0	1	0.0 0.9	0.0 1.1
19,200	12,000,000	39	0	1	0.0 0.0	0.0 0.2
19,200	16,000,000	52	0	1	−0.4 0.0	−0.4 0.1
38,400	4,000,000	6	0	8	−1.8 0.0	−2.2 0.4
38,400	8,000,000	13	0	0	−1.8 0.0	−1.9 0.2
38,400	12,000,000	19	0	8	−1.8 0.0	−1.8 0.1
38,400	16,000,000	26	0	1	0.0 0.9	0.0 1.1
57,600	1,000,000	1	7	0	−34.4 0.0	−33.4 0.0
57,600	4,000,000	4	5	3	−3.5 3.2	−1.8 6.4
57,600	8,000,000	8	0	11	0.0 0.9	0.0 1.6
57,600	12,000,000	13	0	0	−1.8 0.0	−1.9 0.2
57,600	16,000,000	17	0	6	0.0 0.9	−0.1 1.0
115,200	4,000,000	2	3	2	−2.1 4.8	−2.5 7.3
115,200	8,000,000	4	5	3	−3.5 3.2	−1.8 6.4
115,200	12,000,000	6	0	8	−1.8 0.0	−2.2 0.4
115,200	16,000,000	8	0	11	0.0 0.9	0.0 1.6
230,400	4,000,000	1	7	0	−34.4 0.0	−33.4 0.0
230,400	8,000,000	2	3	2	−2.1 4.8	−2.5 7.3
230,400	12,000,000	3	0	4	−1.8 0.0	−2.6 0.9
230,400	16,000,000	4	5	3	−3.5 3.2	−1.8 6.4
460,800	8,000,000	1	7	0	−34.4 0.0	−33.4 0.0
460,800	16,000,000	2	3	2	−2.1 4.8	−2.5 7.3

Table 12.12

UART character format.

Start Bit	D0 · · · D6	*D7*	*Address Bit*	*Parity Bit*	Stop Bit	*Second Stop Bit*

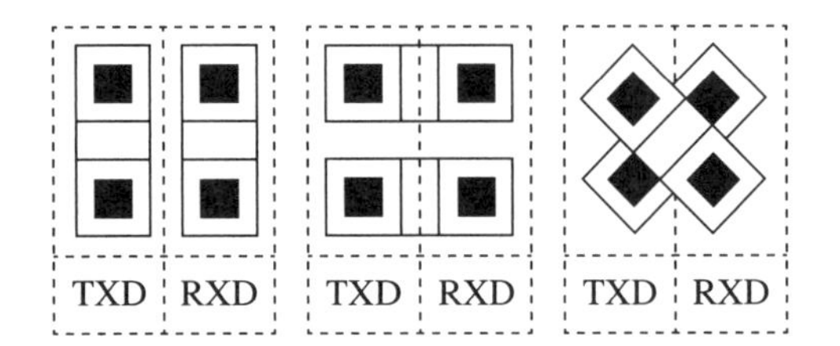

Figure 12.3

Jumper 3 (J3) TXD/RXD connections for UART communication. From left to right: Normal view; UART setting for MSP430 LaunchPad Rev.1.5; UART setting for MSP430 LaunchPad Rev.1.4.

the UART goes to the idle state. Otherwise, each received signal pulse is checked by majority voting. Here, three samples are taken from the pulse. If the number of zeros is more than ones in these samples, then the receive shift register receives a zero. Otherwise, it receives a one. The binary data is shifted in the receive shift register. This operation continues until the stop bit is detected. The final result is transferred to the UCA0RXBUF.

12.2.3 *UART Interrupts*

UART has different interrupt vectors for transmission and reception operations. As given in Table 9.2, for the transmitter the interrupt vector is **USCIAB0TX_VECTOR**. For the receiver, the interrupt vector is **USCIAB0RX_VECTOR**.

The interrupt-based communication operation works as follows in the UART mode. Initially, the UCA0TXIE and UCA0RXIE bits should be set to enable transmission and reception interrupts. These two interrupts are maskable. Therefore, the GIE bit must also be set. In the transmission operation, an interrupt is requested when the UCA0TXBUF is ready for another character. Then the UCA0TXIFG is set. This flag is automatically cleared when a new character is written to the UCA0TXBUF. In the reception operation, an interrupt is requested when a character is loaded to the UCA0RXBUF. Then the UCA0RXIFG is set. This flag is automatically cleared when the data in UCA0RXBUF is read.

12.2.4 *Coding Practices for the UART Mode*

In this section, we provide sample C and assembly codes in the UART communication mode. Before focusing on the code samples, there are important issues to be clarified. First, the MSP430 LaunchPads should be disconnected from the external circuitry before debugging the code. Otherwise, CCS gives a debug error since common grounds are used in the circuitry. Second, the jumper settings of J3 for the MSP430 LaunchPad should be done for the transmit and receive pins as given in Fig. 12.3. Third, we will be using the terminal program under CCS. Please see Sec. 5.8 for its usage.

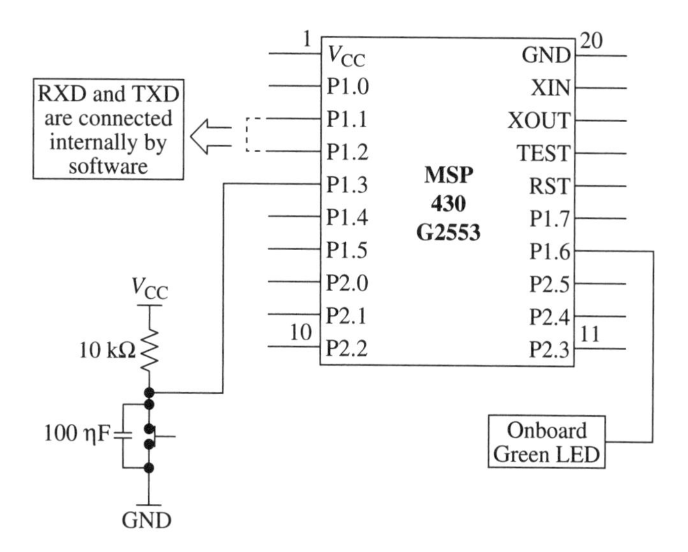

Figure 12.4 The connection diagram for the UART loopback application.

UART in C

In Listing 12.1, the loopback property of the UART mode is used. The connection diagram for this application is given in Fig. 12.4. The C code containing the transmitter and receiver parts are run on the same microcontroller using the loopback property. Hence, the code can be debugged easily. In Listing 12.1, the green LED on the MSP430 LaunchPad is toggled by the button connected to pin P1.3. However, the loopback property is used such that the toggle command is sent and received within the microcontroller.

Listing 12.1 The UART loopback application, in C language.

```
#include <msp430.h>

 int Data = 0;

void main(void)
{
 WDTCTL = WDTPW|WDTHOLD;

 BCSCTL1 = CALBC1_1MHZ;  //Adjust the clock
 DCOCTL = CALDCO_1MHZ;

 P1DIR |= BIT6;  //Adjust pins
 P1OUT = 0x00;
 P1SEL = BIT1|BIT2;
 P1SEL2 = BIT1|BIT2;
 P1IE |= 0x08;
 P1IES |= 0x08;
 P1IFG = 0x00;

 UCA0CTL1 |= UCSSEL_2;  //Setup the UART mode
//Use SMCLK
```

```
 UCA0BR0 = 104;
//Low bit of UCBRx is 104
 UCA0BR1 = 0;
//High bit of UCBRx is 0
 UCA0MCTL = UCBRS_1;
//Second modulation stage select is 1
//Baud Rate = 9600
  UCA0STAT |= UCLISTEN;
//Enable internal loopback
 UCA0CTL1 &= ~UCSWRST;
//Clear SW reset, resume operation
 IE2 |= UCA0RXIE|UCA0TXIE;
//Enable USCI_A0 RX TX interrupt

 _enable_interrupts();

 LPM4;
}

//USCI A transmitter interrupt
#pragma vector=USCIAB0TX_VECTOR
__interrupt void USCIA0TX_ISR(void){
 UCA0TXBUF = Data;
//Load the TX buffer with integer value
}

//USCI A receiver interrupt
#pragma vector=USCIAB0RX_VECTOR
__interrupt void USCIA0RX_ISR(void){
 P1OUT = UCA0RXBUF;
//Write received data to P1OUT
}

#pragma vector=PORT1_VECTOR
__interrupt void Port_1(void){
 Data ^= 0x40;
//Toggle data value
 P1IFG = 0x00;
//Clear interrupt flags
}
```

In Listing 12.2, the MSP430 receives the password through the UART mode from the host computer. The connection diagram for this application is given in Fig. 12.5. If the password is correct, then the green LED on the MSP430 LaunchPad turns on for 5 s, then the code is reset. Otherwise, the red LED on the MSP430 LaunchPad turns on for 2 s, then the MSP430 waits for the new password. Meanwhile, the MSP430 will tell the user to enter the password and will determine whether the entered password is correct or not though the terminal program.

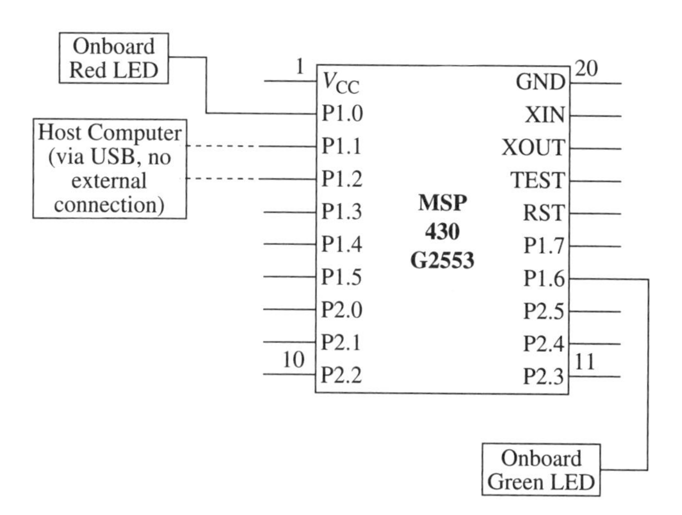

Figure 12.5 The connection diagram for UART password application.

Listing 12.2 The UART password application, in C language.

```
#include <msp430.h>

#define RedLed BIT0
#define GreenLed BIT6

  char password[ ] = "12345";  //The Password
  char enter[ ] = "Enter Your Password\r\n";
  char correct[ ] = "Your password is correct\r\n";
  char incorrect[ ] = "Your password is incorrect\r\n";
  char reenter[ ] = "Please re-enter your password\r\n";
  char input[100];
  int RXByteCtr = 0;
  int cnt = 0;
  int inputlength,passwordlength;
  int difference;

 void transmit(char *str);
 int compare(char *strin, char *strpass);
 int arraylength(char *str);

 void main(void)
{
 WDTCTL = WDTPW|WDTHOLD;

 BCSCTL1 = CALBC1_1MHZ; //Adjust the clock
 DCOCTL = CALDCO_1MHZ;

 P1DIR = RedLed|GreenLed;  //Adjust pins
 P1OUT = 0x00;
 P1SEL = BIT1|BIT2;
 P1SEL2 = BIT1|BIT2;

 UCA0CTL1 |= UCSWRST+UCSSEL_2;  //Setup the UART mode
//Enable SW reset, Use SMCLK
```

```
UCA0BR0 = 104;
//Low bit of UCBRx is 104
UCA0BR1 = 0;
//High bit of UCBRx is 0
UCA0MCTL = UCBRS_1;
//Second modulation stage select is 1
//Baud Rate = 9600
UCA0CTL1 &= ~UCSWRST;
//Clear SW reset, resume operation

transmit(enter);

IE2 |= UCA0RXIE;  //Enable the USCI_A0 RX interrupt

_enable_interrupts();

while(1){
if(cnt == 1){
//Check if cnt is 1
inputlength = arraylength(input);
//Get your input length
passwordlength = arraylength(password);
// Get your password length
difference = compare(input,password);
//Compare the received password with your password
if(difference == 0){
//Check if they match
transmit(correct);
//If they match, transmit correct string
P1OUT = GreenLed;
//Turn on the green LED
__delay_cycles(5000000);
//Wait for 5 seconds
WDTCTL = WDT_MRST_0_064;
//Reset the system
}
else{
//If they do not match
transmit(incorrect);
//Transmit incorrect string
P1OUT = RedLed;
//Turn on the red LED
__delay_cycles(2000000);
//Wait for 2 seconds
P1OUT = 0x00;
//Turn off the red LED
transmit(reenter);
//Transmit reenter string
}
cnt = 0;
//Reset cnt
RXByteCtr = 0;
//Reset Receive Byte counter
```

```
 }}
}

//USCI A receiver interrupt
#pragma vector=USCIAB0RX_VECTOR
__interrupt void USCI0RX_ISR(void){
//Check if the UCA0RXBUF is different from 0x0A
//(Enter key from keyboard)
 if(UCA0RXBUF != 0x0A)
 input[RXByteCtr++] = UCA0RXBUF;
//If it is, load received character
//to current input string element
 else{
 cnt = 1;
//If it is not, set cnt
 input[RXByteCtr] = 0;
//Add null character at the end of input string
 }
}

void transmit(char *str){
 while(*str != 0){
//Do this during current element is not
//equal to null character
 while (!(IFG2&UCA0TXIFG));
//Ensure that transmit interrupt flag is set
 UCA0TXBUF = *str++;
//Load UCA0TXBUF with current string element
//then go to the next element
 }
}

int compare(char *strin, char *strpass){
 int result = 0;
//Clear result
 if(inputlength <= passwordlength){
//Check if passwordlength is greater than or
//equal to inputlength
 while(*strpass != 0){
 result = result + abs((*strin++)-(*strpass++));
//If it is, take the difference between elements of
//strin and strpass until current element of strpass
//is equal to null character, abs() is used to ensure
//that differences do not cancel each other
 }}
 else{
 while(*strin != 0){
 result = result + abs((*strin++)-(*strpass++));
//If it is not, do the same thing until current element
//of strin is equal to null character this time
 }}
 return result;
```

```
//Return result value
}

int arraylength(char *str){
 int length = 0;
//Clear length
 while(*str != 0){
//Until null character is reached
 str++;
//Increase array address
 length++;
//Increase length value
 }
 return length;
//Return length value
}
```

In Listing 12.3, the duty cycle of a PWM signal is obtained from the host computer using the UART mode. The connection diagram for this application is given in Fig. 12.6. Then the PWM signal is used to adjust the brightness of the green LED on the MSP430 LaunchPad. This operation is done continuously.

Listing 12.3 The UART PWM application, in C language.

```
#include <msp430.h>

#define GreenLed BIT6

 char digits[3];
 char enter[] = "Enter Duty Cycle \r\n";
 char enter1[] = "Enter New Duty Cycle \r\n";
 int result = 0;
 int cnt = 0;
 int RXByteCtr = 0;

void transmit(char *str);
void convert(void);

void main(void)
{
 WDTCTL = WDTPW|WDTHOLD;

 BCSCTL1 = CALBC1_1MHZ;  //Adjust the clock
 DCOCTL = CALDCO_1MHZ;

 P1DIR = GreenLed;  //Adjust pins
 P1OUT = 0x00;
 P1SEL = BIT1|BIT2|BIT6;
 P1SEL2 = BIT1|BIT2;

 UCA0CTL1 |= UCSWRST|UCSSEL_2;  //Setup the UART mode
//Enable SW reset, Use SMCLK
 UCA0BR0 = 104;  //Low bit of UCBRx is 104
```

```
 UCA0BR1 = 0;  //High bit of UCBRx is 0
 UCA0MCTL = UCBRS_1;
//Second modulation stage select is 1
//Baud Rate = 9600
 UCA0CTL1 &= ~UCSWRST;
//Clear SW reset, resume operation

 TACCR1 = 0;  //Setup the PWM
 TACCR0 = 999;
 TACCTL1 = OUTMOD_7;
 TACTL = TASSEL_2|MC_1|ID_3;

 transmit(enter);

 IE2 |= UCA0RXIE;  //Enable USCI_A0 RX interrupt

 _enable_interrupts();

 while(1){
 if(cnt == 1){
 convert();  //Convert received character to integer
 CCR1 = 10*result;
 transmit(enter1);
 cnt = 0;
 RXByteCtr = 0;  //Reset the Receive Byte counter
 }}
}

//USCI A receiver interrupt
#pragma vector=USCIAB0RX_VECTOR
__interrupt void USCI0RX_ISR(void){
//Check if the UCA0RXBUF is different from 0x0A
//(Enter key from keyboard)
 if(UCA0RXBUF != 0x0A)
 digits[RXByteCtr++] = UCA0RXBUF;
//If it is, load received character
//to the current string element
//then go to next string element
 else cnt = 1;
//If the received character is 0x0A, set cnt
}

void transmit(char *str){
//While the current element is not equal
//to the null character
 while(*str != 0){
 while (!(IFG2&UCA0TXIFG));
//Ensure the transmit interrupt flag is set
 UCA0TXBUF = *str++;
//Load the UCA0TXBUF with the current string
//element, then go to the next element
 }
}

void convert(void){
```

```
 char hundreds = '0',tens = '0',ones = '0';
 if(RXByteCtr == 1)ones = digits[0];
//If the RXByteCtr equals 1,
//take only ones digit

 if(RXByteCtr == 2){
 ones = digits[1];
 tens = digits[0];
 }
//If the RXByteCtr equals 1,
//take ones and tens digits

 if(RXByteCtr == 3){
 ones = digits[2];
 tens = digits[1];
 hundreds = digits[0];
 }
//If the RXByteCtr equals 1,
//take ones, tens, and hundreds digits

result = ((hundreds-0x30)*100)+((tens-0x30)*10)+\
(ones-0x30);
}
```

In Listings 12.4 and 12.5, the UART mode is used to establish a digital communication between two MSP430 LaunchPads. The connection diagram for this application is given in Fig. 12.7. The C code for the transmitter device is given in Listing 12.4. The C code for the receiver device is given in Listing 12.5. In this application, when the button connected to pin P1.3 of the transmitter device is pressed, the transmitter sends the next PWM constant from the TXData array to control the brightness of the green LED on the receiver. The connection between pin P1.5 of the transmitter device and the RST pin of the receiver device is used for resetting the slave before the communication starts.

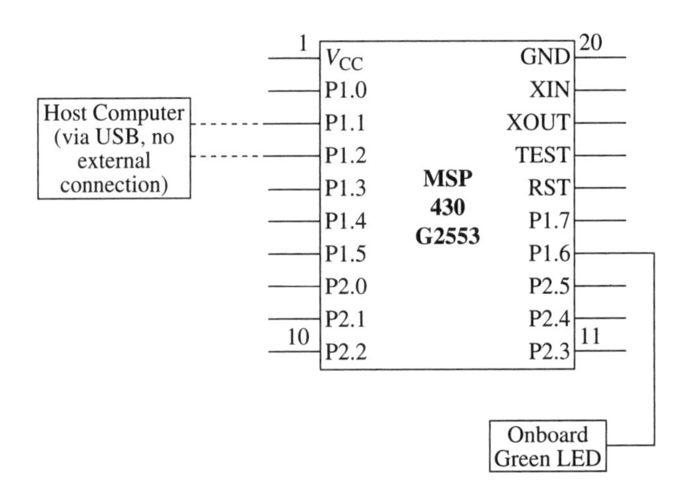

Figure 12.6 The connection diagram for the UART PWM application.

Figure 12.7

The connection diagram for the UART communication between two MSP430 LaunchPads.

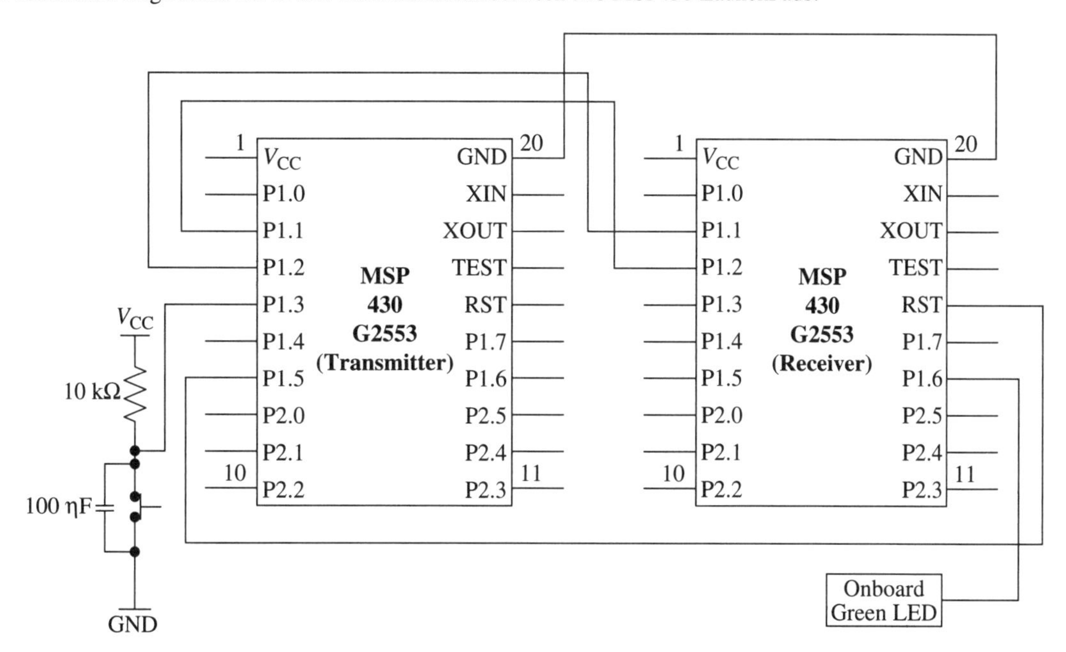

Listing 12.4 The transmitter part of the UART communication between two MSP430 LaunchPads, in C language.

```
#include <msp430.h>

 unsigned int *PTXData = 0;
 unsigned int TXByteCtr;
 unsigned int TXData[] = {0x0000, 0x00FA, 0x01F4, \
 0x02EE, 0x03E8};  //TACCR1 values to be transmitted
 unsigned int High = 0, Low = 0;
 int cntr=0;

void main(void)
{
 WDTCTL = WDTPW|WDTHOLD;

 BCSCTL1 = CALBC1_1MHZ;  //Adjust the clock
 DCOCTL = CALDCO_1MHZ;

 P1DIR |= BIT5;  //Adjust pins
//Assing P1.5 as output for resetting the slave
 P1SEL = BIT1|BIT2;
 P1SEL2 = BIT1|BIT2;
 P1IE |= 0x08;
 P1IES |= 0x08;
 P1IFG = 0x00;
```

```
 UCA0CTL1 |= UCSWRST|UCSSEL_2;  //Setup the UART mode
//Enable SW reset, Use SMCLK
 UCA0BR0 = 104;
//Low bit of UCBRx is 104
 UCA0BR1 = 0;
//High bit of UCBRx is 0
 UCA0MCTL = UCBRS_1;
//Second modulation stage select is 1
//Baud Rate = 9600
 UCA0CTL1 &= ~UCSWRST;
//Clear SW reset, resume operation
 PTXData = TXData;
//Equate TXData array's start address to PTXData

 P1OUT &= ~BIT5;  //Reset UART slave
 P1OUT |= BIT5;

 _enable_interrupts();

 LPM4;
}

//USCI A transmitter interrupt
#pragma vector=USCIAB0TX_VECTOR
__interrupt void USCIA0TX_ISR(void){
 if((TXByteCtr%2) == 0){
 High = *PTXData;
//Write the incoming array element to high integer
 UCA0TXBUF=(High>>8);
//Shift the high byte of High integer then load
//the TX buffer with it
 }
 if((TXByteCtr%2) == 1){
 Low = *PTXData++;
//Write the incoming array element to low integer,
//then increase the PTXData
 UCA0TXBUF = Low;
//Load the TX buffer with low byte of array element
 cntr++;
 if(cntr == 5){
 PTXData = PTXData-5;
 cntr = 0;
//If cntr equals 5, return to array start address
//and reset cntr
 }
 IE2 &= ~UCA0TXIE;
//Disable the transmit interrupt
 }
 TXByteCtr--;
//Decrease TX Byte Counter
}
```

```
#pragma vector=PORT1_VECTOR
__interrupt void Port_1(void)
{
 TXByteCtr = 2;
//Load TX Byte Counter with 2
 IE2 |= UCA0TXIE;
//Enable the transmit interrupt
 P1IFG = 0x00;
}
```

Listing 12.5 The receiver part of the UART communication between two MSP430 LaunchPads, in C language.

```
#include <msp430.h>

 unsigned int RXData;
 unsigned int RXByteCtr = 0;

void main(void)
{
 WDTCTL = WDTPW|WDTHOLD;

 BCSCTL1 = CALBC1_1MHZ;  //Adjust the clock
 DCOCTL = CALDCO_1MHZ;

 P1DIR |= BIT6;  //Adjust pins
 P1SEL = BIT1|BIT2|BIT6;
 P1SEL2 = BIT1|BIT2;
 P1OUT = 0x00;

 UCA0CTL1 |= UCSWRST|UCSSEL_2;  //Setup the UART mode
//Enable SW reset, Use SMCLK
 UCA0BR0 = 104;
//Low bit of UCBRx is 104
 UCA0BR1 = 0;
//High bit of UCBRx is 0
 UCA0MCTL = UCBRS_1;
//Second modulation stage select is 1
//Baud Rate = 9600
 UCA0CTL1 &= ~UCSWRST;
//Clear SW reset, resume operation
 IE2 |= UCA0RXIE;
//Enable the USCI_A0 RX interrupt

 TACCR1 = 0;  //Setup the PWM
 TACCR0 = 999;
 TACCTL1 = OUTMOD_7;
 TACTL = TASSEL_2 | MC_1 | ID_3;

 _enable_interrupts();

  LPM0;
```

```
}
//USCI A receiver interrupt
#pragma vector = USCIAB0RX_VECTOR
__interrupt void USCIAB0RX_ISR(void){
 if((RXByteCtr%2) == 0){
 RXData = UCA0RXBUF;
//Move the received data (high byte) to low byte
//of RXData
 RXData = (RXData<<8);
//Shift the low byte of RXData to high byte
 }
 if((RXByteCtr%2) == 1){
 RXData |= UCA0RXBUF;
//Move received data (low byte) to low byte
//of RXData
 TACCR1 = RXData;
//Move RXData to TACCR1
 }
 RXByteCtr++;
//Increase RX Byte Counter
 if(RXByteCtr == 2) RXByteCtr = 0;
}
```

UART in Assembly

In the first assembly code, given in Listing 12.6, the "Hello World" string is transmitted to the host computer when the button connected to pin P1.3 is pressed. The connection diagram for this application is given in Fig. 12.8.

Listing 12.6 The UART "Hello World" application, in assembly language.

```
 .cdecls C,LIST,"msp430.h"

 .text
 .retain
 .retainrefs

RESET
 mov.w #WDTPW|WDTHOLD,WDTCTL
 mov.w #__STACK_END,SP

;String starts
 mov.b #'H',&0200h
 mov.b #'e',&0201h
 mov.b #'l',&0202h
 mov.b #'l',&0203h
 mov.b #'o',&0204h
 mov.b #' ',&0205h
 mov.b #'W',&0206h
```

```
 mov.b #'o',&0207h
 mov.b #'r',&0208h
 mov.b #'l',&0209h
 mov.b #'d',&020Ah
;String ends

 mov.b #0Ah,&020Bh ;New Line
 mov.b #0Dh,&020Ch
;Cursor returns to the beginning of the line
 mov.b #0h,&020Dh ;Null character

 mov.b &CALBC1_1MHZ,BCSCTL1 ;Adjust the clock
 mov.b &CALDCO_1MHZ,DCOCTL

 mov.b #6h,P1SEL ;Adjust pins
 mov.b #6h,P1SEL2
 bis.b #08h,P1IE
 bis.b #08h,P1IES
 clr.b P1IFG

 bis.b #UCSWRST+UCSSEL_2,UCA0CTL1 ;Adjust the UART mode
;Enable SW reset, Use SMCLK
 mov.b #68h,UCA0BR0 ;Low bit of UCBRx is 104
 mov.b #0h,UCA0BR1 ;High bit of UCBRx is 0
 mov.b #UCBRS_1,UCA0MCTL
;Second modulation stage select is 1
;Baud Rate = 9600
 bic.b #UCSWRST,UCA0CTL1
;Clear SW reset, resume operation

 bis.w #GIE+LPM4,SR

;Transmit subroutine
Transmit:
 mov.w #0200h,R5
Check:
 bit.b #UCA0TXIFG,IFG2
;Ensure the transmit interrupt flag is set
 jeq Check
;If not, jump to check label
 mov.b @R5,R6
 tst.b R6
;Check if new character equals null character
 jeq Finish
;If it is jump to finish
 mov.b @R5+,UCA0TXBUF
;If it is not, Load TX buffer with this
;character then go to new character
 jmp Check
;Jump to check label to send new character
Finish:
 ret
```

```
;----------------------
P1_ISR
;----------------------

 call #Transmit
 clr.b P1IFG
 reti

;------------------------
;Stack Pointer definition
;------------------------
 .global __STACK_END
 .sect .stack

;-------------------
;Interrupt Vectors
;-------------------

 .sect RESET_VECTOR
 .short RESET
 .sect  PORT1_VECTOR
 .short P1_ISR
 .end
```

In the second assembly code, given in Listing 12.7, the red and green LEDs on the MSP430 LaunchPad are controlled by the host computer. The connection diagram for this application is the same as given in Fig. 12.5. In this application, the red LED on the MSP430 LaunchPad turns on when the ‘r’ key is pressed on the keyboard of the host computer. The green LED on the MSP430 LaunchPad turns on when the ‘g’ key is pressed on the keyboard of the host computer. Both LEDs turn off when a different key is pressed.

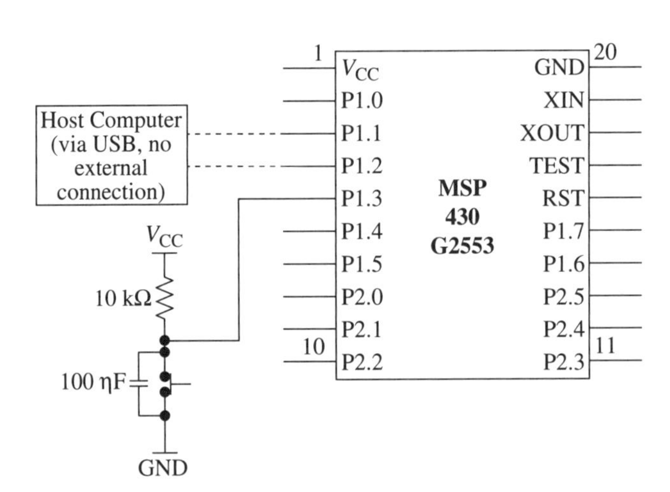

Figure 12.8

The connection diagram for the UART “Hello World” application.

Listing 12.7 The UART LED control application, in assembly language.

```
 .cdecls C,LIST,"msp430.h"

 .text
 .retain
 .retainrefs

RESET
 mov.w #WDTPW|WDTHOLD,WDTCTL
 mov.w #__STACK_END,SP

 mov.b &CALBC1_1MHZ,BCSCTL1 ;Adjust the clock
 mov.b &CALDCO_1MHZ,DCOCTL

 bis.b #41h,P1DIR
 clr.b P1OUT

 mov.b #6h,P1SEL ;Adjust pins
 mov.b #6h,P1SEL2

 bis.b #UCSWRST+UCSSEL_2,UCA0CTL1 ;Adjust the UART mode
;Enable SW reset, Use SMCLK
 mov.b #68h,UCA0BR0 ;Low bit of UCBRx is 104
 mov.b #0h,UCA0BR1 ;High bit of UCBRx is 0
 mov.b #UCBRS_1,UCA0MCTL
;Second modulation stage select is 1
;Baud Rate = 9600
 bic.b #UCSWRST,UCA0CTL1
;Clear SW reset, resume operation
 bis.b #UCA0RXIE,IE2 ;Enable RX interrupt

 bis.w #GIE+LPM0,SR

;----------------------
USCIAB0RX_ISR
;----------------------

 cmp.b #'r',UCA0RXBUF
;Check if the received character is 'r'
 jne Second
;If not, jump Second
 mov.b #01h,P1OUT
;If it is, turn on red LED
 jmp EndISR
Second:
 cmp.b #'g',UCA0RXBUF
;Check if the received character is 'g'
 jne Third
;If not, jump Third
 mov.b #40h,P1OUT
;If it is, turn on green LED
 jmp EndISR
Third:
 mov.b #00h,P1OUT
```

```
;If anything else, turn off LEDs
EndISR:
 reti

;------------------------
;Stack Pointer definition
;------------------------

 .global __STACK_END
 .sect .stack

;-------------------
;Interrupt Vectors
;-------------------

 .sect RESET_VECTOR
 .short RESET
 .sect  USCIAB0RX_VECTOR
 .short USCIAB0RX_ISR
 .end
```

12.3 UART in Grace

Grace can be used to configure the USCI_A0 and USCI_B0 modules. The first module is called USCI_A0: UART/LIN, IRDA, SPI and the second module is called USCI_B0: SPI, I2C as shown in Fig. 5.11. First, the target block should be clicked. Then, the Enable USCI_x0 in my configuration box must be checked to enable it. For both modules, a selection window appears with the Basic User, Power User, and Registers options. For all options, a selection window appears. For the USCI _A0 block, this window will have two buttons, UART and SPI. The same window appears when the USCI_B0 block is chosen. There the buttons will be SPI and I^2C. When a button is clicked in this selection window, the related communication mode appears. The user can return to the previous selection window by clicking the Return to USCI_x0 Mode Selection View button.

In this section, we will focus on the UART mode. Therefore, we should select the USCI_A0 block first. We assume that the user has clicked the UART button in the initial selection window for all user modes explored below.

12.3.1 *The Basic User Mode*

The basic user mode window appears as shown in Fig. 12.9. In this mode, we can enable or disable the UART pins from the related drop-down lists. We can also select the UART baud rate from the Baud drop-down list. Here, we have an option to set a custom baud rate by first selecting the Custom option from the list. Then we can enter the desired value into the Set Custom box. In the basic user mode, we can enable the transmit and receive interrupts by checking the "USCI_A0 UART transmit interrupt enable" and "USCI_A0 UART receive interrupt enable"

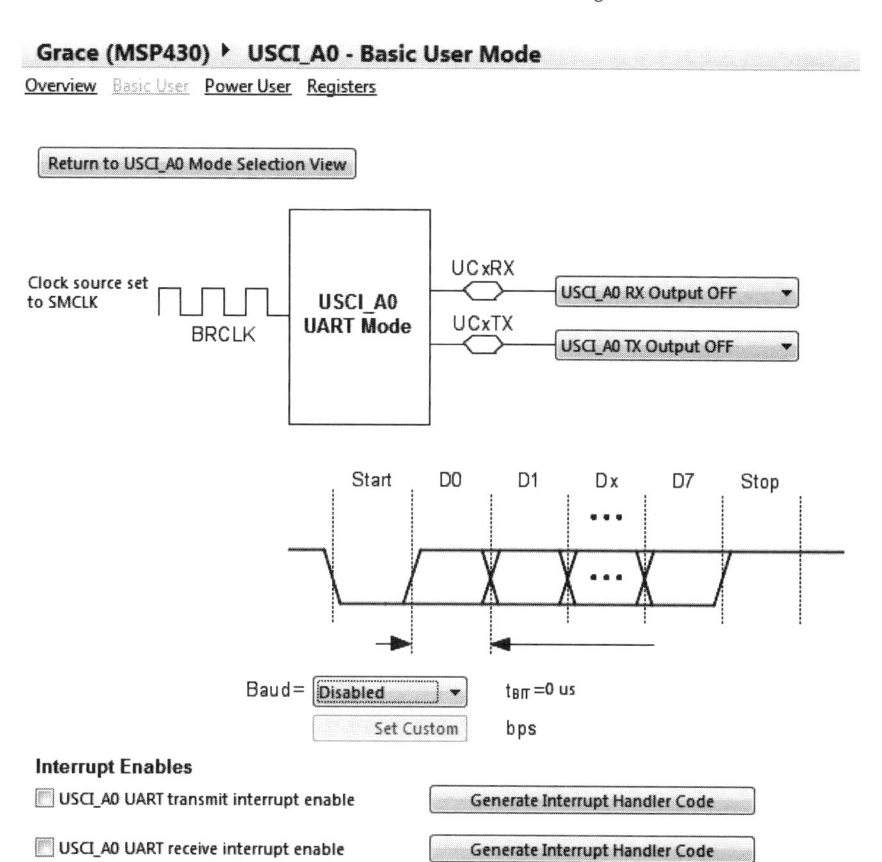

Figure 12.9 The basic user mode for the UART under Grace.

boxes respectively. We can also generate ISRs related to these interrupts using the associated Generate Interrupt Handler Code button.

12.3.2 *The Power User Mode*

The power user mode for the UART is shown in Fig. 12.10. In this mode, the user can adjust the clock source, character length, parity, and stop bits in addition to the arrangements in the basic user mode. All these can be adjusted by using related drop-down list items.

12.3.3 *The Register Controls Mode*

Finally, the UART registers can be adjusted under Grace. The user should select the register controls mode, as shown in Fig. 12.11 for this purpose. Some register entries are not available under Grace. They are labeled R/W. Some entries are only available for reading a value. They are labeled R. The same format applies to the SPI and I^2C under Grace also.

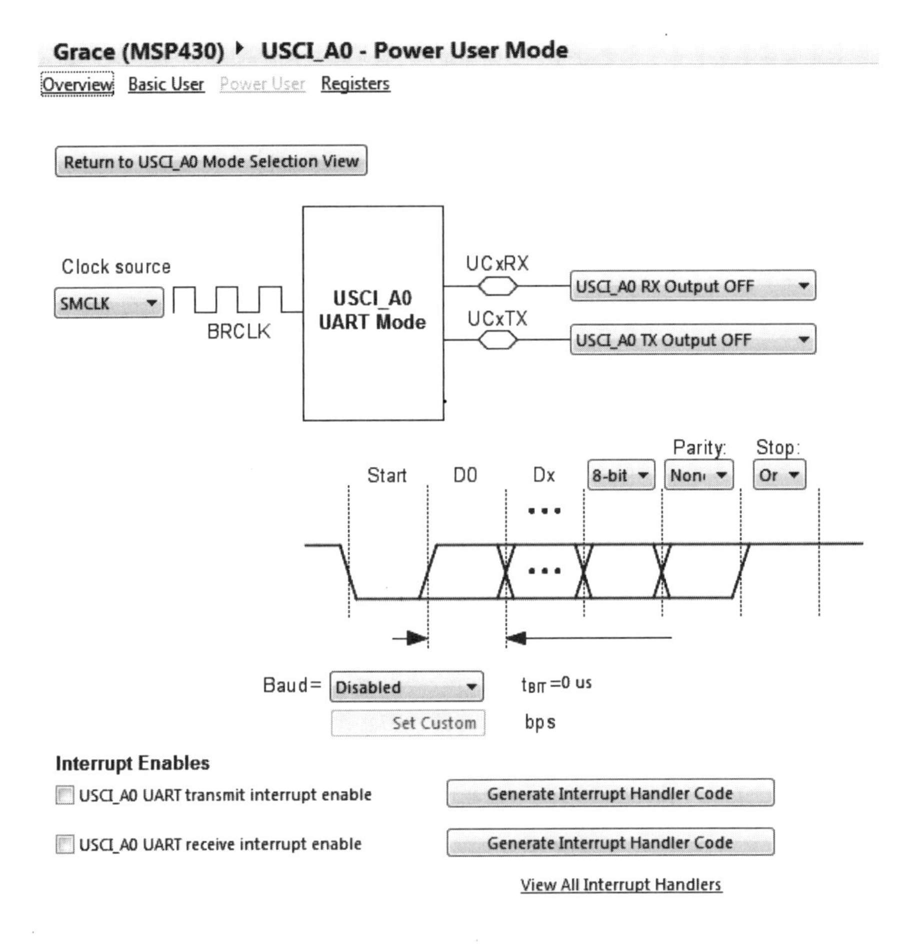

Figure 12.10 The power user mode for the UART under Grace.

12.3.4 *Coding Practices*

In this section, we redo the previous UART-based applications using Grace. We first redo the application given in Listing 12.6 in C language. For this application, we generate an empty Grace project. Then we enable the USCI_A block. We select the UART option from the selection window in the basic user mode. We set the baud rate to 9600 bps from the drop-down list. The main.c file of the Grace project will be as in Listing 12.8 for this application. As we debug and run the program, the "Hello World" string will be transmitted from the MSP430 to the host computer. Do not forget to use the terminal program to see the string in the host computer.

In the second example, we redo the application given in Listing 12.7, now in C language. We first generate an empty Grace project for this application. Then we enable the USCI_A block. We select the UART option from the selection window in the basic user mode. We set the baud rate to 9600 bps from the drop-down list. We also enable the receive interrupt by checking its box. We set the clock to 1 MHz under the BCM+ module. We also set P1.0 and P1.6 as output (both initially turned off) from the GPIO block. For this application, we do not add any codes to the main.c file of the Grace project. The USCI_A ISR under InterruptVectors_init.c

Figure 12.11

The register controls mode for the UART under Grace.

Grace (MSP430) ▸ USCI_A0 - Register Controls

Overview Basic User Power User Registers

UCAxCTL0, USCI_Ax Control Register 0

7	6	5	4	3	2	1	0
UCPEN	UCPAR	UCMSB	UC7BIT	UCSPB	UCMODEx		UCSYNC
					UART Mode		

UCAxCTL1, USCI_Ax Control Register 1

7	6	5	4	3	2	1	0
UCSSELx		UCRXEIE	UCBRKIE	UCDORM	UCTX ADDR	UCTX BRK	UCSW RST
SMCLK							[R/W]

UCAxBR0 7:0 — 0

UCAxBR1 7:0 — 0

UCAxMCTL, USCI_Ax Modulation Control Register

7	6	5	4	3	2	1	0
UCBRFx				UCBRSx			UCOS16
First stage 0				Second stage 0			

UCAxSTAT, USCI_Ax Status Register

7	6	5	4	3	2	1	0
UC LISTEN	UCFE	UCOE	UCPE	UCBRK	UCRX ERR	UCADDR UCIDLE	UCBUSY
[R/W]	[R/W]	[R/W]	[R/W]	[R/W]	[R/W]	[R/W]	[R/W]

UCAxRXBUF 7:0

UCAxTXBUF 7:0

UCAxIRTCTL, USCI_Ax IrDA Transmit Control Register

7	6	5	4	3	2	1	0
UCIRTX PL5	UCIRTX PL4	UCIRTX PL3	UCIRTX PL2	UCIRTX PL1	UCIRTX PL0	UCIR TXCLK	UCIREN

UCAxIRRCTL, USCI_Ax IrDA Receive Control Register

7	6	5	4	3	2	1	0
UCIRRX FL5	UCIRRX FL4	UCIRRX FL3	UCIRRX FL2	UCIRRX FL1	UCIRRX FL0	UCIR RXPL	UCIR RXFE

UCAxABCTL, USCI_Ax Auto Baud Rate Control Register

7	6	5	4	3	2	1	0
Reserved		UC DELIM1	UC DELIM0	UCSTOE	UCBTOE	Reserved	UCABD EN
				[R/W]	[R/W]		

IE2, Interrupt Enable Register 2

7	6	5	4	3	2	1	0
						UCA0 TXIE	UCA0 RXIE

IFG2, Interrupt Flag Register 2

7	6	5	4	3	2	1	0
						UCA0 TXIFG	UCA0 RXIFG
						[R/W]	[R/W]

will be as in Listing 12.9. As we debug and run the program, we can control the red and green LEDs by the keyboard entries of the host computer. Do not forget to use the terminal program in the host computer for this application.

Listing 12.8 The main.c file of the UART "Hello World" application, under Grace.

```
/*
 * ========  Standard MSP430 includes ========
 */
#include <msp430.h>

/*
 * ========  Grace related includes ========
 */
#include <ti/mcu/msp430/Grace.h>

/*
 *  ========  main ========
 */

const char string[] = {"Hello World"};
unsigned int i=0;

int main(void)
{

 Grace_init();
//Activate Grace-generated configuration

 while(string[i] != 0){
 while (!(IFG2&UCA0TXIFG));
 UCA0TXBUF = string[i++];
}

 while(1);
 return(0);
}
```

Listing 12.9 The ISR of the UART LED control application, under Grace.

```
#pragma vector=USCIAB0RX_VECTOR
__interrupt void USCI0RX_ISR_HOOK(void)
{
 if (UCA0RXBUF == 'r') P1OUT = BIT0;
 else if (UCA0RXBUF == 'g') P1OUT = BIT6;
 else P1OUT = 0x00;
}
```

12.4 Serial Peripheral Interface

SPI is a synchronous communication mode. It can be used between multiple masters and one slave. It can also be used for one master and one or more slaves. As in the UART mode, in this chapter we will only focus on the SPI mode between one

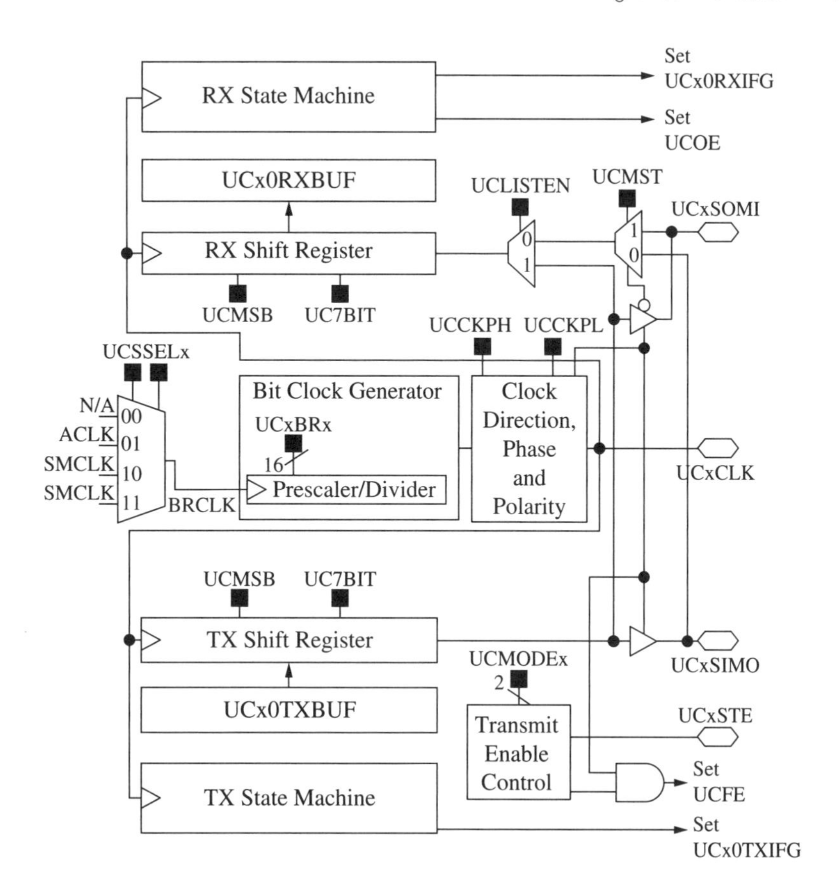

Figure 12.12

Block diagram of the SPI mode.

master and slave. A block diagram of the SPI mode is given in Fig. 12.12. SPI is the only communication mode available in both USCI_A0 and USCI_B0 modules. Therefore, the character x is used in register or variable names to indicate that the same register can be used for USCI_A0 or USCI_B0.

As can be seen in Fig. 12.12, the SPI mode has four pins for communication. These pins are slave in master out (**UCxSIMO**), master in slave out (**UCxSOMI**), SPI clock (**UCxCLK**), and slave transmit enable (**UCxSTE**). UCxSIMO and UCxSOMI pins are used for data transmission. UCxSIMO is the data output line and UCxSOMI is the data input line for the master device. UCxSIMO is the data input line and UCxSOMI is the data output line for the slave device. UCxCLK is the SPI clock generated by the master device. It ensures synchronization between the master and slave devices. UCxSTE is used to enable the chosen master in the multiple master mode or chosen slave in the multiple slave mode. When this pin is used, the SPI mode is called four pin. In single master and slave mode, some slave devices need this pin to start or end the SPI communication. Therefore, four-pin SPI is a necessity for them. If the SPI communication is established between one master and slave (and UCxSTE is not needed), then UCxSIMO, UCxSOMI, and UCxCLK pins will be enough. The UCxSTE pin can be connected to the ground in this setting. This SPI mode is called three pin.

Table 12.13

USCI_x0 control register 0 (UCx0CTL0).

Bits	7	6	5	4	3	2 - 1	0
	UCCKPH	UCCKPL	UCMSB	UC7BIT	UCMST	UCMODEx	UCSYNC

Table 12.14

SPI clock modes.

Mode	UCCKPL	UCCKPH
0	0	1
1	0	0
2	1	1
3	1	0

The SPI mode is configured by the USCI_x0 Control Register 0 (**UCx0CTL0**) and USCI_x0 Control Register 1 (**UCx0CTL1**). As a reminder, the UCA0CTL0 and UCA0CTL1 registers are also used in the UART mode. Here they are used for SPI with different entries. The reader should be aware of this overlap. The UCx0CTL0 and UCx0CTL1 register entries are given in Tables 12.13 and 12.15.

In Table 12.13, **UCCKPH** and **UCCKPL** bits are used together to adjust the SPI clock modes. The **UCCKPL** bit is used to set the clock polarity. When this bit is reset, the clock is kept low in the idle state. When it is set, the clock is kept high in the idle state. The UCCKPL bit does not affect the transmission format. The UCCKPH bit, on the other hand, has a direct effect on the transmission format. When this bit is reset, data is sent on the first clock edge and read on the next edge. When it is set, data is read on the first clock edge and sent on the next edge. UCCKPL and UCCKPH bits must be same for both master and slave devices to set up an SPI communication between them. The clock modes for the SPI are shown in Table 12.14. Here, modes 0 and 3 are the most commonly used ones. In these, data is read on the rising edge and sent on the falling edge of the clock. Mode 0 needs the UCxSTE pin. Therefore, it is preferred in the four-pin SPI mode. Unlike mode 0, mode 3 does not need the UCxSTE pin. Therefore, it is used in the three-pin SPI mode.

The **UCMSB** bit in Table 12.13 is used to choose the start bit for the data transfer. If this bit is reset, the transmission starts from the LSB. If it is set, the transmission starts from the MSB. The **UC7BIT** bit is used to select the data length. When this bit is reset, the data length is taken as 8 bits. When it is set, the data length is taken as 7 bits. The **UCMST** bit is used to decide on the usage type of the device. When this bit is reset, the device is used as a slave. When the bit is set, it is used as a master. **UCMODEx** bits are used to select the synchronization mode. Constants for these bits are as follows: **UCMODE_0** (three-pin SPI mode), **UCMODE_1** (four-pin SPI mode with UCxSTE active high), **UCMODE_2**

Table 12.15

USCI_x0 control register 1 (UCx0CTL1).

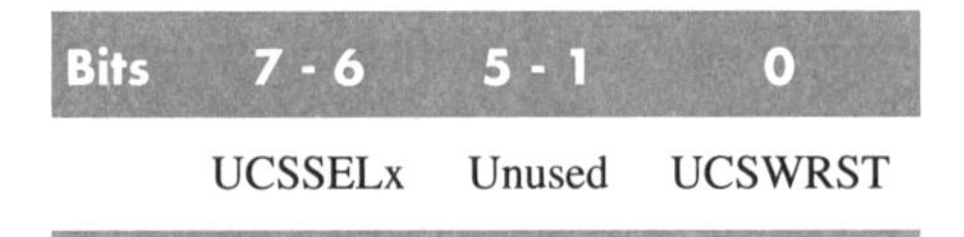

Bits	7 - 6	5 - 1	0
	UCSSELx	Unused	UCSWRST

(four-pin SPI mode with UCxSTE active low), and **UCMODE_3** (I^2C mode). Finally, the **UCSYNC** bit is used to select the communication mode. When this bit is reset, asynchronous mode is selected. When it is set, synchronous mode is selected. Therefore, this bit must be set for the SPI mode.

In Table 12.15, **UCSSELx** bits are used to select the SPI clock source. Constants for these bits are **UCSSEL_0** (not available), **UCSSEL_1** (for ACLK), and **UCSSEL_2** and **UCSSEL_3** (for SMCLK). After the clock source is selected, it can be divided by the 16-bit coefficient **UCBRx** as explained in Sec. 12.1.2. The SPI mode does not use modulation for clock generation. Therefore, the UCA0MCTL register must be cleared when the USCLA0 module is used for the SPI mode. The **UCSWRST** bit is used to reset the USCI module. When this bit is set, the USCI module is reset. When it is reset, the USCI module will be ready for operation.

The SPI mode also has a status register called **UCx0STAT**. It is specifically used to observe the changes in the system. The entries of this register are given in Table 12.16. In this table, the **UCLISTEN** bit is used to generate an internal loop between the transmitter and receiver on the same device. When this bit is set, the loopback is enabled. When it is reset, the loopback is disabled. The **UCFE** bit is the framing error flag. This bit is set when a bus conflict occurs in the four-pin SPI mode. It is not used in the three-pin SPI mode. The **UCOE** bit is the overrun error flag. This bit is set when a new character is sent to the receive buffer register (UCx0RXBUF) before the previous one is read. This bit is cleared automatically when the UCx0RXBUF is read. Therefore, the user should not try to clear it by software. The **UCBUSY** bit shows whether the USCI module is in process or not. This bit is set when the transmit or receive operation is performed. It is reset when the system is inactive.

12.4.1 *SPI Transmit/Receive Operations*

Transmission and reception must be carried out simultaneously in the SPI mode. Therefore, data must be received from the slave or transmitted from the master (or vice versa) even if it is completely redundant. Next, we provide transmit/receive operations for the master and slave modes separately.

Master Mode

One transmit-receive cycle for the SPI master mode is as follows: The USCI module is enabled. Transfer starts when data is written to the UCx0TXBUF. Then, data is transferred to the transfer shift register from the UCx0TXBUF. UCx0TXIFG is set to indicate that the UCx0TXBUF is ready to accept new data. Data in the transfer shift register is sent to the UCxSIMO pin starting with MSB or LSB order (based on the UCMSB bit setting). Meanwhile, the received data is kept waiting at the UCxSOMI pin until the next clock edge. Data in the UCxSOMI pin is moved to the receive shift register with the next clock edge. Then data is transferred to the UCx0RXBUF from the receive shift register. This operation is

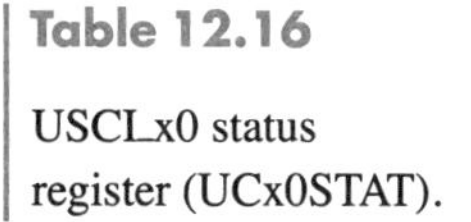

Table 12.16

USCLx0 status register (UCx0STAT).

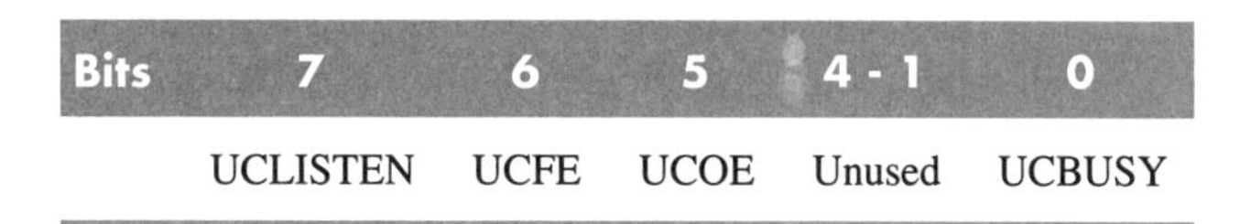

Bits	7	6	5	4 - 1	0
	UCLISTEN	UCFE	UCOE	Unused	UCBUSY

repeated until the 7 or 8 bits (depending on the setting of the UC7BIT) are transferred. As the transfer is completed, the UCx0RXIFG bit is set to indicate that the transmit-receive cycle is completed.

As mentioned before, four-pin SPI master mode is extensively used for the multimaster SPI communication. Here, the desired master is set active by the UCMODEx and UCxSTE bits. When a master is set inactive, UCxSIMO and UCxCLK pins are reconfigured as input. Receive-transmit operations are reset. Any ongoing shift operation is terminated. The UCFE bit is set to show that bus conflict on the system is handled by the user. If there is no ongoing shift operation when the master is set inactive, data in the UCx0TXBUF is transmitted after the master is set active again. But if there is a transmission in process when the master is set inactive, this data is lost and must be rewritten to the UCx0TXBUF.

Slave Mode

One transmit-receive cycle for the slave mode is as follows: The UCxCLK supplied by the master is used to start the data transfer. Before this clock is enabled, data is transferred to the transmit shift register from UCx0TXBUF. UCx0TXIFG is also set to indicate that UCx0TXBUF is ready to accept new data. Data in the transmit shift register is sent to the UCxSOMI pin starting with MSB or LSB order (based on the UCMSB bit setting). This data waits until the clock is activated. Data kept in the UCxSOMI pin is sent to output as the clock is activated. Meanwhile, the received data is kept waiting at the UCxSIMO pin until the next clock edge. Data in the UCxSIMO pin is moved to the receive shift register with the next clock edge. Then, data is transferred to the UCx0RXBUF from the receive shift register. This operation is repeated until the 7 or 8 bits (depending on the setting of the UC7BIT) are transferred. As the transfer is completed, the UCx0RXIFG bit is set to indicate that the transmit-receive cycle is completed.

Four-pin SPI slave mode is used for multislave SPI communication. Here, the desired slave is set active by the UCMODEx and UCxSTE bits. When a slave is set inactive, its UCxSOMI pin is reconfigured as input. Any receive operation in progress in the UCxSIMO pin is stopped. Until the slave is set active by reconfiguring the UCxSTE bit, ongoing shift operations are also stopped.

The four-pin SPI also provides an option to disable the slave in a single master slave setup. To do so, we should have a connection between a digital I/O pin of the master to the STE pin of the slave device. Then we can set/reset the digital I/O pin to enable/disable the slave device. We provide such an example in Sec. 12.4.3.

12.4.2 *SPI Interrupts*

The SPI mode shares the same interrupt vectors with the UART mode for the transmission and reception operations. As a reminder, for the transmitter the interrupt vector is **USCIAB0TX_VECTOR**. For the receiver, the interrupt vector is **USCIAB0RX_VECTOR**. Also, the SPI interrupt–related registers are the same as the UART ones. These are the interrupt enable register 2 (IE2) and interrupt flag register 2 (IFG2) given in Tables 12.3 and 12.4.

The interrupt operations for the transmitter and the receiver are similar in the SPI mode. More specifically, the interrupt-based communication operation works as follows: Initially, the UCx0TXIE and UCx0RXIE bits should be set in the master and slave devices to enable transmission and reception interrupts. These

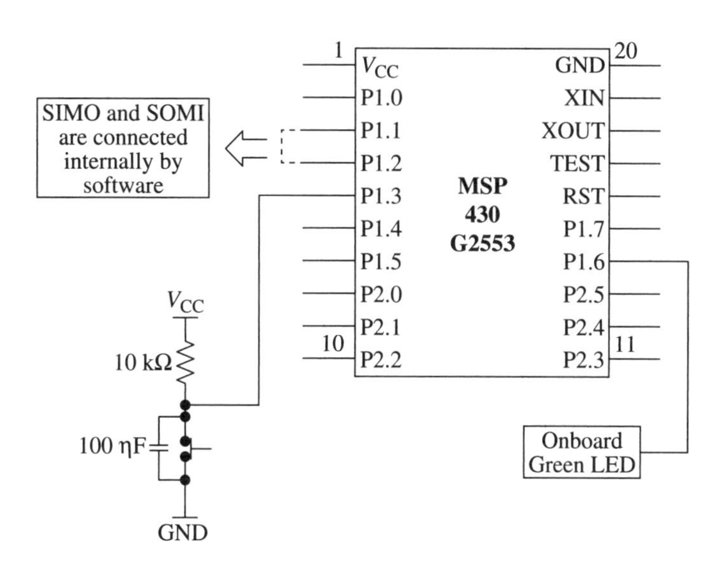

Figure 12.13 The connection diagram for the SPI loopback application.

two interrupts are maskable. Therefore, the GIE bit must also be set in both devices. In the transmitter, an interrupt is requested when the UCA0TXBUF is ready for another character. Then the UCA0TXIFG is set. This flag is automatically cleared when a new character is written to the UCA0TXBUF. In the receiver, an interrupt is requested when a character is loaded to the UCA0RXBUF. Then the UCA0RXIFG is set. This flag is automatically cleared when the data in UCA0RXBUF is read.

12.4.3 Coding Practices for the SPI Mode

In this section, we provide sample C and assembly codes on the SPI communication mode. The problems mentioned for the UART mode are also applicable here. Therefore, please see Sec. 12.2.4 first.

SPI in C

In Listing 12.10, the loopback property of the SPI mode is used. The connection diagram for this application is given in Fig. 12.13. Here, the green LED on the MSP430 LaunchPad is toggled by the button connected to pin P1.3. However, the loopback property is used such that the toggle command is sent and received within the microcontroller.

Listing 12.10 The SPI loopback application, in C language.

```
#include <msp430.h>
 int Data = 0;
void main(void)
{
 WDTCTL = WDTPW|WDTHOLD;
 P1DIR |= BIT6;  //Adjust pins
 P1OUT = 0x00;
 P1SEL = BIT1|BIT2;
 P1SEL2 = BIT1|BIT2;
 P1IE |= 0x08;
```

```
 P1IES |= 0x08;
 P1IFG = 0x00;
 UCA0CTL1 |= UCSWRST;  //Setup the SPI mode
//Enable SW reset
 UCA0CTL0 |= UCCKPL | UCMSB | UCMST | UCSYNC;
//Clock Mode 3, MSB first, 8-bit SPI master,
//three pin mode
 UCA0CTL1 = UCSSEL_2|UCSWRST;
//Use SMCLK, keep SW reset
 UCA0BR0 |= 0x02;
//Low bit of UCBRx is 2
 UCA0BR1 = 0;
//High bit of UCBRx is zero,
//fSCL = SMCLK/2 = ~600kHz
 UCA0MCTL = 0;
//No modulation
 UCA0STAT |= UCLISTEN;
//Enable internal loopback
 UCA0CTL1 &= ~UCSWRST;
//Clear SW reset, resume operation
 IE2 |= UCA0RXIE;
//Enable USCI0 RX interrupt
 _enable_interrupts();
 LPM4;
}
//USCI A transmitter interrupt
#pragma vector=USCIAB0TX_VECTOR
__interrupt void USCIA0TX_ISR(void){
 UCA0TXBUF = Data;
//Load the TX buffer with integer value
 IE2 &= ~UCA0TXIE;
//Disable transmit interrupt
}
//USCI A receiver interrupt
#pragma vector=USCIAB0RX_VECTOR
__interrupt void USCIA0RX_ISR(void){
 P1OUT = UCA0RXBUF;
//Write received data to P1OUT
}
#pragma vector=PORT1_VECTOR
__interrupt void Port_1(void){
 Data ^= 0x40;
//Toggle data value
 IE2 |= UCA0TXIE;
//Enable transmit interrupt
 P1IFG = 0x00;
//Clear interrupt flags
}
```

Figure 12.14

The connection diagram for the SPI PWM application in four-pin mode.

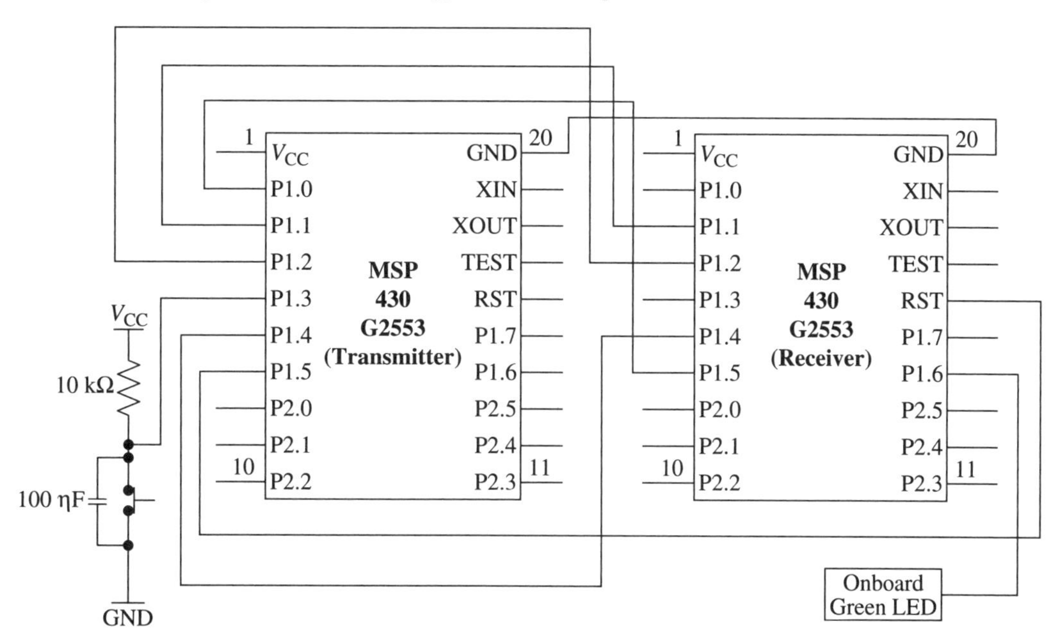

In Listings 12.11 and 12.12, the four-pin SPI mode is used to establish a digital communication between two MSP430 LaunchPads. The connection diagram for this application is given in Fig. 12.14. The C code for the master device (used as transmitter) is given in Listing 12.11. The C code for the slave device (used as receiver) is given in Listing 12.12. In this application, when the button connected to pin P1.3 of the master device is pressed, it sends the next PWM constant from the TXData array to control the brightness of the green LED on the receiver. The connection between pin P1.5 of the master device and the RST pin of the slave device is used for resetting the slave before the communication starts.

Listing 12.11 The SPI PWM application in four-pin mode, the master transmitter code in C language.

```
#include <msp430.h>

#define SlaveActive (P1OUT &= ~0x01)
#define SlaveInactive (P1OUT |= 0x01)
//Define outputs to activate
//or deactivate the slave

unsigned int *PTXData = 0;
unsigned int TXData[] = {0x0000, 0x00FA, 0x01F4,\
0x02EE, 0x03E8}; //TACCR1 values to be transmitted
int Receive = 0;
int cntr = 0;
unsigned int High = 0, Low = 0;
```

```
unsigned int TXByteCtr;
void main(void)
{
WDTCTL = WDTPW|WDTHOLD;
P1DIR |= BIT0|BIT5;  //Adjust pins
//Outputs to enable and reset the slave
P1OUT |= 0x01;
//Slave is initially inactive
P1SEL = BIT1|BIT2|BIT4;
P1SEL2 = BIT1|BIT2|BIT4;
P1IE |= 0x08;
P1IES |= 0x08;
P1IFG = 0x00;
UCA0CTL1 |= UCSWRST;  //Setup the SPI mode
//Enable SW reset
UCA0CTL0 |= UCCKPH | UCMSB | UCMST | UCSYNC;
//Clock Mode 0, MSB first, 8-bit SPI master,
//three pin mode
UCA0CTL1 = UCSSEL_2 | UCSWRST;
//Use SMCLK, keep SW reset
UCA0BR0 |= 0x02;
//Low bit of UCBRx is 2
UCA0BR1 = 0;
//High bit of UCBRx is zero,
//fSCL = SMCLK/2 = ~600kHz
UCA0MCTL = 0;
//No modulation
UCA0CTL1 &= ~UCSWRST;
//Clear SW reset, resume operation
PTXData = TXData;
//Equate TXData array's start address
//to PTXData
IE2 |= UCA0RXIE;  //Enable USCI0 RX interrupt
P1OUT &= ~BIT5;  //Reset SPI slave
P1OUT |= BIT5;
_enable_interrupts();
while(1){
TXByteCtr = 2;  //Load TX byte counter with two bytes
LPM0;
//Enter LPM0 (until two byte SPI communication
//is done)
}
}
//SPI master transmit interrupt service routine
#pragma vector=USCIAB0TX_VECTOR
__interrupt void USCIA0TX_ISR(void){
if (TXByteCtr){
if((TXByteCtr%2) == 0){High = *PTXData;
//Write the incoming array element to high integer
```

```
 UCA0TXBUF = (High>>8);
//Shift the high byte of High integer then
//load the TX buffer with it
 }
 if((TXByteCtr%2) == 1){
 Low = *PTXData++;
//Write the incoming array element to low
//integer, then increase PTXData
 UCA0TXBUF = Low;
//Load the TX buffer with low byte of array element
 cntr++;
 }
 TXByteCtr--;
 }
 else{
 if(cntr == 5){
 PTXData = TXData;
 cntr = 0;
//If cntr equals to 5, return to array start address
//and reset cntr
 }
 IE2 &= ~UCA0TXIE;  //Disable the TX interrupt
}}
//SPI master receive interrupt service routine
#pragma vector=USCIAB0RX_VECTOR
__interrupt void USCIA0RX_ISR(void){
 Receive = UCA0RXBUF;  //Receive data from slave
 if(TXByteCtr == 0){
 LPM0_EXIT;
 SlaveInactive;
 }
}
#pragma vector=PORT1_VECTOR
__interrupt void Port_1(void){
 SlaveActive;
 IE2 |= UCA0TXIE;  //Enable TX interrupt
 P1IFG = 0x00;
}
```

Listing 12.12 The SPI PWM application in four-pin mode, the slave receiver code in C language.

```
#include <msp430.h>

 unsigned int RXData;
 unsigned int RXByteCtr=0;
 int Null=0x00;

void main(void)
{
```

```
 WDTCTL=WDTPW|WDTHOLD;

 P1DIR |= BIT6;  //Adjust pins
 P1SEL = BIT1|BIT2|BIT4|BIT5|BIT6;
 P1SEL2 = BIT1|BIT2|BIT4|BIT5;

 UCA0CTL1 |= UCSWRST;  //Setup the SPI mode
//Enable SW reset
 UCA0CTL0 |= UCCKPH | UCMSB | UCSYNC | UCMODE_2;
//Clock Mode 0, MSB first, 8-bit SPI slave,
//four pin mode (STE active low)
 UCA0CTL1 &= ~UCSWRST;
//Clear SW reset, resume operation
 IE2 |= UCA0RXIE | UCA0TXIE;
//Enable USCI0 RX and TX interrupts

 TACCR1 = 0;  //Setup the PWM
 TACCR0 = 999;
 TACCTL1 = OUTMOD_7;
 TACTL = TASSEL_2 | MC_1 | ID_3;

 _enable_interrupts();

 while(1)
 LPM0;
}

//SPI slave transmit interrupt
#pragma vector=USCIAB0TX_VECTOR
__interrupt void USCI0TX_ISR (void){
 UCA0TXBUF = Null;  //Send null byte
 IE2 &= ~UCA0TXIE;  //Disable the TX interrupt
}

//SPI slave receive interrupt
#pragma vector=USCIAB0RX_VECTOR
__interrupt void USCI0RX_ISR (void){
 if((RXByteCtr%2)==0){
 RXData=UCA0RXBUF;
//Move received data (high byte) to low byte of RXData
 RXData=(RXData<<8);
//Shift low byte of RXData to high byte
 }
 else{
 RXData|=UCA0RXBUF;
//Move received data (low byte) to low byte of RXData
 TACCR1=RXData;
 }
 RXByteCtr++;
 if(RXByteCtr==2) RXByteCtr=0;
 IE2 |= UCA0TXIE;  //Enable TX interrupt
}
```

In Listings 12.13 and 12.14, the SPI PWM application is implemented in three-pin mode. Therefore, the previous application is redone. The connection diagram for this application is given in Fig. 12.15. The C code for the master device (used as transmitter) is given in Listing 12.13. The C code for the slave device (used as receiver) is given in Listing 12.14.

Listing 12.13 The SPI PWM application in three-pin mode, the master transmitter code in C language.

```
#include <msp430.h>

 unsigned int *PTXData = 0;
 unsigned int TXData[] = {0x0000, 0x00FA, 0x01F4,\
 0x02EE, 0x03E8};  //TACCR1 values to be transmitted
 int Receive = 0;
 int cntr = 0;
 unsigned int High = 0, Low = 0;
 unsigned int TXByteCtr;

void main(void)
{
 WDTCTL = WDTPW|WDTHOLD;

 P1DIR |= BIT5;  //Adjust pins
 P1OUT = 0x00;
 P1SEL = BIT1|BIT2|BIT4;
 P1SEL2 = BIT1|BIT2|BIT4;
 P1IE |= 0x08;
 P1IES |= 0x08;
 P1IFG = 0x00;

 UCA0CTL1 |= UCSWRST;  //Setup the SPI mode
//Enable SW reset
 UCA0CTL0 |= UCCKPL | UCMSB | UCMST | UCSYNC;
//Clock Mode 3, MSB first, 8-bit SPI master,
//three pin mode
 UCA0CTL1 = UCSSEL_2 | UCSWRST;
//Use SMCLK, keep SW reset
 UCA0BR0 |= 0x02;
//Low bit of UCBRx is 2
 UCA0BR1 = 0;
//High bit of UCBRx is zero,
//fSCL = SMCLK/2 = ~600kHz
 UCA0MCTL = 0;
//No modulation
 UCA0CTL1 &= ~UCSWRST;
//Clear SW reset, resume operation
 PTXData = TXData;
//Equate TXData array's start address to PTXData
 IE2 |= UCA0RXIE;
```

```
//Enable USCI0 RX interrupt

 P1OUT &= ~BIT5;  //Reset SPI slave
 P1OUT |= BIT5;

 _enable_interrupts();

while(1){
 TXByteCtr = 2;  //Load TX byte counter with two bytes
 LPM0;
//Enter LPM0 (until two byte SPI communication
//is done)
 }
}

//SPI master transmit interrupt
#pragma vector=USCIAB0TX_VECTOR
__interrupt void USCIA0TX_ISR(void){
 if (TXByteCtr){
 if((TXByteCtr%2) == 0){
 High = *PTXData;
//Write the incoming array element to high integer
 UCA0TXBUF = (High>>8);
//Shift the high byte of High integer then load TX
//buffer with this
 }
 if((TXByteCtr%2) == 1){
 Low = *PTXData++;
//Write the incoming array element to low integer,
//then increase PTXData
 UCA0TXBUF = Low;
 cntr++;
 }
 TXByteCtr--;
 }
 else{
 if(cntr == 5){
 PTXData = TXData;
 cntr = 0;
//If cntr equals to 5, return to array start address
//and reset cntr
 }
 IE2 &= ~UCA0TXIE;  //Disable TX interrupt
 }
}

//SPI master receive interrupt
#pragma vector=USCIAB0RX_VECTOR
__interrupt void USCIA0RX_ISR(void){
 Receive = UCA0RXBUF;  //Receive data from slave
 if(TXByteCtr == 0)LPM0_EXIT;
}

#pragma vector=PORT1_VECTOR
```

```
__interrupt void Port_1(void){
 IE2 |= UCA0TXIE;  //Enable TX interrupt
 P1IFG = 0x00;
}
```

Listing 12.14 The SPI PWM application in three-pin mode, the slave receiver code in C language.

```
#include <msp430.h>

 unsigned int RXData;
 unsigned int RXByteCtr = 0;
 int Null = 0x00;

void main(void)
{
 WDTCTL = WDTPW|WDTHOLD;

 P1DIR |= BIT6;  //Adjust pins
 P1SEL = BIT1|BIT2|BIT4|BIT6;
 P1SEL2 = BIT1|BIT2|BIT4;

 UCA0CTL1 |= UCSWRST;  //Setup the SPI mode
//Enable SW reset
 UCA0CTL0 |= UCCKPL | UCMSB | UCSYNC;
//Clock Mode 3, MSB first, 8-bit SPI slave,
//three pin mode
 UCA0CTL1 &= ~UCSWRST;
//Clear SW reset, resume operation
 IE2 |= UCA0RXIE | UCA0TXIE;
//Enable USCI0 RX and TX interrupts

 TACCR1 = 0;  //Setup the PWM
 TACCR0 = 999;
 TACCTL1 = OUTMOD_7;
 TACTL = TASSEL_2|MC_1|ID_3;

 _enable_interrupts();

 while (1){
 RXByteCtr=0;
 LPM0;
//Enter LPM0 and wait until transmit
//or receive interrupts occur
 }
}

//SPI slave transmit interrupt
#pragma vector=USCIAB0TX_VECTOR
__interrupt void USCI0TX_ISR (void){
 UCA0TXBUF = Null;
 IE2 &= ~UCA0TXIE;  //Disable TX interrupt
```

```
}
//SPI slave receive interrupt
#pragma vector=USCIAB0RX_VECTOR
__interrupt void USCI0RX_ISR (void){
 if((RXByteCtr%2) == 0){
 RXData = UCA0RXBUF;
//Move received data (high byte) to
//low byte of RXData
 RXData = (RXData<<8);
//Shift low byte of RXData to high byte
 }
 else{
 RXData |= UCA0RXBUF;
//Move received data (low byte) to low byte of RXData
 TACCR1 = RXData;
//Move RXData to TACCR1
 LPM0_EXIT;

 }
 RXByteCtr++;
 IE2 |= UCA0TXIE;  //Enable TX interrupt
}
```

Figure 12.15

The connection diagram for the SPI PWM application in three-pin mode.

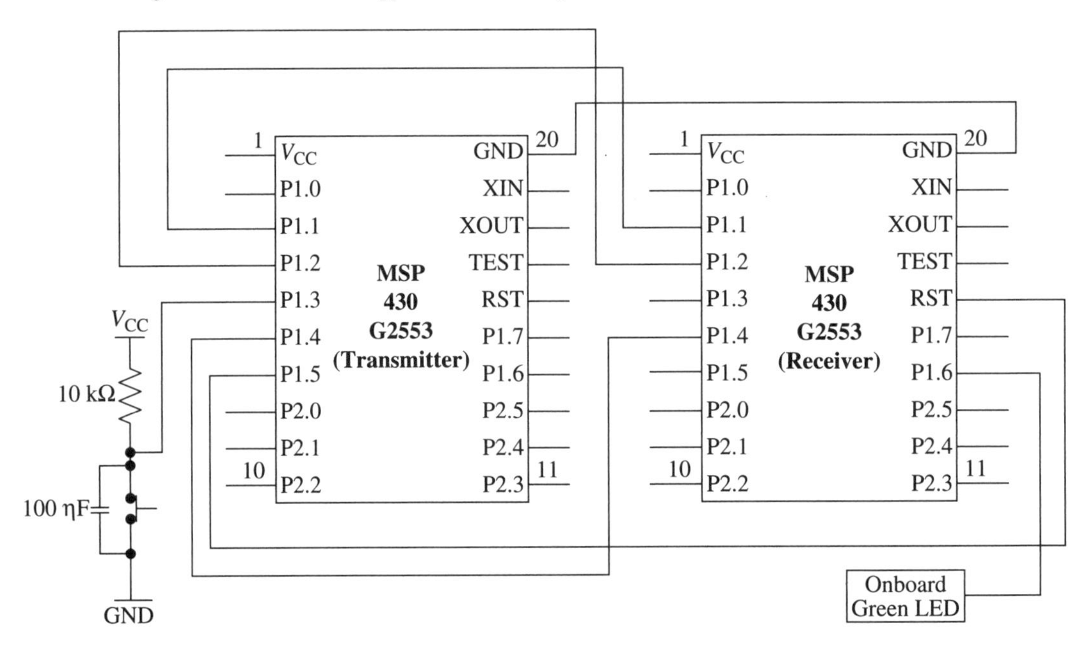

SPI in Assembly

The assembly codes, given in Listings 12.15 and 12.16, perform the same operation done in Listings 12.11 and 12.12. The connection diagram for this application is also the same as that given in Fig. 12.14. The assembly code for the master device (used as transmitter) is given in Listing 12.15. The assembly code for the slave device (used as receiver) is given in Listing 12.16.

Listing 12.15 The SPI PWM application in four-pin mode, the master transmitter code in assembly language.

```
 .cdecls C,LIST,"msp430.h"

 .text
 .retain
 .retainrefs

RESET
 mov.w #WDTPW|WDTHOLD,WDTCTL
 mov.w #__STACK_END,SP

; These five values are to be sent
 mov.w #0000h,&0200h
 mov.w #00FAh,&0202h
 mov.w #01F4h,&0204h
 mov.w #02EEh,&0206h
 mov.w #03E8h,&0208h

 bis.b #21h,P1DIR ;Adjust pins
 bis.b #01h,P1OUT
 bis.b #16h,P1SEL
 bis.b #16h,P1SEL2
 bis.b #08h,P1IE
 bis.b #08h,P1IES
 clr.b P1IFG

 bis.b #UCSWRST,UCA0CTL1 ;Adjust the SPI mode
;Enable SW reset
 bis.b #UCCKPH + UCMSB + UCMST + UCSYNC,UCA0CTL0
;Clock Mode 0, MSB first, 8-bit SPI master,
;four pin mode
 mov.b #UCSSEL_2+UCSWRST,UCA0CTL1
;Use SMCLK, keep SW reset
 mov.b #02h,UCA0BR0 ;Low bit of UCBRx is 2
 mov.b #00h,UCA0BR1 ;High bit of UCBRx is zero,
;fSCL = SMCLK/12 = ~100kHz
 clr.b UCA0MCTL
;No Modulation
 bic.b #UCSWRST,UCA0CTL1
;Clear SW reset, resume operation
 bis.b #UCA0RXIE,IE2 ;Enable RX interrupt
 mov.w #0200h,R5
;Write the start address of TX data into R5
 clr R10
;Clear R10 used for checking which element
```

```
;is transmitted

 bic.b #20h,P1OUT ;Reset slave
 bis.b #20h,P1OUT

 bis.w #GIE,SR

Mainloop:
 mov.w #2,R7
 bis.w #LPM0,SR
 jmp Mainloop

;-----------------------------------------------
USCIAB0TX_ISR
;SPI master transmit interrupt service routine
;-----------------------------------------------
 tst.w R7 ;Test if R7 is zero
 jeq AllBytesTransmitted
;If it is, jump AllBytesTransmitted
 bit.w #1h,R7
;Check if R7 is odd
 jne OddByte
;If it is, jump OddByte
 mov.w @R5,R8
;Write the incoming array element to R8
 swpb R8
 mov.b R8,UCA0TXBUF
;Load TX buffer with high byte of incoming element
OddByte:
 cmp.w #1h,R7 ;Check if R7 is 1
 jne DecrementByteNumber
 mov.w @R5+,R9
 mov.b R9,UCA0TXBUF
;Load TX buffer with low byte of incoming element
 inc.w R10
;Increase R10 to indicate that one element is sent
DecrementByteNumber:
 dec.w R7
;Decrease byte number by one
 jmp EndISR
AllBytesTransmitted:
 cmp.w #5h,R10
;Check if all five elements are sent
 jne DisableTX
 sub.w #10h,R5
;If it is 5, reload PTXData pointer with
;address of the first element
 clr R10
DisableTX:
 bic.b #UCA0TXIE,IE2 ;Disable TX interrupt
EndISR:
 reti
```

```
;-------------------------------------------
USCIAB0RX_ISR
;SPI slave receive interrupt service routine
;-------------------------------------------
 mov.b UCA0RXBUF,R6
;Move received data from slave to R6
 tst.w R7
;Check if R7 is zero
 jne EndISR2
;If it is not, jump EndISR2
 bic.w #LPM0,0(SP)
 bis.b #01h,P1OUT
;Exit LPM0 and deactivate slave
EndISR2:
 reti
;----------------------
P1_ISR
;----------------------
 bic.b #01h,P1OUT ;Activate slave
 bis.b #UCA0TXIE,IE2 ;Enable the TX interrupt
 clr.b P1IFG
 reti
;------------------------
;Stack Pointer definition
;------------------------
 .global __STACK_END
 .sect .stack

;-------------------
;Interrupt Vectors
;-------------------
 .sect RESET_VECTOR
 .short RESET
 .sect  USCIAB0TX_VECTOR
 .short USCIAB0TX_ISR
 .sect  USCIAB0RX_VECTOR
 .short USCIAB0RX_ISR
 .sect  PORT1_VECTOR
 .short P1_ISR
 .end
```

Listing 12.16 The SPI PWM application in four-pin mode, the slave receiver code in assembly language.

```
 .cdecls C,LIST,"msp430.h"

 .text
 .retain
 .retainrefs

RESET
```

```
 mov.w #WDTPW|WDTHOLD,WDTCTL
 mov.w #__STACK_END,SP

 bis.b #40h,P1DIR ;Adjust pins
 bis.b #076h,P1SEL
 bis.b #036h,P1SEL2

 bis.b #UCSWRST,UCA0CTL1 ;Adjust the SPI mode
;Enable SW reset
 mov.b #UCCKPH + UCMSB + UCSYNC + UCMODE_2,UCA0CTL0
;Clock Mode 0, MSB first, 8-bit SPI slave,
;four pin mode (STE active low)
 bic.b #UCSWRST,UCA0CTL1
;Clear SW reset, resume operation
 bis.b #UCA0RXIE + UCA0TXIE,IE2
;Enable USCI0 RX and TX interrupts

 clr.w TACCR1 ;Adjust the PWM
 mov.w #03E7h,TACCR0
 mov.w #OUTMOD_7,TACCTL1
 mov.w #TASSEL_2+MC_1+ID_3,TACTL

 clr.w R7
;Clear R7 used for counting the received byte

 bis.w #GIE,SR

Mainloop:
 clr.w R7
 bis.w #LPM0,SR
 jmp Mainloop

;-----------------------------------------------
USCIAB0TX_ISR
;SPI slave transmit interrupt service routine
;-----------------------------------------------
 mov.b #0h,UCA0TXBUF ;Send Null character
 bic.b #UCA0TXIE,IE2 ;Disable TX interrupt
 reti

;-----------------------------------------------
USCIAB0RX_ISR
;SPI slave receive interrupt service routine
;-----------------------------------------------
 bit.w #1h,R7 ;Check if R7 is odd
 jne OddByte
;If it is, jump OddByte
 mov.b UCA0RXBUF,R8
;Move the received data (low byte) to low byte of R8
 jmp IncrementByteNumber
OddByte:
 mov.b UCA0RXBUF,R9
;Move the received data (low byte) to R9
 swpb R8
 and.w #0F00h,R8
```

```
;And R8 with 0x0F00; for the high byte
;numbers are located in these four bits

 bis.w R8,R9 ;Or R8 and R9
 mov.w R9,TACCR1
;Move the received two byte data to TACCR1
  bic.w #LPM0,0(SP)
;Exit from LPM0
IncrementByteNumber:
 inc.w R7 ;Increase byte number by one
 bis.b #UCA0TXIE,IE2 ;Enable TX interrupt
 reti

;------------------------
;Stack Pointer definition
;------------------------
 .global __STACK_END
 .sect .stack

;-------------------
;Interrupt Vectors
;-------------------
 .sect RESET_VECTOR
 .short RESET
 .sect  USCIAB0TX_VECTOR
 .short USCIAB0TX_ISR
 .sect  USCIAB0RX_VECTOR
 .short USCIAB0RX_ISR
 .end
```

The assembly codes, given in Listings 12.17 and 12.18, perform the same operation done in Listings 12.13 and 12.14. The connection diagram for this application is also the same as that given in Fig. 12.15. The assembly code for the master device (used as transmitter) is given in Listing 12.17. The assembly code for the slave device (used as receiver) is given in Listing 12.18.

Listing 12.17 The SPI PWM application in three-pin mode, the master transmitter code in assembly language.

```
 .cdecls C,LIST,"msp430.h"

 .text
 .retain
 .retainrefs

RESET
 mov.w #WDTPW|WDTHOLD,WDTCTL
 mov.w #__STACK_END,SP

;These five values are to be sent
 mov.w #0000h,&0200h
 mov.w #00FAh,&0202h
```

```
 mov.w #01F4h,&0204h
 mov.w #02EEh,&0206h
 mov.w #03E8h,&0208h

 bis.b #20h,P1DIR

 bis.b #16h,P1SEL ;Adjust pins
 bis.b #16h,P1SEL2
 bis.b #08h,P1IE
 bis.b #08h,P1IES
 clr.b P1IFG

 bis.b #UCSWRST,UCA0CTL1 ;Adjust the SPI mode
;Enable SW reset
 bis.b #UCCKPL + UCMSB + UCMST + UCSYNC,UCA0CTL0
;Clock Mode 3, MSB first, 8-bit SPI master
;three pin mode
 mov.b #UCSSEL_2+UCSWRST,UCA0CTL1
;Use SMCLK, keep SW reset
 mov.b #02h,UCA0BR0 ;Low bit of UCBRx is 2
 mov.b #00h,UCA0BR1 ;High bit of UCBRx is zero
;fSCL = SMCLK/12 = ~100kHz
 clr.b UCA0MCTL
;No Modulation
 bic.b #UCSWRST,UCA0CTL1
;Clear SW reset, resume operation
 bis.b #UCA0RXIE,IE2 ;Enable RX interrupt
 mov.w #0200h,R5
; Write the start address of TX data to R5
 clr R10
;Clear R10 used for checking which element
;is transmitted

 bic.b #20h,P1OUT ;Reset I2C slave
 bis.b #20h,P1OUT

 bis.w #GIE,SR

Mainloop:
 mov.w #2h,R7 ;Load R7 with two bytes
 bis.w #LPM0,SR
 jmp Mainloop

;-------------------------------------------------
USCIAB0TX_ISR
;SPI master transmit interrupt service routine
;-------------------------------------------------
 tst.w R7 ;Test if R7 is zero
 jeq AllBytesTransmitted
;If it is, jump to AllBytesTransmitted
 bit.w #1h,R7 ;Check if R7 is odd
 jne OddByte
;If it is, jump to OddByte
 mov.w @R5,R8
```

```
;Write the incoming array element to R8
 swpb R8
 mov.b R8,UCA0TXBUF
;Load TX buffer with high byte of incoming element
OddByte:
 cmp.w #1h,R7 ;Check if R7 is 1
 jne DecrementByteNumber
 mov.w @R5+,R9
 mov.b R9,UCA0TXBUF
;Load TX buffer with the low byte of the incoming
;element
 inc.w R10
;Increase R10 to indicate that one element is sent
DecrementByteNumber:
 dec.w R7
;Decrease byte number by one
 jmp EndISR
AllBytesTransmitted:
 cmp.w #5h,R10 ;Check if all five elements are sent
 jne DisableTX
 sub.w #10h,R5
;If it is 5, reload PTXData pointer with
;address of the first element
 clr R10 ;Clear R10
DisableTX:
 bic.b #UCA0TXIE,IE2 ;Disable TX interrupt
EndISR:
 reti

;-----------------------------------------
USCIAB0RX_ISR
;SPI slave receive interrupt service routine
;-----------------------------------------
 mov.b UCA0RXBUF,R6
;Move the received data from slave to R6
 tst.w R7 ;Check if R7 is zero
 jne EndISR2 ;If it is not, jump to EndISR2
 bic.w #LPM0,0(SP) ;Exit LPM0
EndISR2:
 reti

;---------------------
P1_ISR
;---------------------
 bis.b #UCA0TXIE,IE2 ;Enable TX interrupt
 clr.b P1IFG
 reti

;----------------------
;Stack Pointer definition
;----------------------
 .global __STACK_END
```

```
 .sect .stack
;-------------------
;Interrupt Vectors
;-------------------
 .sect RESET_VECTOR
 .short RESET
 .sect  USCIAB0TX_VECTOR
 .short USCIAB0TX_ISR
 .sect  USCIAB0RX_VECTOR
 .short USCIAB0RX_ISR
 .sect  PORT1_VECTOR
 .short P1_ISR
 .end
```

Listing 12.18 The SPI PWM application in three-pin mode, the slave receiver code in assembly language.

```
 .cdecls C,LIST,"msp430.h"

 .text
 .retain
 .retainrefs

RESET
 mov.w #WDTPW|WDTHOLD,WDTCTL
 mov.w #__STACK_END,SP

 bis.b #40h,P1DIR ;Adjust pins
 bis.b #056h,P1SEL
 bis.b #016h,P1SEL2

 bis.b #UCSWRST,UCA0CTL1 ;Adjust the SPI mode
;Enable SW reset
 mov.b #UCCKPL + UCMSB + UCSYNC,UCA0CTL0
;Clock Mode 3, MSB first, 8-bit SPI slave,
;three pin mode
 bic.b #UCSWRST,UCA0CTL1
;Clear SW reset, resume operation
 bis.b #UCA0RXIE + UCA0TXIE,IE2
;Enable USCI0 RX and TX interrupts

 clr.w TACCR1 ;Adjust the PWM
 mov.w #03E7h,TACCR0
 mov.w #OUTMOD_7,TACCTL1
 mov.w #TASSEL_2+MC_1+ID_3,TACTL

 clr.w R7
;Clear R7 used for counting the received byte

 bis.w #GIE,SR

Mainloop:
 clr.w R7
```

```
 bis.w #LPM0,SR
 jmp Mainloop
;------------------------------------------------
USCIAB0TX_ISR
;SPI slave transmit interrupt service routine
;------------------------------------------------
 mov.b #0h,UCA0TXBUF ;Send Null character
 bic.b #UCA0TXIE,IE2 ;Disable TX interrupt
 reti
;-----------------------------------------------
USCIAB0RX_ISR
;SPI slave receive interrupt service routine
;-----------------------------------------------
 bit.w #1h,R7 ;Check if R7 is odd
 jne OddByte ;If it is, jump OddByte
 mov.b UCA0RXBUF,R8
;Move the received data (low byte) to low byte of R8
 jmp IncrementByteNumber
OddByte:
 mov.b UCA0RXBUF,R9
;Move the received data (low byte) to R9
 swpb R8
 and.w #0F00h,R8
;And R8 with 0x0F00, high byte
;numbers are located only these four bits
 bis.w R8,R9 ;Or R8 and R9
 mov.w R9,TACCR1
;Move the received two bytes of data to TACCR1
 bic.w #LPM0,0(SP)
;Exit from LPM0
IncrementByteNumber:
 inc.w R7 ;Increase byte number by one
 bis.b #UCA0TXIE,IE2 ;Enable TX interrupt
 reti
;-----------------------
;Stack Pointer definition
;-----------------------
 .global __STACK_END
 .sect .stack

;------------------
;Interrupt Vectors
;------------------
 .sect RESET_VECTOR
 .short RESET
 .sect  USCIAB0TX_VECTOR
 .short USCIAB0TX_ISR
 .sect  USCIAB0RX_VECTOR
 .short USCIAB0RX_ISR
 .end
```

12.5 SPI in Grace

SPI is available in both USCI_A0 and USCI_B0 blocks under Grace. However, they are used in the same manner. Therefore, we only provide the SPI mode under the USCI_A0 block in this section. We assume that the user clicked the SPI button in the first selection window for all user modes to be explored below.

12.5.1 *The Basic User Mode*

The basic user mode for the SPI is given in Fig. 12.16. Here, we can set the device as master or slave from the drop-down list in the USCI_A0 SPI Mode block. We can configure the SPI pins from the related drop-down lists. We can select the bit rate from the Bitrate drop-down list. We can also select a custom bit rate value. First, we should select the Custom option from the drop-down list. Then we can enter the desired bit rate in the Set Custom box. We can also set the clock phase and polarity values from the related drop-down lists. We can enable the transmit and receive interrupts by checking the "USCI_A0 SPI transmit interrupt enable" and "USCI_A0 SPI receive interrupt enable" boxes respectively. We can also generate an ISR related to these interrupts using the associated Generate Interrupt Handler Code button.

12.5.2 *The Power User Mode*

The power user mode for the SPI is given in Fig. 12.17. In addition to the basic user mode, we can select the three- or four-pin mode from the drop-down list in the USCI_A0 SPI Mode block. We can also select the bit order (MSB or LSB first) and the character length using the related drop-down lists.

12.5.3 *The Register Controls Mode*

Finally, the SPI registers can be adjusted under Grace. The user should select the register controls mode, as given in Fig. 12.18, for this purpose. As in the UART mode, some registers are not available here. Some register entries are also read only in this mode.

12.6 Inter Integrated Circuit

Inter integrated circuit (I^2C) is the second synchronous communication mode supported by the MSP430. It can be used between multiple masters and slaves. The master and slave devices are represented by address values in I^2C. In addition to this, a simple protocol establishes an effective communication between multiple master and slave devices. A block diagram of the I^2C mode is given in Fig. 12.19.

As can be seen in Fig. 12.19, the I^2C mode has two bidirectional pins for communication. These are the serial data pin (**SDA**) and serial clock pin (**SCL**). These pins must be connected to the positive supply voltage (V_{CC}) via pull-up resistors. The pull-up resistors should be external. Their values should be around 10 kΩ. Unlike SPI, here the transmission and reception operations are done on a single line. This saves pins, but it slows down the communication speed. More information on I^2C can be found in [7].

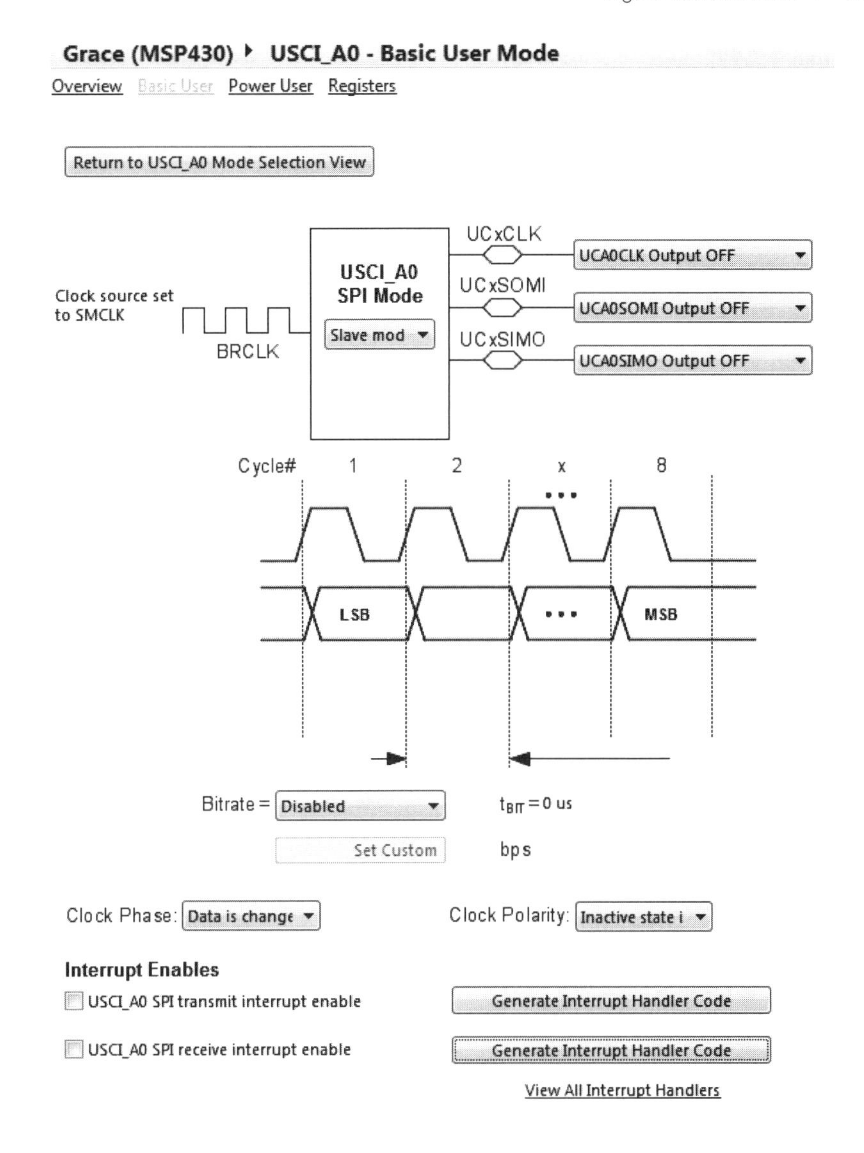

Figure 12.16

The basic user mode for the SPI under Grace.

The I^{2}C mode is configured by the USCIB0 control register 0 (**UCB0CTL0**) and USCIB0 control register 1 (**UCB0CTL1**). As a reminder, the same registers are also used in the SPI mode. Here they are used for the I^{2}C with different entries. The reader should be aware of this overlap. The UCB0CTL0 and UCB0CTL1 register entries are given in Tables 12.17 and 12.18.

In Table 12.17, the **UCA10** bit is used to select the own-address length of the device. When this bit is reset, 7-bit address is used. When it is set, 10-bit address is used. The **UCSLA10** bit is used to set the slave address length similar to the UCA10 bit settings. The **UCMM** bit is used to choose the master number. This bit should be reset if there is only one master in the system. Otherwise, it should

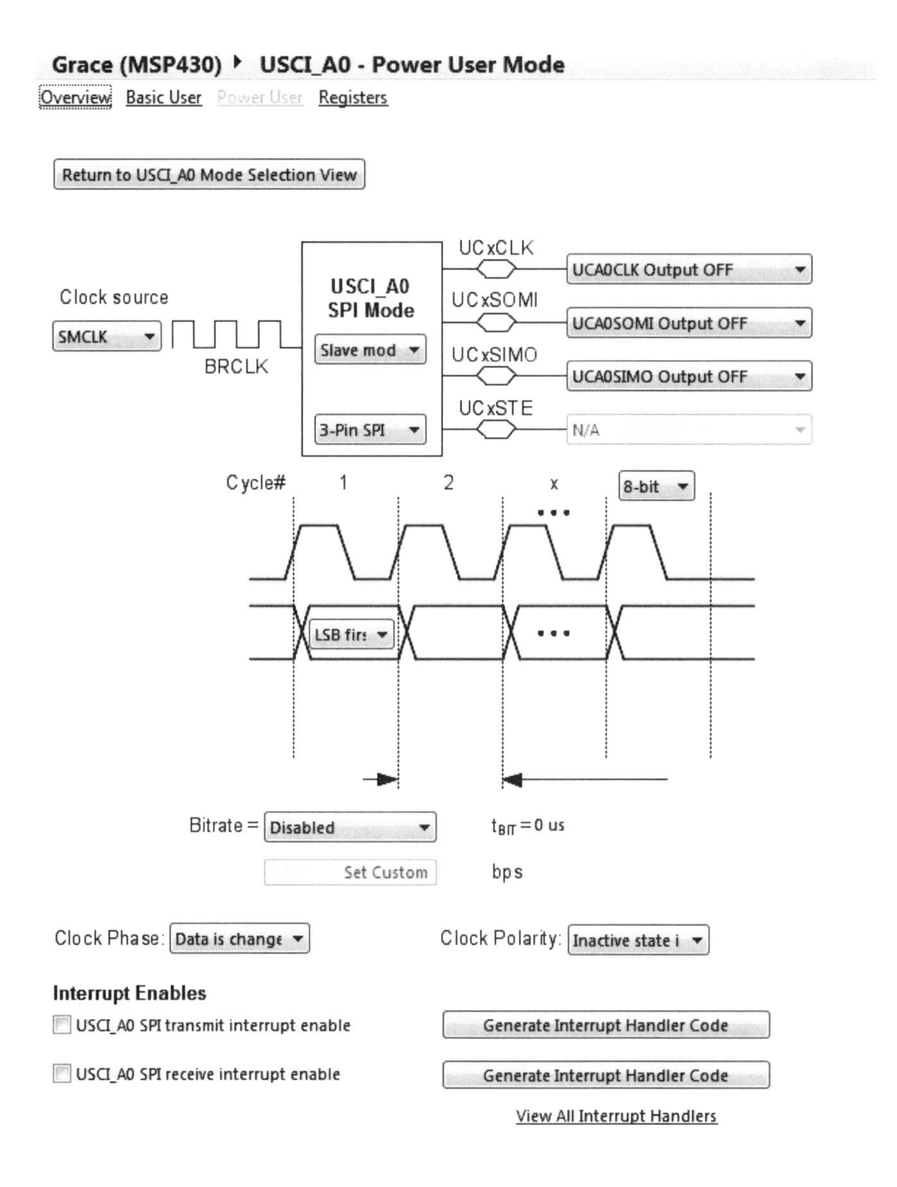

Figure 12.17 The power user mode for the SPI under Grace.

be set to indicate that more than one master device will be used in communication. The **UCMST** bit is used to decide on whether the device is master or slave. When this bit is reset, the device will be used as slave. When it is set, the device will be used as master. The **UCMODEx** bits are used to select the synchronous communication mode. For I^2C, they should be set to the constant UCMODE_3. Finally, the **UCSYNC** bit is used to choose the communication mode. When this bit is reset, asynchronous mode is chosen. When it is set, synchronous mode is chosen. Therefore, this bit must be set for the I^2C mode.

In Table 12.18, **UCSSELx** bits are used to select the I^2C clock source. Constants for these bits are UCSSEL_0 (for UC1CLK), UCSSEL_1 (for ACLK), and

Figure 12.18

The register controls mode for the SPI under Grace.

Grace (MSP430) ▸ USCI_A0 - Register Controls

UCAxCTL0, USCI_Ax Control Register 0

7	6	5	4	3	2	1	0
UCCKPH	UCCKPL	UCMSB	UC7BIT	UCMST	UCMODEx		UCSYNC
☐	☐	☐	☐	☐	3-Pin SPI		☑

UCAxCTL1, USCI_Ax Control Register 1

7	6	5	4	3	2	1	0
UCSSELx		Unused					UCSWRST
SMCLK							[R/W]

UCAxBR0

7:0
0

UCAxBR1

7:0
0

UCAxSTAT, USCI_Ax Status Register

7	6	5	4	3	2	1	0
UCLISTEN	UCFE	UCOE	Unused	Unused	Unused	Unused	UCBUSY
[R/W]	[R/W]	[R/W]					[R/W]

UCAxRXBUF

7:0

UCAxTXBUF

7:0

IE2, Interrupt Enable Register 2

7	6	5	4	3	2	1	0
						UCA0TXIE	UCA0RXIE
						☐	☐

IFG2, Interrupt Flag Register 2

7	6	5	4	3	2	1	0
						UCA0TXIFG	UCA0RXIFG
						[R/W]	[R/W]

UCSSEL_2 and UCSSEL_3 (for SMCLK). After the clock source is selected, it can be divided by the 16-bit coefficient UCBRx as in the SPI mode. The **UCTR** bit is used to select whether the device is a transmitter or a receiver. When this bit is set, the device becomes a transmitter. When it is reset, the device becomes a receiver. The **UCTXNACK** bit is used to adjust the not-acknowledge (NACK) bit settings. When this bit is reset, sending an ACK bit occurs normally. When it is set, the NACK bit is generated. The UCTXNACK bit is reset automatically after the NACK bit is sent. More information on these settings can be found in [7]. **UCTXSTT** and **UCTXSTP** bits are used to transmit start and stop conditions respectively. When these bits are set, start or stop conditions are generated. These can be produced only by the master device. Therefore, they are not used in the

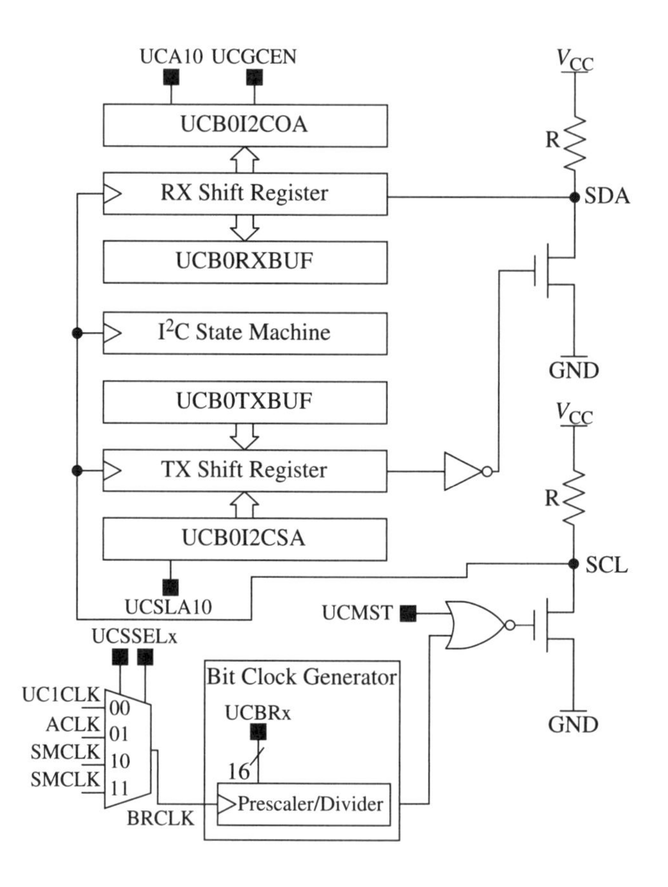

Figure 12.19 Block diagram of the I^2C mode.

Table 12.17 USCLB0 control register 0 (UCB0CTL0).

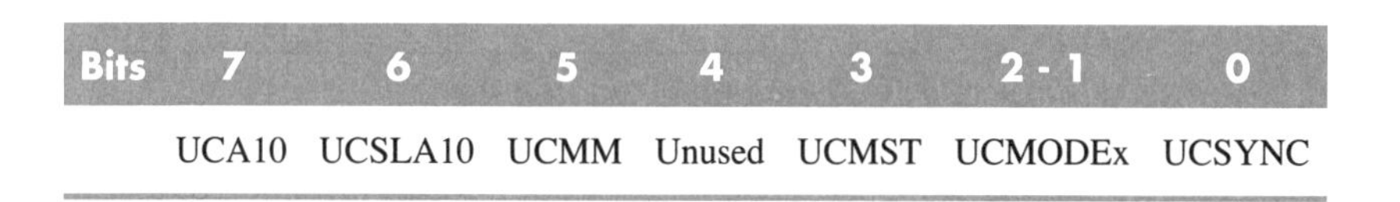

Bits	7	6	5	4	3	2 - 1	0
	UCA10	UCSLA10	UCMM	Unused	UCMST	UCMODEx	UCSYNC

slave mode. When these bits are set in the master receiver mode to generate a repeated start or stop condition, they are followed by a NACK bit. The UCTXSTT bit is reset automatically after the start condition and the address data is sent. The UCTXSTP bit is reset automatically after the stop condition is generated. The **UCSWRST** bit is used to reset the USCI module. When this bit is set, the USCI module is reset. When it is reset, the USCI module will be ready for operation.

The I^2C mode also has a status register called **UCB0STAT**. It is specifically used to observe the changes in the system. The entries of this register are given in Table 12.19. In this table, the **UCSCLLOW** bit is used to check the condition of the SCL. When this bit is set, the SCL is held low. When it is reset, the SCL is held high (default case). The **UCGC** bit is used to control whether the general call address is received or not. The general call address is used when more than

Table 12.18

USCI_B0 control register 1 (UCB0CTL1).

Bits	7 - 6	5	4	3	2	1	0
	UCSSELx	Unused	UCTR	UCTXNACK	UCTXSTP	UCTXSTT	UCSWRST

Table 12.19

USCI_B0 status register (UCB0STAT).

Bits	7	6	5	4	3	2	1	0
	Unused	UCSCLLOW	UCGC	UCBBUSY	UCNACKIFG	UCSTPIFG	UCSTTIFG	UCALIFG

two devices are used in communication. Please see the I^2C data sheet for detailed information [7]. When the UCGC bit is reset, this means the general call address is not received. When it is set, this means the address information is received. The **UCBBUSY** bit is used to indicate that whether the I^2C bus is busy or not. When this bit is reset, it means the bus is inactive. When it is set, it means that the bus is active. **UCNACKIFG**, **UCSTPIFG**, **UCSTTIFG** and **UCALIFG** bits are used to observe interrupts for NACK, stop, start, and arbitration lost conditions respectively. When these bits are reset, this means that there is no interrupt pending. When these bits are set, it means there is an interrupt pending from the related source. UCNACKIFG and UCSTPIFG bits are reset automatically after the start condition is received. UCSTTIFG is reset automatically after the stop condition is received. Start and stop conditions are received by the slave device. Therefore, the UCSTPIFG and UCSTTIFG bits are related only with the slave device.

The I^2C mode has two additional address registers. These are called the USCI_B0 I^2C own address register (**UCB0I2COA**) and USCI_B0 I^2C slave address register (**UCB0I2CSA**). UCB0I2COA keeps the device's own address. It is used when the device is used as a slave. Also the fifteenth bit of this register (**UCGCEN**) is used to respond to a general call. When this bit is set, the device responds to a general call. When it is reset, the device does not respond to any general calls. UCB0I2CSA keeps the address of the slave device to be connected by the master. Therefore, it is used when the device is in the master mode. When the master wants to communicate with another slave, this register must be changed to the address of the new slave.

12.6.1 I^2C Transmit/Receive Operations

Data transfer in the I^2C mode is carried out byte by byte. Every bit of the byte is transferred during one SCL pulse. Communication starts when the master sends the start condition to the slave. This is done by generating a high to low transition on the SDA while the SCL is high. Then, the slave address is transmitted by the master in the next 1 or 2 bytes according to the addressing mode.

In the 7-bit addressing mode, the address information is sent in 1 byte. In this byte, the first 7 bits represent the slave address. The 8 bit is $R/\overline{W}$. When the $R/\overline{W}$ bit is 0, it means that the master will transmit data to the slave. When the $R/\overline{W}$ bit is 1, it means that the master will receive data from the slave. After this byte is transmitted by the master, the slave sends an acknowledge (ACK) bit to the master to indicate that the address information is received. Actually, this

ACK bit is sent by the receiver (master or slave) after each received byte throughout the communication to show that the transmitted byte is received.

In the 10-bit addressing mode, the address information is sent in 2 bytes. The first byte is formed by the constant number 11110b, the first 2 bits of the slave address, and the $R/\overline{W}$ bit. The second byte contains the remaining 8 bits of the slave address. After receiving each byte, the slave sends an ACK bit. After the address information is acknowledged by the slave, data is transmitted or received byte by byte according to the $R/\overline{W}$ bit. As in the 7-bit addressing mode, the ACK bit is sent by the receiver (master or slave) after receiving each byte.

The UCBBUSY bit is set to indicate that the bus is busy during the communication period. As the data transfer is completed, the communication halts. This is done by the master device by sending the stop condition (a low to high transition on the SDA while the SCL is high).

Sometimes, the direction of data transfer has to be changed during I^2C communication. This can be achieved by sending a start condition followed by the address information and new $R/\overline{W}$ bit after an ACK bit anywhere in the data transfer. This way, the direction can be changed without stoping the data transfer since no stop condition is generated.

There are four transmit/receive operation options for the I^2C mode. These are slave transmitter, slave receiver, master transmitter, and master receiver. In the following sections, we explore each in detail.

Slave Transmitter Mode

In this mode, first the device is set as slave by setting UCSYNC and resetting UCMST bits. Then, the slave address is written to the UCB0I2COA register. This address can be either seven or 10 bits long based on the UCA10 bit value. After a start condition is detected by the slave, its own address is compared with the received one coming from the master (from the UCB0I2CSA register). If both addresses match, UCSTTIFG is set and UCSTPIFG is reset. The slave must be set as receiver first by resetting the UCTR bit. This is done to get the address information from the master. Then, if the master is configured as the receiver, the $R/\overline{W}$ bit is set and the slave is configured as transmitter automatically. The UCTR and UCB0TXIFG bits are also set automatically in this step. Then, the first data bit is written to the UCB0TXBUF and an ACK bit is sent by the slave to indicate that the address information is acknowledged. Afterwards, UCB0TXIFG is reset and the data byte is transmitted. UCB0TXIFG is set again as soon as data in UCB0TXBUF is transferred to the transmit shift register. After data is transmitted to the master, there are three options for the system. First, the master can send an ACK bit, a new data byte is transmitted, and the transfer proceeds. Second, the master can send a NACK bit followed by the stop condition to end data transfer. Here, the UCB0TXIFG bit is reset after the NACK bit. The UCSTPIFG is set and UCSTTIFG is reset after the stop condition is received by the slave. Third, the master can also send a NACK bit followed by the restart condition to restart the data transfer. Then, the data transfer cycle returns to the step where the start condition and address information is received by the slave.

Slave Receiver Mode

This mode has the same configuration steps as the previous one. Only in the slave receiver mode, the master is configured as transmitter and the $R/\overline{W}$ bit is reset. Then, the slave is configured as receiver automatically. The UCTR bit is also reset automatically in this step. UCB0RXIFG is set automatically after the first data byte is received. Then, the received data is read from UCB0RXBUF and an ACK bit is sent by the slave. There are four options for the system after the ACK bit is sent.

First, the master can transfer a new data byte and the transfer proceeds. Second, the master can send a stop condition. Here, UCSTPIFG is set and UCSTTIFG is reset after the stop condition. Third, the master can send a restart condition. Then, the data transfer cycle returns to the step where the start condition and address information are received by the slave. Fourth, the slave device can also send a NACK bit instead of an ACK bit to the master if the UCTXNACK bit is set during the last data cycle. The master device must respond to this by generating a stop or restart condition. If a NACK is transmitted before the last data in UCB0RXBUF is read, new data is written to UCB0RXBUF and the last data is lost. In order to prevent this, data in UCB0RXBUF must be read before UCTXNACK is set. After the NACK is transmitted, the UCTXNACK bit is reset automatically. UCSTPIFG is set and UCSTTIFG is reset after the stop condition. If a restart condition occurs, the data transfer cycle returns to the step where the start condition and address information are received by the slave. The fourth option can be used if the slave wants to stop the communication.

Master Transmitter Mode

In this mode, first the device is set as master by setting the UCSYNC and UCMST bits. Then, the target slave address is written to the UCB0I2CSA register in accordance with the UCSLA10 bit (7- or 10-bit addressing modes). Also the UCTR bit must be set to indicate that the master is used as the transmitter. The master generates a start condition to initiate the communication if the UCTXSTT bit is set by software. When this start condition is generated, UCB0TXIFG is set to show that UCB0TXBUF is ready for new data. Then, the slave address is transmitted with the $R/\overline{W}$ bit being 0. An ACK bit is expected from the slave as the first data byte is written to UCB0TXBUF. The UCTXSTT bit is reset automatically after the ACK bit is received. Also, UCB0TXIFG is set again as soon as data in UCB0TXBUF is transferred to the transmit shift register. Then the data byte is transmitted from the master to the slave. After this transmission, there are four options.

First, the slave can send an ACK bit, a new data byte is transmitted, and transfer proceeds. Second, the master can generate a stop condition after the last ACK bit is received from the slave if UCTXSTP is set. When the data is transferred from UCB0TXBUF to the transmit shift register, UCB0TXIFG is set to show that data transmission has started and the UCTXSTP bit may be set. UCB0TXIFG must be reset by the user when UCTXSTP is set. UCTXSTP is reset automatically after the stop condition is generated. Third, the master can generate a restart condition after the last ACK bit is received from the slave if UCTXSTT is set. Then the data transfer cycle returns to the step where the start condition and address information are received by the slave. If desired, UCTR and UCB0I2CSA can be changed here. Fourth, the slave can send a NACK bit. This sets the UCNACKIFG bit. The master must respond to this by generating a stop or restart condition. Data in the UCB0TXBUF is discarded here. If this data needs to be transmitted after a restart condition, it must be rewritten to the UCB0TXBUF.

In the first address transmission operation by the master, the following scenario may occur. If the address information cannot be acknowledged by the slave, it sends a NACK bit to the master and UCNACKIFG is set. The master device must respond to this by generating a stop or restart condition. This is also the case for the master receiver mode to be explained next.

Master Receiver Mode

This mode has the same configuration steps as the previous one. Only the UCTR bit must be reset to indicate that the master is used as a receiver. Here, the master generates a start condition to initiate the communication if the UCTXSTT bit is set by software. Then the slave address is transmitted with $R/\overline{W} = 1$. As the ACK bit is received from the slave (for the address information), the UCTXSTT bit is reset automatically and the first data byte can be received. After this byte is received, UCB0RXIFG is set to indicate that data is loaded to UCB0RXBUF. After the UCB0RXIFG bit is set, there are three options for the system.

First, the master can send an ACK bit, a new data byte is received, and transfer proceeds. Second, the master can generate a stop condition by setting UCTXSTP and sending a NACK bit. Here, UCTXSTP is reset automatically after the stop condition is generated. Third, the master can generate a restart condition by setting UCTXSTT and sending a NACK bit. Here, the data transfer cycle returns to the step where the start condition and the address information are received by the slave. If desired, UCTR and UCB0I2CSA can be changed here.

12.6.2 I²C Interrupts

The I²C mode shares the same interrupt vectors with the UART and SPI modes for the transmit and receive operations. However, they are used in a different way in the I²C mode. The **USCIAB0TX_VECTOR** is used for both transmit and receive interrupts. The **USCIAB0RX_VECTOR** is used for checking UCNACKIFG, UCSTPIFG, UCSTTIFG, and UCALIFG flags (in UCB0STAT) generated by related interrupts. In order to use these flags, related interrupt enable bits of the USCI_B0 I²C interrupt enable register (**UCB0I2CIE**) must be set. The entries of this register are given in Table 12.20. The interrupt enable register 2 (IE2) and interrupt flag register 2 (IFG2) given in Tables 12.3 and 12.4 are also used here as with the SPI mode.

Table 12.20 USCI_B0 I²C interrupt enable register (UCB0I2CIE).

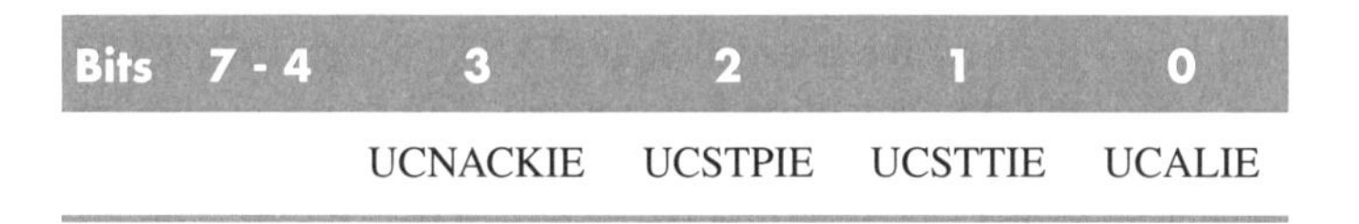

Bits	7 - 4	3	2	1	0
		UCNACKIE	UCSTPIE	UCSTTIE	UCALIE

The interrupt-based communication operation in the I^2C is the same as in the SPI mode. The only difference is the UCB0TXIFG. This bit is reset if a NACK bit is received in addition to writing a new character to UCB0TXBUF.

12.6.3 Coding Practices for the I^2C Mode

In this section, we provide sample C and assembly codes on the I^2C communication mode. The problems mentioned for the UART mode are also applicable here. Therefore, please see Sec. 12.2.4 first.

I^2C in C

In Listings 12.19 and 12.20, the I^2C mode is used to establish a digital communication between two MSP430 LaunchPads. The connection diagram for this application is given in Fig. 12.20. The C code for the master transmitter device is given in Listing 12.19. The C code for the slave receiver device is given in Listing 12.20. In this application, when the button connected to pin P1.3 of the master device is pressed, it sends the next PWM constant from the TXData array to control the brightness of the LED connected to pin P1.2 of the slave device. The connection between pin P1.5 of the master device and the RST pin of the slave device is used for resetting the slave before the communication starts.

Listing 12.19 The I^2C PWM application, master transmitter code in C language.

```
#include <msp430.h>

 unsigned int *PTXData = 0;
 unsigned int TXByteCtr;
 unsigned int TXData[] = {0x0000, 0x00FA, 0x01F4, \
0x02EE, 0x03E8}; //TACCR1 values to be transmitted
 unsigned int High = 0, Low = 0;
 int StartEnable = 0, cntr = 0;

void main(void)
{
 WDTCTL = WDTPW|WDTHOLD;

 P1DIR |= BIT5;  //Adjust pins
 P1SEL |= BIT6|BIT7;
 P1SEL2 |= BIT6|BIT7;
 P1IE |= 0x08;
 P1IES |= 0x08;
 P1IFG = 0x00;

 UCB0CTL1 |= UCSWRST;  //Setup the I2C mode
//Enable SW reset
 UCB0CTL0 = UCMST | UCMODE_3 | UCSYNC;
//I2C Master, synchronous mode
 UCB0CTL1 = UCSSEL_2 | UCSWRST;
```

```
//Use SMCLK, keep SW reset
 UCB0BR0 = 12;
//Low byte of UCBRx is 12
 UCB0BR1 = 0;
//High byte of UCBRx is zero,
//fSCL = SMCLK/12 = ~100kHz
 UCB0I2CSA = 0x48;  //Slave Address
 UCB0CTL1 &= ~UCSWRST;
//Clear SW reset, resume operation
 IE2 |= UCB0TXIE;  //Enable TX interrupt
 PTXData = TXData;
//Equate TXData array's start address to PTXData

 P1OUT &= ~BIT5;  //Reset the I2C slave
 P1OUT |= BIT5;

 _enable_interrupts();

 while (1) {
 TXByteCtr = 2;
//Load TX byte counter with two bytes
 while (UCB0CTL1 & UCTXSTP);
//Ensure that the stop condition is sent
 if(StartEnable == 1)  //Check for the button press
 {
 UCB0CTL1 |= UCTR | UCTXSTT;
//I2C Transmitter, start condition
 LPM0;
 }}
}

//I2C transmit interrupt
#pragma vector = USCIAB0TX_VECTOR
__interrupt void USCIAB0TX_ISR(void){
 if (TXByteCtr){
 if((TXByteCtr%2) == 0){
 High = *PTXData;
//Write the incoming array element to high integer
 UCB0TXBUF = (High>>8);
//Shift the high byte of High integer then load
//TX buffer with it
 }
 if((TXByteCtr%2) == 1){
 Low = *PTXData++;
//Write incoming array element to low integer,
//then increase PTXData
 UCB0TXBUF = Low;
//Load TX buffer with low byte of array element
 cntr++;
 }
 TXByteCtr--;
 }
 else
```

```
 {
 if(cntr == 5){
 PTXData = PTXData-5;
//Reload PTXData pointer with address
//of the first element
 cntr = 0;
 }
 StartEnable = 0;
//Reset StartEnable to stop the I2C communication
//until the button is pressed again
 UCB0CTL1 |= UCTXSTP;  //I2C stop condition
 IFG2 &= ~UCB0TXIFG;  //Clear the USCI_B0 TX int. flag
 LPM0_EXIT;
 }
}

#pragma vector=PORT1_VECTOR
__interrupt void Port_1(void){
 StartEnable = 1;
//Set StartEnable variable to start the
//I2C communication
 P1IFG = 0x00;
}
```

Listing 12.20 The I^2C PWM application, slave receiver code in C language.

```
#include <msp430.h>

 unsigned int RXData;
 unsigned int RXByteCtr = 0;

void main(void)
{
 WDTCTL = WDTPW|WDTHOLD;

 P1DIR |= BIT2;  //Adjust pins
 P1SEL |= BIT6|BIT7|BIT2;
 P1SEL2 |= BIT6|BIT7;

 UCB0CTL1 |= UCSWRST;  //Setup the I2C mode
//Enable SW reset
 UCB0CTL0 = UCMODE_3 | UCSYNC;
//I2C Slave, synchronous mode
 UCB0I2COA = 0x48; // Own Address
 UCB0CTL1 &= ~UCSWRST;
//Clear SW reset, resume operation
 UCB0I2CIE |= UCSTPIE | UCSTTIE;
//Enable STT and STP interrupt
 IE2 |= UCB0RXIE;  //Enable the RX interrupt

 TACCR1 = 0;  //Setup the PWM
```

```
  TACCR0 = 999;
  TACCTL1 = OUTMOD_7;
  TACTL = TASSEL_2|MC_1|ID_3;
  _enable_interrupts();
  while (1){
  RXByteCtr = 0;
  LPM0;
  }
}
//I2C transmit interrupt
#pragma vector = USCIAB0TX_VECTOR
__interrupt void USCIAB0TX_ISR(void){
  if((RXByteCtr%2) == 0){
  RXData = UCB0RXBUF;
//Move the received data (high byte) to low byte
//of RXData
  RXData = (RXData<<8);
//Shift the low byte of RXData to high byte
  }
  if((RXByteCtr%2) == 1){
  RXData |= UCB0RXBUF;
//Move the received data (low byte) to low byte
//of RXData
  TACCR1 = RXData; // Move the RXData to TACCR1
  }
  RXByteCtr++;
}
//I2C receive interrupt
#pragma vector = USCIAB0RX_VECTOR
__interrupt void USCIAB0RX_ISR(void){
  UCB0STAT &= ~(UCSTPIFG | UCSTTIFG);
//Clear interrupt flags
  if (RXByteCtr) LPM0_EXIT;
//Exit LPM0 if data is received
}
```

In Listings 12.21 and 12.22, the I^2C mode is again used to establish a digital communication between two MSP430 LaunchPads. The connection diagram for this application is given in Fig. 12.21. However, this time the slave becomes the transmitter and the master becomes the receiver. The C code for the slave transmitter device is given in Listing 12.21. The C code for the master receiver device is given in Listing 12.22. In this application, when the button connected to pin P1.3 of the master device is pressed, the slave sends the next PWM constant from the TXData array to control the brightness of the LED connected to pin P1.2 of the master device. The connection between pin P1.5 of the master device and the RST pin of the slave device is used for resetting the slave before the communication starts.

Figure 12.20

The connection diagram for the I²C PWM application (master transmitter and slave receiver).

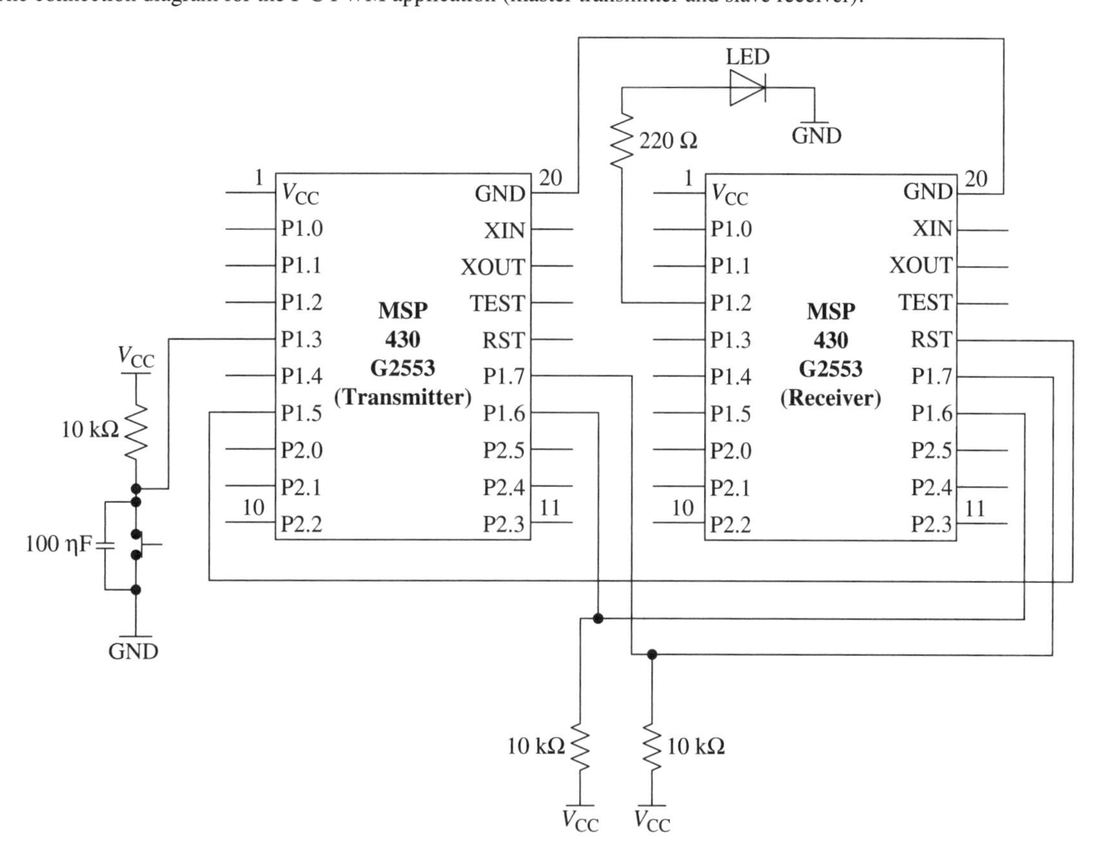

Listing 12.21 The I²C PWM application, slave transmitter code in C language.

```
#include <msp430.h>

 unsigned int *PTXData = 0;
 unsigned int TXByteCtr = 2;
 unsigned int TXData[] = {0x0000, 0x00FA, 0x01F4, \
 0x02EE, 0x03E8}; //TACCR1 values to be transmitted
 unsigned int High = 0, Low = 0;
 int cntr = 0;

void main(void)
{
 WDTCTL = WDTPW|WDTHOLD;

 P1SEL |= BIT6|BIT7;   //Adjust pins
 P1SEL2 |= BIT6|BIT7;

 UCB0CTL1 |= UCSWRST;  //Setup the I2C mode
//Enable SW reset
 UCB0CTL0 = UCMODE_3 | UCSYNC;
```

```
//I2C Slave, synchronous mode
 UCB0I2COA = 0x48;  //Own Address
 UCB0CTL1 &= ~UCSWRST;
//Clear SW reset, resume operation
 UCB0I2CIE |= UCSTPIE + UCSTTIE;
//Enable STT and STP interrupt
 IE2 |= UCB0TXIE;  //Enable TX interrupt
 PTXData = TXData;

 _enable_interrupts();

 while (1){
 if(cntr == 5){
 PTXData = PTXData - 5;
//Reload PTXData pointer with the address of
//the first element
 cntr = 0;
 }
 TXByteCtr = 0;
 LPM0;
 }
}

//I2C transmit interrupt
#pragma vector = USCIAB0TX_VECTOR
__interrupt void USCIAB0TX_ISR(void){
 if((TXByteCtr%2) == 0){
 High = *PTXData;
//Write the incoming array element to high integer
 UCB0TXBUF=(High>>8);
//Shift high byte of High integer then load TX buffer
//with it
 }
 if((TXByteCtr%2) == 1){
 Low = *PTXData++;
//Write the incoming array element to low integer,
//then increase PTXData
 UCB0TXBUF=Low;
//Load TX buffer with the low byte of the array element
 cntr++;
 }
 TXByteCtr++;
}

//I2C receive interrupt
#pragma vector = USCIAB0RX_VECTOR
__interrupt void USCIAB0RX_ISR(void){
 UCB0STAT &= ~(UCSTPIFG | UCSTTIFG);
//Clear interrupt flags
 if (TXByteCtr) LPM0_EXIT;
//Exit LPM0 if data is transmitted
}
```

Listing 12.22 The I²C PWM application, master receiver code in C language.

```
#include <msp430.h>

 unsigned int RXData;
 unsigned int RXByteCtr;
 int StartEnable = 0;

void main(void)
{
 WDTCTL = WDTPW|WDTHOLD;

 P1DIR |= BIT2|BIT5;  //Adjust pins
//Assign P1.5 as output to reset the slave
 P1SEL |= BIT6|BIT7|BIT2;
 P1SEL2 |= BIT6|BIT7;
 P1IE |= 0x08;
 P1IES |= 0x08;
 P1IFG = 0x00;

 UCB0CTL1 |= UCSWRST;  //Setup the I2C mode
//Enable SW reset
 UCB0CTL0 = UCMST | UCMODE_3 | UCSYNC;
//I2C Master, synchronous mode
 UCB0CTL1 = UCSSEL_2 | UCSWRST;
//Use SMCLK, keep SW reset
 UCB0BR0 = 12;
//Low byte of UCBRx is 12
 UCB0BR1 = 0;
//High byte of UCBRx is 0,
//fSCL = SMCLK/12 = ~100kHz
 UCB0I2CSA = 0x48;  //Slave Address
 UCB0CTL1 &= ~UCSWRST;
//Clear SW reset, resume operation
 IE2 |= UCB0RXIE;   //Enable RX interrupt

 TACCR1 = 0;  //Setup the PWM
 TACCR0 = 999;
 TACCTL1 = OUTMOD_7;
 TACTL = TASSEL_2 | MC_1 | ID_3;

 P1OUT &= ~BIT5;  //Reset the I2C slave
 P1OUT |= BIT5;

 _enable_interrupts();

 while (1){
 RXByteCtr = 2;
 while (UCB0CTL1 & UCTXSTP);
//Ensure that the stop condition is sent
```

```
 if(StartEnable == 1){
 UCB0CTL1 |= UCTXSTT;
 LPM0;
 }}
}

//I2C transmit interrupt
#pragma vector = USCIAB0TX_VECTOR
__interrupt void USCIAB0TX_ISR(void){
 RXByteCtr--;
 if(RXByteCtr){
 RXData = UCB0RXBUF;
//Move the received data (high byte) to
//low byte of RXData
 RXData = (RXData<<8);
//Shift low byte of RXData to high byte
 if(RXByteCtr == 1)
 UCB0CTL1 |= UCTXSTP;
//Generate I2C stop condition
 }
 else{
 RXData |= UCB0RXBUF;
//Move the received data (low byte) to
//low byte of RXData
 TACCR1 = RXData;
//Load TACCR1 with RXData
 StartEnable = 0;
//Reset StartEnable to stop I2C communication
//until the button is pressed again
 LPM0_EXIT;
 }
}

#pragma vector=PORT1_VECTOR
__interrupt void Port_1(void){
 StartEnable = 1;
//Set StartEnable to start the I2C communication
 P1IFG = 0x00;
}
```

I^2C in Assembly

The assembly codes, given in Listings 12.23 and 12.24, perform the same operation done in Listings 12.19 and 12.20. The connection diagram for this application is also the same as that shown in Fig. 12.20. The assembly code for the master transmitter device is given in Listing 12.23. The assembly code for the slave receiver device is given in Listing 12.24.

Figure 12.21

The connection diagram for the I^{2}C PWM application (master receiver and slave transmitter).

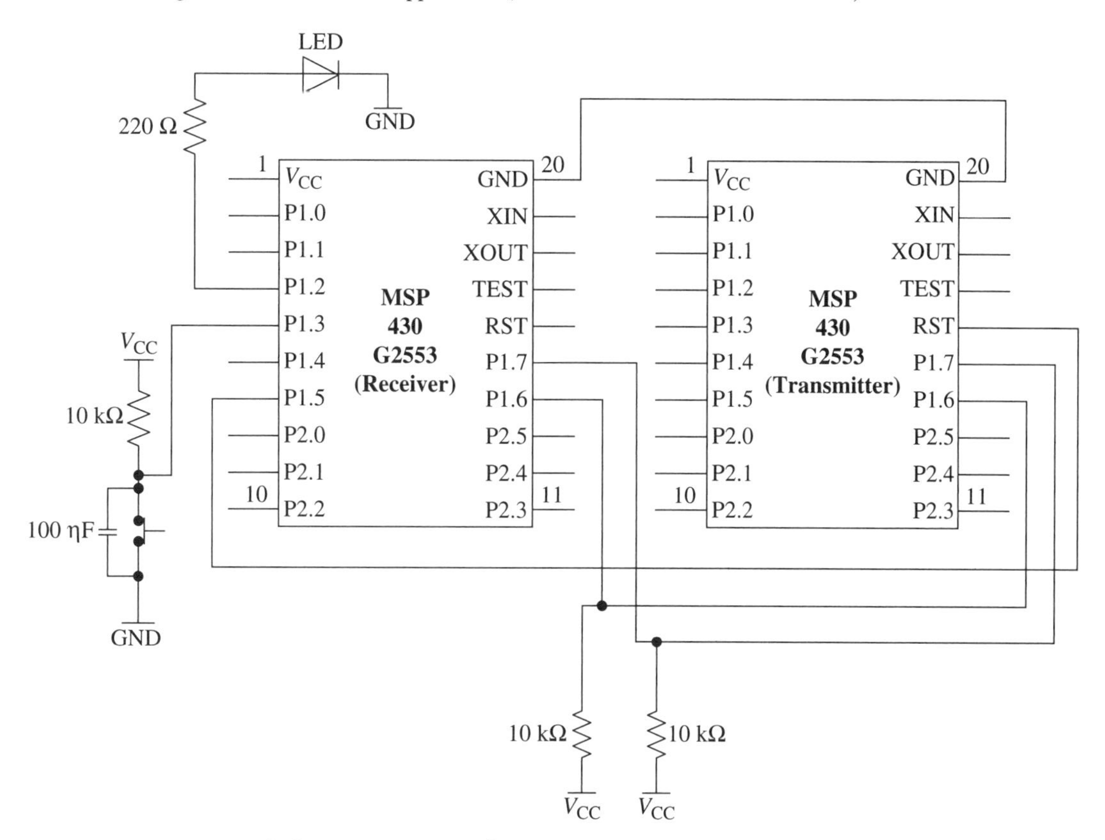

Listing 12.23 The I^{2}C PWM application, master transmitter code in assembly language.

```
    .cdecls C,LIST,"msp430.h"

    .text
    .retain
    .retainrefs

RESET
    mov.w #WDTPW|WDTHOLD,WDTCTL
    mov.w #__STACK_END,SP

;These five values are to be sent
    mov.w #0000h,&0200h
    mov.w #00FAh,&0202h
    mov.w #01F4h,&0204h
    mov.w #02EEh,&0206h
    mov.w #03E8h,&0208h

    bis.b #20h,P1DIR ;Adjust pins
    bis.b #0C0h,P1SEL
```

```
 bis.b #0C0h,P1SEL2
 bis.b #08h,P1IE
 bis.b #08h,P1IES
 clr.b P1IFG

 bis.b #UCSWRST,UCB0CTL1 ;Adjust the I2C mode
;Enable SW reset
 mov.b #UCMST+UCMODE_3+UCSYNC,UCB0CTL0
;I2C Master, synchronous mode
 mov.b #UCSSEL_2+UCSWRST,UCB0CTL1
;Use SMCLK, keep SW reset
 mov.b #0Ch,UCB0BR0  ;Low bit of UCBRx is 12
 mov.b #00h,UCB0BR1  ;High bit of UCBRx is zero
;fSCL = SMCLK/12 = ~100kHz
 mov.w #48h,UCB0I2CSA ;Slave Address
 bic.b #UCSWRST,UCB0CTL1
;Clear SW reset, resume operation
 bis.b #UCB0TXIE,IE2 ;Enable TX interrupt

 mov.w #0200h,R5
;Write the start address of TX data to R5
 mov.w #0h,R6
;Reset R6 used for controlling I2C start/stop
 clr R10
;Clear R10 used for checking which element is
;transmitted

 bic.b #20h,P1OUT ;Reset I2C slave
 bis.b #20h,P1OUT

 bis.w #GIE,SR

Mainloop:
 mov.w #2h,R7
;Load R7 with desired byte number to be transmitted
StopConditionLoop:
 bit.b #UCTXSTP,UCB0CTL1
 jne StopConditionLoop
;Ensure the stop condition is sent
 cmp.w #1h,R6
 jne Mainloop

 bis.b #UCTR+UCTXSTT,UCB0CTL1
;I2C Transmitter, start condition
 bis.w #LPM0,SR
 jmp Mainloop

;-----------------------------------------
USCIAB0TX_ISR
;I2C transmit interrupt service routine to
;control transmit operation
;------------------------------------------
 tst.w R7 ;Test if R7 is zero
 jeq AllBytesTransmitted
```

```
;If it is, jump AllBytesTransmitted
 bit.w #1h,R7 ;Check if R7 is odd
 jne OddByte ;If it is, jump OddByte

 mov.w @R5,R8
; Write the incoming array element to R8
 swpb R8
 mov.b R8,UCB0TXBUF
;Load TX buffer with high byte of incoming element
OddByte:
 cmp.w #1,R7
 jne DecrementByteNumber
 mov.w @R5+,R9
 mov.b R9,UCB0TXBUF
;Load TX buffer with low byte of incoming element
 inc.w R10
;Increase R10 to indicate that one element is sent
DecrementByteNumber:
 dec.w R7
 jmp EndISR
AllBytesTransmitted:
 cmp.w #5h,R10 ;Check if all five elements are sent
 jne SendStop
 sub.w #10h,R5
;If they are sent, reload PTXData pointer with
;the address of the first element
 clr R10
SendStop:
 clr R6
 bis.b #UCTXSTP,UCB0CTL1 ;I2C stop condition
 bic.b #UCB0TXIFG,IFG2 ;Clear USCI_B0 TX int flag
 bic.w #LPM0,0(SP)
EndISR:
 reti

;----------------------
P1_ISR
;----------------------
 mov.w #1h,R6
 clr.b P1IFG
 reti

;------------------------
;Stack Pointer definition
;------------------------
 .global __STACK_END
.sect .stack

;-------------------
;Interrupt Vectors
;-------------------
 .sect RESET_VECTOR
 .short RESET
```

```
 .sect  USCIAB0TX_VECTOR
 .short USCIAB0TX_ISR
 .sect  PORT1_VECTOR
 .short P1_ISR
 .end
```

Listing 12.24 The I²C PWM application, slave receiver code in assembly language.

```
 .cdecls C,LIST,"msp430.h"

 .text
 .retain
 .retainrefs

RESET
 mov.w #WDTPW|WDTHOLD,WDTCTL
 mov.w #__STACK_END,SP

 bis.b #04h,P1DIR ;Adjust pins
 bis.b #0C4h,P1SEL
 bis.b #0C0h,P1SEL2

 bis.b #UCSWRST,UCB0CTL1 ;Adjust the I2C mode
;Enable SW reset
 mov.b #UCMODE_3+UCSYNC,UCB0CTL0
;I2C Slave, synchronous mode
 mov.w #48h,UCB0I2COA ;Own Address
 bic.b #UCSWRST,UCB0CTL1
;Clear SW reset, resume operation
 bis.b #UCSTPIE+UCSTTIE,UCB0I2CIE
;Enable STT and STP interrupts
 bis.b #UCB0RXIE,IE2 ;Enable RX interrupt

 clr.w TACCR1 ;Adjust the PWM
 mov.w #03E7h,TACCR0
 mov.w #OUTMOD_7,TACCTL1
 mov.w #TASSEL_2+MC_1+ID_3,TACTL

 bis.w #GIE,SR

Mainloop:
 clr.w R7
 bis.w #LPM0,SR
 jmp Mainloop

;-----------------------------------------
USCIAB0TX_ISR
;I2C transmit interrupt service routine to
;control receive operation
;-----------------------------------------
 bit.w #1h,R7 ;Check if R7 is odd
 jne OddByte ;If it is, jump OddByte
 mov.b UCB0RXBUF,R8
;Move the received data (low byte) to low byte of R8
```

```
OddByte:
 cmp.w #1h,R7 ;Check if R7 is 1
 jne IncrementByteNumber
; If it is not, jump IncrementByteNumber
 mov.b UCB0RXBUF,R9
;Move the received data (low byte) to R9
 swpb R8
 and.w #0F00h,R8
;And R8 with 0x0F00, high byte
;numbers are located only these four bits
 bis.w R8,R9
 mov.w R9,TACCR1 ;Move received 2 byte data to TACCR1
IncrementByteNumber:
 inc.w R7
 reti
;----------------------------------------------
USCIAB0RX_ISR
;I2C receive interrupt service routine to check
;restart or stop conditions
;----------------------------------------------
 mov.b #UCSTPIFG + UCSTTIFG,R10
;Move stop and start interrupt flags to R10
 inv.b R10
 and.b R10,UCB0STAT ;Clear interrupt flags
 tst.w R7
 jeq EndISR
 bic.w #LPM0,0(SP)
EndISR:
 reti
;-----------------------
;Stack Pointer definition
;-----------------------
 .global __STACK_END
 .sect .stack
;------------------
;Interrupt Vectors
;------------------
 .sect RESET_VECTOR
 .short RESET
 .sect  USCIAB0TX_VECTOR
 .short USCIAB0TX_ISR
 .sect  USCIAB0RX_VECTOR
 .short USCIAB0RX_ISR
 .end
```

The assembly codes given in Listings 12.25 and 12.26, perform the same operation done in Listings 12.21 and 12.22. The connection diagram for this application is shown in Fig. 12.21. The assembly code for the slave transmitter device is given in Listing 12.25. The assembly code for the master receiver device is given in Listing 12.26.

Listing 12.25 The I²C PWM application, slave transmitter code in assembly language.

```
 .cdecls C,LIST,"msp430.h"

 .text
 .retain
 .retainrefs

RESET
 mov.w #WDTPW|WDTHOLD,WDTCTL
 mov.w #__STACK_END,SP

;These five values are to be sent
 mov.w #0000h,&0200h
 mov.w #00FAh,&0202h
 mov.w #01F4h,&0204h
 mov.w #02EEh,&0206h
 mov.w #03E8h,&0208h

 bis.b #0C0h,P1SEL ;Adjust pins
 bis.b #0C0h,P1SEL2

 bis.b #UCSWRST,UCB0CTL1 ;Adjust the I2C mode
;Enable SW reset
 mov.b #UCMODE_3+UCSYNC,UCB0CTL0
;I2C Slave, synchronous mode
 mov.w #48h,UCB0I2COA ;Own Address
 bic.b #UCSWRST,UCB0CTL1
;Clear SW reset, resume operation
 bis.b #UCSTPIE+UCSTTIE,UCB0I2CIE
;Enable STT and STP interrupts
 bis.b #UCB0TXIE,IE2 ;Enable TX interrupt

 mov.w #0200h,R5
;Write the start address of TX data to R5
 clr.w R10
;Clear R10 used for checking which element
;is transmitted

 bis.w #GIE,SR

Mainloop:
 cmp.w #5h,R10 ;Check if all five elements are sent
 jne Subloop
;If they are sent, reload R10 with address of the
;first element
 sub.w #10,R5
 clr R10
Subloop:
 clr.w R7
 bis.w #LPM0,SR
 jmp Mainloop

;----------------------------------------
USCIAB0TX_ISR
```

```
;I2C transmit interrupt service routine to
;control transmit operation
;-------------------------------------------
 bit.w #1h,R7 ;Check R7 is odd
 jne OddByte ;If it is, jump OddByte
 mov.w @R5,R8
;Write the incoming array element to R8
 swpb R8
 mov.b R8,UCB0TXBUF
;Load TX buffer with the high byte of
;the incoming element
OddByte:
 cmp.w #1h,R7
 jne IncrementByteNumber
 mov.w @R5+,R9
 mov.b R9,UCB0TXBUF
;Load TX buffer with low byte of incoming element
 inc.w R10
;Increase R10 to indicate that one element is sent
IncrementByteNumber:
 inc.w R7
 reti

;------------------------------------------
USCIAB0RX_ISR
;I2C receive interrupt service routine to
;check restart or stop conditions
;------------------------------------------
 mov.b #UCSTPIFG + UCSTTIFG,R11
;Move stop and start interrupt flags to R11
 inv.b R11 ;Invert R11
 and.b R11,UCB0STAT ;Clear interrupt flags
 tst.w R7 ;Check if R7 is 0
 jeq EndISR ;If it is, jump EndISR
 bic.w #LPM0,0(SP)

EndISR:
 reti

;------------------------
;Stack Pointer definition
;------------------------
 .global __STACK_END
 .sect .stack

;------------------
;Interrupt Vectors
;------------------
 .sect RESET_VECTOR
 .short RESET
 .sect  USCIAB0TX_VECTOR
 .short USCIAB0TX_ISR
 .sect  USCIAB0RX_VECTOR
```

```
.short USCIAB0RX_ISR
.end
```

Listing 12.26 The I^{2}C PWM application, master receiver code in assembly language.

```
.cdecls C,LIST,"msp430.h"

.text
.retain
.retainrefs

RESET
mov.w #WDTPW|WDTHOLD,WDTCTL
mov.w #__STACK_END,SP

bis.b #24h,P1DIR ;Adjust pins
bis.b #0C4h,P1SEL
bis.b #0C0h,P1SEL2
 bis.b #08h,P1IE
bis.b #08h,P1IES
clr.b P1IFG

bis.b #UCSWRST,UCB0CTL1 ;Adjust the I2C mode
;Enable SW reset
mov.b #UCMST+UCMODE_3+UCSYNC,UCB0CTL0
;I2C Master, synchronous mode
mov.b #UCSSEL_2+UCSWRST,UCB0CTL1
;Use SMCLK, keep SW reset
mov.b #0Ch,UCB0BR0 ;Low bit of UCBRx is 12
mov.b #00h,UCB0BR1 ;High bit of UCBRx is zero,
;fSCL = SMCLK/12 = ~100kHz
mov.w #48h,UCB0I2CSA ;Slave Address
bic.b #UCSWRST,UCB0CTL1
;Clear SW reset, resume operation
bis.b #UCB0RXIE,IE2 ;Enable RX interrupt

clr.w TACCR1 ;Adjust the PWM
mov.w #03E7h,TACCR0
mov.w #OUTMOD_7,TACCTL1
mov.w #TASSEL_2+MC_1+ID_3,TACTL

mov.w #0h,R6
;Reset StartEnable used for controlling
;I2C start/stop
 clr.w R8
;Clear R8 used for getting the high byte of data
clr.w R9
;Clear R9  used for getting the low byte of data

bic.b #20h,P1OUT
bis.b #20h,P1OUT ;Reset I2C slave

bis.w #GIE,SR
```

```
Mainloop:
 mov.w #2h,R7
;Load R7 with the desired byte number to be received
StopConditionLoop:
 bit.b #UCTXSTP,UCB0CTL1
 jne StopConditionLoop
;Ensure the stop condition is sent
 cmp.w #1h,R6
;Check if R6 is set
 jne Mloop
 bis.b #UCTXSTT,UCB0CTL1
 ;I2C Receiver, start condition
 bis.w #LPM0,SR
 jmp Mloop

;--------------------------------------
USCIAB0TX_ISR
;I2C transmit interrupt service routine
;to control receive operation
;---------------------------------------
 dec.w R7
 tst.w R7
 jeq LastByte
 mov.b UCB0RXBUF,R8 ;Load R8 with UCB0RXBUF
 cmp.w #1h,R7
 jne EndISR
 bis.b #UCTXSTP,UCB0CTL1 ;I2C stop condition
 jmp EndISR
LastByte:
 mov.b UCB0RXBUF,R9
;Move the received data (low byte) to R9
 swpb R8
 and.w #0F00h,R8
;And R8 with 0x0F00, high byte
; numbers are located on these four bits
 bis.w R8,R9
 mov.w R9,TACCR1
;Move received 2 bytes of data to TACCR1
 clr.w R6 ;Clear StartEnable
 clr.w R8
 clr.w R9
 bic.w #LPM0,0(SP)
EndISR:
 reti

;----------------------
P1_ISR
;----------------------
 mov.w #1h,R6 ;Set StartEnable
 clr.b P1IFG
 reti
```

```
;------------------------
;Stack Pointer definition
;------------------------
 .global __STACK_END
 .sect .stack

;-------------------
;Interrupt Vectors
;-------------------
 .sect RESET_VECTOR
 .short RESET
 .sect  USCIAB0TX_VECTOR
 .short USCIAB0TX_ISR
 .sect  PORT1_VECTOR
 .short P1_ISR
 .end
```

12.7 I²C in Grace

The I²C mode can also be configured under Grace. Since I²C is present only under the USCI_B0 module, we should enable it first. Then we should click the I²C button from the selection window for all user modes to be explored below.

12.7.1 *The Basic User Mode*

The basic user mode for I²C is shown in Fig. 12.22. In this mode, we can set the device as master or slave from the drop-down list in the USCI B0 I2C block. We can enable or disable I²C pins. We can select the bit rate from the Bitrate drop-down list. We can enter the slave and own addresses to the I²C Slave Address and I²C Own Address boxes respectively.

We can enable the I²C interrupts by checking the appropriate "USCI_B0 I2C transmit interrupt enable," "USCI_B0 I2C receive interrupt enable," "Start condition interrupt enable," or "Stop condition interrupt enable" box respectively. We can also generate an ISR related to these interrupts using the associated Generate Interrupt Handler Code button.

12.7.2 *The Power User Mode*

The power user mode for I²C is shown in Fig. 12.23. In addition to the basic user mode, we can set the clock source here. We can also set two additional interrupts by checking the "Not-acknowledge interrupt enable" and "Arbitration lost interrupt enable" boxes.

12.7.3 *The Register Controls Mode*

Finally, the I²C registers can be adjusted under Grace. The user should select the register controls mode, as shown in Fig. 12.24, for this purpose. As in the UART and SPI modes, some registers are not available here. Some register entries are also read only in this mode.

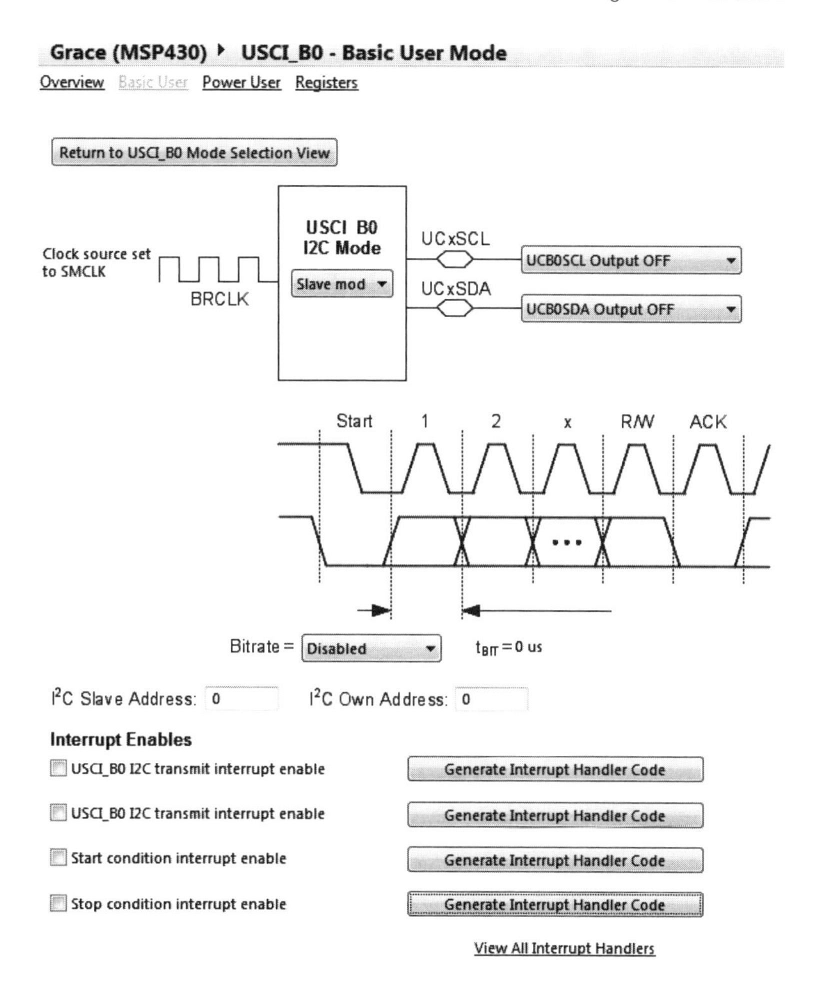

Figure 12.22 The basic user mode for I^2C under Grace.

12.8 Digital Communication Application

In this section, we provide a generic application different from the previous chapters. Our aim here is using the UART and I^2C modes together to form a communication link between two MSP430 LaunchPads and a host computer. The user should be aware of the hardware issues (related to digital communication) mentioned in the previous sections.

12.8.1 *Equipment List*

Following is a list of the equipment to be used in this application.

- Two MSP430 LaunchPads
- One LED
- One 220 Ω resistor
- Two 10 kΩ resistors

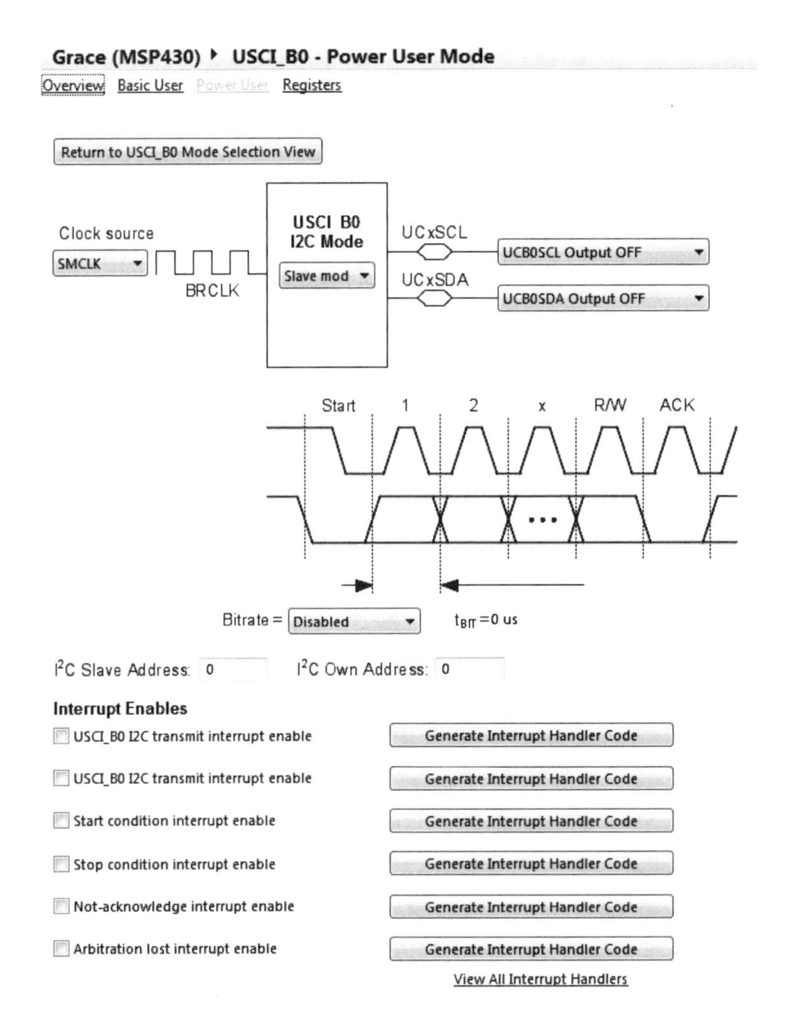

Figure 12.23 The power user mode for I²C under Grace.

12.8.2 *Layout*

The layout of this application is given in Fig. 12.25.

12.8.3 *System Design Specifications*

In this application, we use two MSP430 LaunchPads and a host computer for communication. The host computer communicates with the first MSP430 using UART mode. The first MSP430 communicates with the second MSP430 using I²C. The aim here is to change the brightness of the LED connected to the second MSP430 by data coming from the host computer.

First, UART communication between the host computer and the first MSP430 must be constructed. A variable in the first MSP430 is used to keep the duty cycle value. This variable should change between 0 and 100. When the '+' key is pressed on the keyboard of the host computer, this variable will increase by one. When the '−' key is pressed, the same variable will decrease by one. If these keys are held

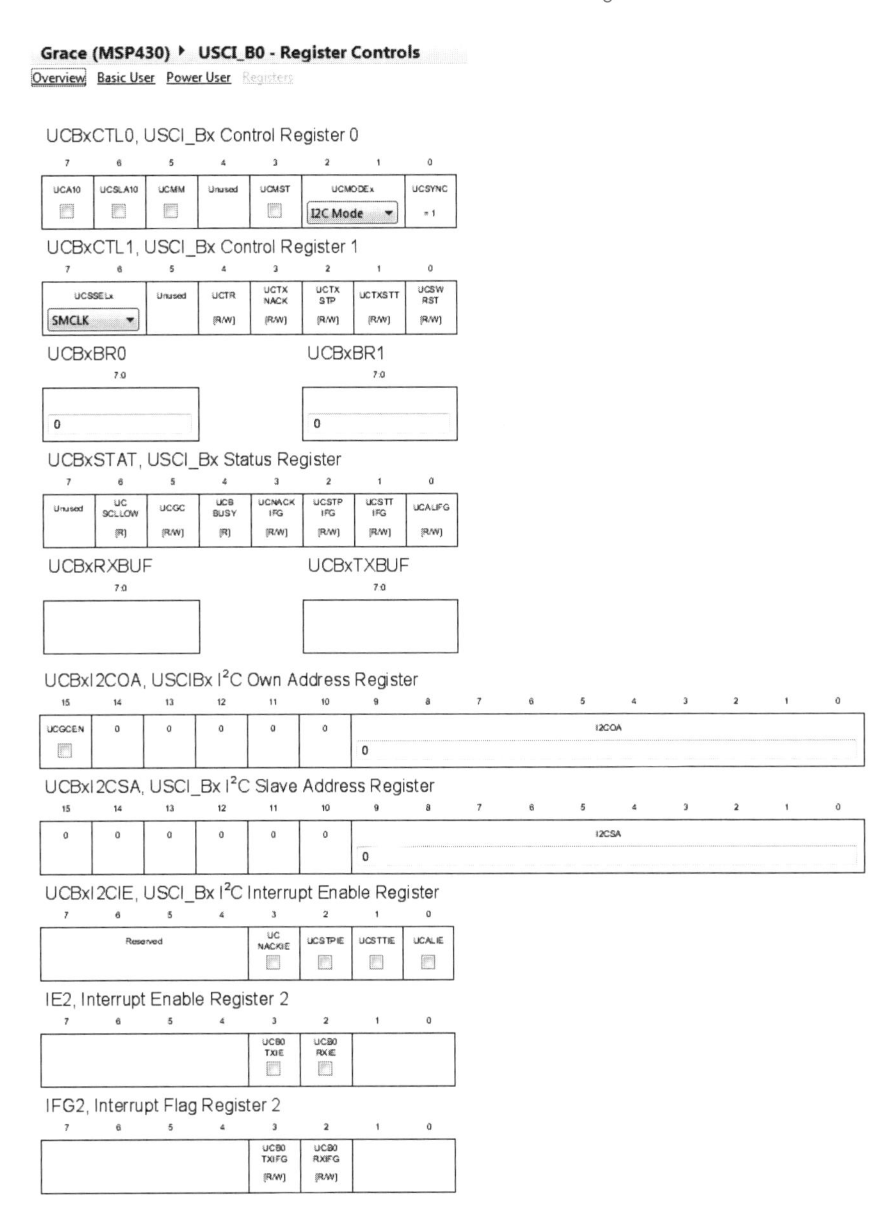

Figure 12.24

The register controls mode for I^2C under Grace.

down, the data is transmitted continuously. But the amount of data that can be sent in one second is limited (to 30–40) because of the BIOS setup restrictions.

While the duty cycle variable is changing, the first MSP430 will check this value. If the duty cycle equals 0, the MSP430 will send the Minimum Duty Cycle Value warning to the host computer. If the duty cycle equals 100, the MSP430 will send the Maximum Duty Cycle Value warning to the host computer. If the duty cycle is between 0 and 100, the MSP430 will send the Duty Cycle is Changing warning to the host computer.

Figure 12.25

Layout of the digital communication application.

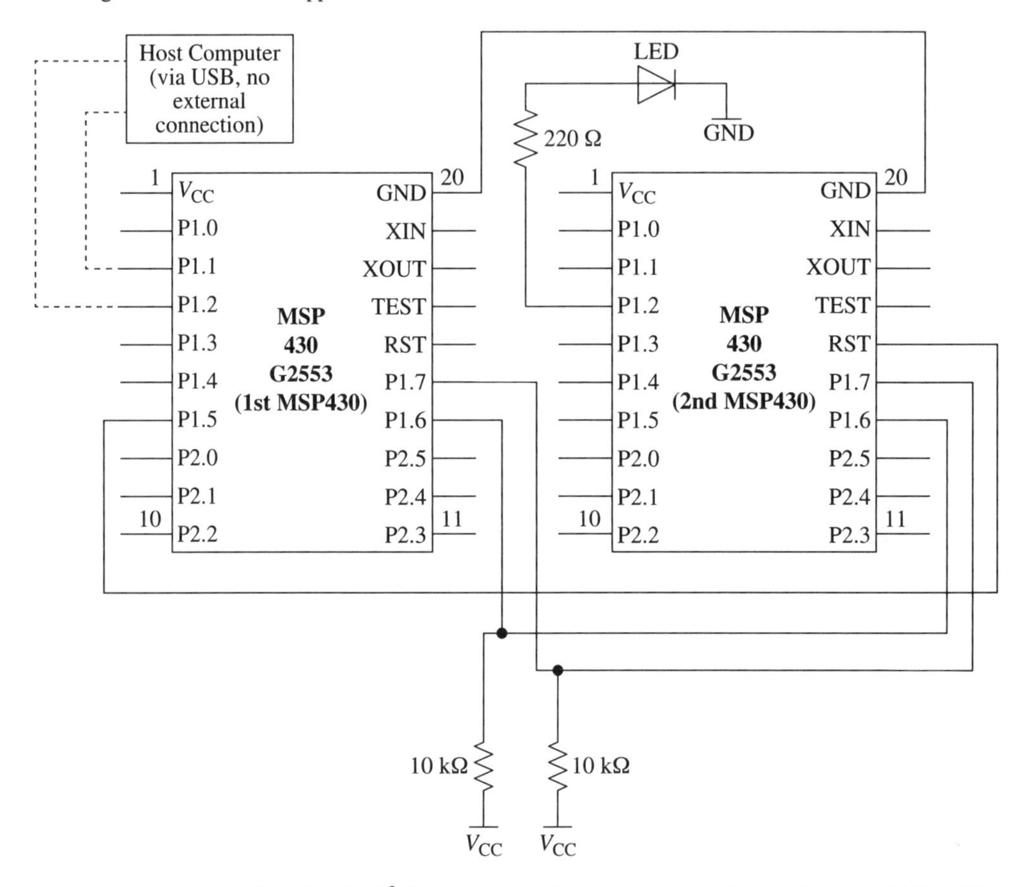

Finally, the I^2C communication between the first and second MSP430 must be constructed. When the duty cycle variable has changed on the first MSP430, it transmits this data via I^2C to the second MSP430. Then the second MSP430 receives this data. It uses the received data to change the duty cycle of the PWM signal connected to the LED. Hence, the brightness of the LED changes accordingly. Initially, the brightness of the LED is at minimum.

12.8.4 *The C Codes for the System*

The first MSP430 communicates with the host computer using UART mode. It is also the master transmitter device for the I^2C communication between the first and second MSP430. The second MSP430 is only used as the slave receiver device for I^2C communication.

The C Code for the First MSP430

In Listing 12.27, global variables are defined as the first part of the code. Here, `Min`, `Max`, and `Mid` arrays hold the text information to be sent to the host computer. The `DutyCycle` variable is used to get the duty cycle from the host computer and send it to the second MSP430. The `StartEnable` variable is used to control the start of the I^{2}C communication. The `TXByteCtr` variable is used as a counter for the transmitted bytes.

Listing 12.27 Digital communication application, the transmitter code part I.

```
char Min[] = "Minimum Duty Cycle Value\r\n";
char Max[] = "Maximum Duty Cycle Value\r\n";
char Mid[] = "Duty Cycle is Changing\r\n";

int DutyCycle = 0;
int StartEnable = 0;
unsigned int TXByteCtr;
```

In the second part of the code, given in Listing 12.28, the hardware configurations for the digital I/O, timer, UART, and I^{2}C modules are done. In this code block, configuration for each hardware module is done in a separate function.

In the `PinConfig()` function, pin directions are assigned to `P1DIR=BIT5` in the first line to obtain the reset output from P1.5. In the second and third lines: P1.1 is set as UART receive data input (RXD); P1.2 is set as UART transmit data output (TXD); P1.6 is set as the I^{2}C clock pin (SCL); and P1.7 is set as the I^{2}C data pin (SDA). In the fourth and fifth lines, a low to high transition is given from P1.5 with `P1OUT &= ~BIT5` and `P1OUT |= BIT5` to reset the slave at the beginning of communication.

In the `TimerConfig()` function, the watchdog timer is disabled in the first line. In the second and third lines, DCO is calibrated to 1 MHz.

In the `UARTConfig()` function, software reset is enabled and the clock source for a baud rate generator is selected as SMCLK with `UCA0CTL1 |= UCSWRST | UCSSEL_2` in the first line. In the second, third, and fourth lines the baud rate is set to 9600 bps with `UCA0BR0 = 104`, `UCA0BR1 = 0`, and `UCA0MCTL = UCBRS_1`. These values are obtained from Table 12.10. In the fifth line, software reset is disabled with `UCA0CTL1 &= ~UCSWRST` to resume the USCI operation. In the sixth line, the receive interrupt is enabled with `IE2 |= UCA0RXIE`.

In the `I2CConfig()` function, the software reset is enabled with `UCB0CTL1 |= UCSWRST` in the first line to start the configuration. In the second line, the synchronous communication mode is selected by setting the `UCSYNC` bit. The device is set as master by setting the `UCMST` bit. Also, `UCMODE_3` is used to select the I^{2}C mode. In the third line, the clock source for the bit rate generator is selected as SMCLK while the software reset is kept enabled with `UCA0CTL1 = UCSSEL_2 | UCSWRST`. In the fourth and fifth lines, f_{BITCLK} is set to 100 kHz with `UCB0BR0 = 10` and `UCA0BR1 = 0`. In the sixth line, the slave address is set as 0x48. In the seventh line, software reset is disabled with `UCB0CTL1 &= ~UCSWRST` to resume the USCI operation. Finally, the transmit interrupt is enabled with `IE2 |= UCB0TXIE` in the last line.

Listing 12.28 Digital communication application, the transmitter code part II.

```
void PinConfig(void){
 P1DIR |= BIT5;
 P1SEL = BIT1|BIT2|BIT6|BIT7;
 P1SEL2 = BIT1|BIT2|BIT6|BIT7;
 P1OUT &= ~BIT5;
 P1OUT |= BIT5;
}

void TimerConfig(void){
 WDTCTL = WDTPW|WDTHOLD;
 BCSCTL1 = CALBC1_1MHZ;
 DCOCTL = CALDCO_1MHZ;
}

void UARTConfig(void){
 UCA0CTL1 |= UCSWRST|UCSSEL_2;
 UCA0BR0 = 104;
 UCA0BR1 = 0;
 UCA0MCTL = UCBRS_1;
 UCA0CTL1 &= ~UCSWRST;
 IE2 |= UCA0RXIE;
}

void I2CConfig(void){
 UCB0CTL1 |= UCSWRST;
 UCB0CTL0 = UCMST | UCMODE_3 | UCSYNC;
 UCB0CTL1 = UCSSEL_2 | UCSWRST;
 UCB0BR0 = 10;
 UCB0BR1 = 0;
 UCB0I2CSA = 0x48;
 UCB0CTL1 &= ~UCSWRST;
 IE2 |= UCB0TXIE;
}
```

In the third part of the code, given in Listing 12.29, the ISR settings for the UART and I^{2}C modes are done as follows: In the UART receive ISR, the `DutyCycle` variable is changed according to the incoming character data. If this character equals '+', the `DutyCycle` variable is increased by one. If it equals '−', the `DutyCycle` variable is decreased by one. In order to generate a duty cycle range between 0 and 100, the `DutyCycle` variable is set to 0 if it is less than 0. The duty cycle is set to 100 if it is greater than 100. Also, the `StartEnable` variable is set to 1 to trigger the start of I^{2}C communication. In the I^{2}C transmit ISR, the most recent `DutyCycle` variable is loaded to the transmit buffer and `TXByteCtr` is decreased by one when the ISR is called the first time. Then the stop condition is sent with `UCB0CTL1 |= UCTXSTP`. The `StartEnable` variable is reset to halt the transmission until the next data is received from the host computer. The transmit interrupt flag is cleared manually with `IFG2 &= ~UCB0TXIFG`.

Listing 12.29 Digital communication application, the transmitter code part III.

```
#pragma vector=USCIAB0RX_VECTOR
__interrupt void USCIA0RX_ISR(void){
 if (UCA0RXBUF == '+')DutyCycle++;
 if (UCA0RXBUF == '-')DutyCycle--;

 if(DutyCycle > 100)DutyCycle = 100;
 if(DutyCycle < 0)DutyCycle = 0;

 StartEnable = 1;
}

#pragma vector = USCIAB0TX_VECTOR
__interrupt void USCIB0TX_ISR(void){
 if (TXByteCtr)
 {
 UCB0TXBUF = DutyCycle;
 TXByteCtr--;
 }
 else
 {
 UCB0CTL1 |= UCTXSTP;
 StartEnable = 0;
 IFG2 &= ~UCB0TXIFG;
 }
}
```

Finally, the C code for the transmitter part is given in Listing 12.30. In a while loop, `TXByteCtr` is loaded with 1 first. Then the system is halted with `while (UCB0CTL1 & UCTXSTP)` until the `UCTXSTP` bit is reset. The system waits for the new received data to get `StartEnable=1` after the bit `UCTXSTP` is reset. Then the device can start the I^2C communication as a transmitter with `UCB0CTL1 |= UCTR | UCTXSTT`. Also in this step, an update for the `DutyCycle` is sent to the host computer with UART communication when `TXByteCtr` is 0. If the `DutyCycle` equals zero, the Minimum Duty Cycle Value message is sent. If the `DutyCycle` equals 100, the Maximum Duty Cycle Value message is sent. Finally, if the `DutyCycle` is between 0 and 100, the Duty Cycle is Changing message is sent. The `Transmit` function is used here to send the desired message to the host computer. In this function, a string is taken as input. All characters in the string are sent one by one until a null character is reached. Also, before sending the next character, the system is halted with `while(!(IFG2&UCA0TXIFG))` until the transmit interrupt flag is cleared.

Listing 12.30 Digital communication application, the transmitter code.

```
#include <msp430.h>

 char Min[] = "Minimum Duty Cycle Value\r\n";
 char Max[] = "Maximum Duty Cycle Value\r\n";
```

```
 char Mid[] = "Duty Cycle is Changing\r\n";

 int DutyCycle = 0;
 int StartEnable = 0;
 unsigned int TXByteCtr;

void transmit(char *str);
void PinConfig(void);
void TimerConfig(void);
void UARTConfig(void);
void I2CConfig(void);

void main(void)
{
 PinConfig();
 TimerConfig();
 UARTConfig();
 I2CConfig();

 _enable_interrupts();

 while(1){
 TXByteCtr = 1;
 while (UCB0CTL1 & UCTXSTP);
 if(StartEnable == 1){
 UCB0CTL1 |= UCTR | UCTXSTT;
 if(TXByteCtr == 0){
 if(DutyCycle == 0)transmit(Min);
 else if (DutyCycle == 100)transmit(Max);
 else transmit(Mid);
 }}}
}

#pragma vector=USCIAB0RX_VECTOR
__interrupt void USCIA0RX_ISR(void){
 if (UCA0RXBUF == '+')DutyCycle++;
 if (UCA0RXBUF == '-')DutyCycle--;

 if(DutyCycle > 100)DutyCycle = 100;
 if(DutyCycle < 0)DutyCycle = 0;

 StartEnable = 1;
}

#pragma vector = USCIAB0TX_VECTOR
__interrupt void USCIB0TX_ISR(void){
 if (TXByteCtr){
 UCB0TXBUF = DutyCycle;
 TXByteCtr--;
 }
 else
 {
 UCB0CTL1 |= UCTXSTP;
 StartEnable = 0;
 IFG2 &= ~UCB0TXIFG;
```

```
    }
}

void transmit(char *str){
 while(*str != 0){
 while (!(IFG2&UCA0TXIFG));
 UCA0TXBUF = *str++;
 }
}

void PinConfig(void){
 P1DIR |= BIT5;
 P1SEL = BIT1|BIT2|BIT6|BIT7;
 P1SEL2 = BIT1|BIT2|BIT6|BIT7;
 P1OUT &= ~BIT5;
 P1OUT |= BIT5;
}

void TimerConfig(void){
 WDTCTL = WDTPW|WDTHOLD;
 BCSCTL1 = CALBC1_1MHZ;
 DCOCTL = CALDCO_1MHZ;
}

void UARTConfig(void){
 UCA0CTL1 |= UCSWRST|UCSSEL_2;
 UCA0BR0 = 104;
 UCA0BR1 = 0;
 UCA0MCTL = UCBRS_1;
 UCA0CTL1 &= ~UCSWRST;
 IE2 |= UCA0RXIE;
}

void I2CConfig(void){
 UCB0CTL1 |= UCSWRST;
 UCB0CTL0 = UCMST | UCMODE_3 | UCSYNC;
 UCB0CTL1 = UCSSEL_2 | UCSWRST;
 UCB0BR0 = 10;
 UCB0BR1 = 0;
 UCB0I2CSA = 0x48;
 UCB0CTL1 &= ~UCSWRST;
 IE2 |= UCB0TXIE;
}
```

The C Code for the Second MSP430

In the first part of the receiver code, given in Listing 12.31, the hardware configurations for the digital I/O, timer, and I^2C modules are done. In this code block, configurations for each hardware module are done in a separate function.

In the `PinConfig()` function, pin directions are assigned by `P1DIR=BIT2` in the first line since the LED is connected to pin P1.2. In the second and third lines: P1.2 is set as PWM output; P1.6 is set as the I^2C clock pin (SCL); and P1.7 is set as the I^2C data pin (SDA).

In the `TimerConfig()` function, the watchdog timer is disabled in the first line. In the second and third lines, DCO is calibrated to 1 MHz. In the fourth line, the timer configurations are done. SMCLK is chosen as the clock source in up mode, and it is divided by 8 so that f_{CLK} equals 125 kHz. In the fifth and sixth lines, TACCR0 and TACCR1 are set for PWM generation. TACCR1 is set to 999 so that f_{PWM} is 100 Hz and TACCR0 is set to 0. Therefore, the initial duty cycle is 0. The reset/set mode is chosen for the PWM in the last line.

In the `I2CConfig()` function, software reset is enabled by `UCB0CTL1 |= UCSWRST` in the first line to enable the configuration change. In the second line, synchronous communication mode is selected by setting the `UCSYNC` bit. Also `UCMODE_3` is used to select the I^2C mode. In the third line, own address is set as 0x48. In the fourth line, software reset is disabled with `UCB0CTL1 &= ~UCSWRST` to resume the USCI operation. In the fifth line, start and stop interrupts are enabled by `UCB0I2CIE |= UCSTPIE | UCSTTIE`. Finally, the receive interrupt is enabled with `IE2 |= UCB0RXIE` in the last line.

Listing 12.31 Digital communication application, the receiver code part I.

```
void PinConfig(void){
 P1DIR = BIT2;
 P1SEL |= BIT6|BIT7|BIT2;
 P1SEL2 |= BIT6|BIT7;
}

void TimerConfig(void){
 WDTCTL = WDTPW|WDTHOLD;
 BCSCTL1 = CALBC1_1MHZ;
 DCOCTL = CALDCO_1MHZ;
 TACTL = TASSEL_2|MC_1|ID_3;
 TACCR1 = 0;
 TACCR0 = 999;
 TACCTL1 = OUTMOD_7;
}

void I2CConfig(void){
 UCB0CTL1 |= UCSWRST;
 UCB0CTL0 = UCMODE_3 | UCSYNC;
 UCB0I2COA = 0x48;
 UCB0CTL1 &= ~UCSWRST;
 UCB0I2CIE |= UCSTPIE | UCSTTIE;
 IE2 |= UCB0RXIE;
}
```

In the second part of the code, given in Listing 12.32, the ISR settings for the I^2C mode are done as follows: In transmit ISR, the received data is written to the `ReceivedDutyCycle`. Then this value is multiplied by 10 and written to the TACCR1 to obtain the desired duty cycle. In receive ISR, start and stop interrupt flags are cleared manually by `UCB0STAT &= ~(UCSTPIFG | UCSTTIFG)` to resume the communication. Finally, the C code for the receiver part is given in Listing 12.33.

Listing 12.32 Digital communication application, the receiver code part II.

```
#pragma vector = USCIAB0TX_VECTOR
__interrupt void USCIAB0TX_ISR(void){
 ReceivedDutyCycle = UCB0RXBUF;
 TACCR1 = ReceivedDutyCycle*10;
}

#pragma vector = USCIAB0RX_VECTOR
__interrupt void USCIAB0RX_ISR(void){
 UCB0STAT &= ~(UCSTPIFG | UCSTTIFG);
}
```

Listing 12.33 Digital communication application, the receiver code.

```
#include <msp430.h>

 unsigned int ReceivedDutyCycle = 0;

void PinConfig(void);
void TimerConfig(void);
void I2CConfig(void);

void main(void)
{
 PinConfig();
 TimerConfig();
 I2CConfig();

 _enable_interrupts();
 LPM0;
}

#pragma vector = USCIAB0TX_VECTOR
__interrupt void USCIAB0TX_ISR(void){
 ReceivedDutyCycle = UCB0RXBUF;
 TACCR1 = ReceivedDutyCycle*10;
}

#pragma vector = USCIAB0RX_VECTOR
__interrupt void USCIAB0RX_ISR(void){
 UCB0STAT &= ~(UCSTPIFG | UCSTTIFG);
}

void PinConfig(void){
 P1DIR = BIT2;
 P1SEL |= BIT6|BIT7|BIT2;
 P1SEL2 |= BIT6|BIT7;
}

void TimerConfig(void){
 WDTCTL = WDTPW|WDTHOLD;
 BCSCTL1 = CALBC1_1MHZ;
 DCOCTL = CALDCO_1MHZ;
 TACTL = TASSEL_2|MC_1|ID_3;
```

```
  TACCR1 = 0;
  TACCR0 = 999;
  TACCTL1 = OUTMOD_7;
}

void I2CConfig(void){
  UCB0CTL1 |= UCSWRST;
  UCB0CTL0 = UCMODE_3 | UCSYNC;
  UCB0I2COA = 0x48;
  UCB0CTL1 &= ~UCSWRST;
  UCB0I2CIE |= UCSTPIE | UCSTTIE;
  IE2 |= UCB0RXIE;
}
```

12.9 Summary

The MSP430 has digital communication capabilities. In this chapter, we explored these in detail. We started with the USCI_A and USCI_B modules available on the MSP430G2553. USCI_A supports UART and SPI communication modes. UART is the only asynchronous communication mode available on the MSP430G2553. SPI, on the other hand is a synchronous and fast communication mode. USCI_B supports SPI and I^{2}C communication modes. Therefore, SPI is supported by the two USCI modules. I^{2}C is also synchronous. We explored each mode in detail in this chapter. We also provided sample C and assembly codes for the three communication modes. We benefit from Grace to configure and use UART, SPI, and I^{2}C. Finally, we provided a generic digital communication application jointly using UART and I^{2}C communication modes.

12.10 Problems

In this section, we will not offer new problems. Instead, we will ask the reader to solve problems given in previous chapters using two MSP430 LaunchPad boards and establishing a digital communication link between them. Some sample problems are given below. In all below questions, please use available Port 2 pins for LED connections.

12.1 Solve Prob. 8.9 using two MSP430 LaunchPad boards. The first MSP430 LaunchPad will be used for the push button. The second will be used for the LEDs. Establish a digital communication link between these two boards using UART, SPI, and I^{2}C communication modes.

12.2 Repeat Prob. 12.1 in assembly language.

12.3 Solve Prob. 8.12 using two MSP430 LaunchPad boards. The first MSP430 LaunchPad will be used for the push button. The second will be used for the LEDs. Establish a digital communication link between these two boards using UART, SPI, and I^{2}C communication modes.

12.4 Repeat Prob. 12.3 in assembly language.

12.5 Solve Prob. 9.4 using two MSP430 LaunchPad boards. The first MSP430 LaunchPad will be used for the push button. The second will be used for the LEDs. Establish a digital communication link between these two boards using UART, SPI, and I^2C communication modes.

12.6 Repeat Prob. 12.5 in assembly language.

12.7 Solve Prob. 11.2 using two MSP430 LaunchPad boards. The first MSP430 LaunchPad will be used for the comparator operation. The second will be used for the LEDs. Establish a digital communication link between these two boards using UART, SPI, and I^2C communication modes.

12.8 Repeat Prob. 12.7 in assembly language.

12.9 Solve Prob. 11.4 using two MSP430 LaunchPad boards. The first MSP430 LaunchPad will be used for the ADC operation. The second will be used for the LEDs. Establish a digital communication link between these two boards using UART, SPI, and I^2C communication modes.

12.10 Repeat Prob. 12.9 in assembly language.

13 Flash Memory

Chapter Outline

MSP430 memory is divided into two parts, flash and RAM. The flash memory is the main topic of this chapter. We will see how to program the flash using C and assembly languages.

13.1 MSP430 Flash Memory

A flash memory cell is composed of a MOS transistor with an additional floating gate under the control gate. Its working principle is based on charging and discharging this floating gate. More information on this operation and the physical characteristics of a flash cell can be found in [9, 10].

The flash is nonvolatile. It can keep the saved data even when energy is not provided. Therefore, it can be taken as another form of ROM. However, the flash can easily be programmed by feeding a suitable voltage to it. The MSP430 has circuitry to program its flash memory.

MSP430 flash memory is divided into two sections, main and information. The executable code and the constant values are kept in the main section. The calibration data, serial number, and similar factory settings are kept in the information section. To note here, there is no physical difference between these two sections.

The main section of the flash memory for the MSP430G2553 is 16 kB. It spans the memory addresses between FFC0h and C000h. This space is divided into segments, each being 512 bytes. Therefore, there are 12 segments. The information section is divided into four segments. These are called A, B, C, and D. Here, each segment is 64 bytes. Segment A holds the calibration data. Therefore, it is protected. Although this protection can be bypassed during programming, it should be done with caution.

13.2 Flash Memory Programming

There are three options to program the flash memory. The first one is using the JTAG port. For more detail on this option, please see [15]. The second option to program the flash memory is using the bootstrap loader. For more detail on this option, please see [14]. The last option is using a custom solution. In this option, we rely on the CPU's ability to program its own flash memory. In this book, we will only focus on this property.

The CPU can write to a single byte or word location of the flash memory. In the erasing operation, it can only operate on a segment level. The CPU can use *flash to flash* or *RAM to flash* programming. In this book, we will only deal with the former solution. For the RAM to flash programming, please see [17]. Moreover, segment A in the information memory is handled separately in programming the flash memory. To see how it is handled, please see [17].

13.2.1 *The Flash Memory Controller*

The MSP430 has an internal flash memory controller for erasing and writing operations. This controller has three registers, a timing generator, and a programming voltage generator. The registers can be used to configure the writing and erasing operations. The writing and erasing operations are controlled by the flash timing generator. The properties of this timing generator are also configured through the registers. The voltage generator is used to generate necessary voltage values for writing and erasing operations.

In order to program the flash memory, the timing generator frequency must be set within the 257 to 476 kHz range. The supply voltage should be between 2.2 and 3.6 V. If these values are not satisfied during writing or erasing, the result of the operation will be unpredictable. For more detail on these limits, please see [16].

There are specific options when the CPU is used to write or erase the flash memory. Byte, word, or block levels can be used in writing to flash. When erasing the flash memory, segment level, mass erase (to erase all main memory segments), and all erase (to erase all segments) options can be selected.

13.2.2 *Flash Memory Registers*

The operations in the flash memory controller are configured by three registers. These are called the flash memory control register 1 (FCTL1), flash memory control register 2 (FCTL2), and flash memory control register 3 (FCTL3).

The FCTL1 register, given in Table 13.1, is for selecting the erasing mode. In Table 13.1, the **FRKEY** will be read as 96h and **FWKEY** should be written as A5h. These are the **FCTLx** passwords. The **BLKWRT** bit should be set for the block write mode. The **WRT** bit should be set to select any write mode. The **MERAS** and **ERASE** bits are used together for mass erase and erase modes. There will be no erasing when MERAS=0 and ERASE=0. Individual segments can be erased when MERAS=0 and ERASE=1. All main memory segments can be erased when MERAS=1 and ERASE=0. The main and information sections of the flash memory can be erased when MERAS=1, ERASE=1, and LOCKA=0 (in FCTL3). Only the main section of the flash memory can be erased when MERAS=1, ERASE=1, and LOCKA=1.

The FCTL2 register, given in Table 13.2, is for configuring the flash controller clock source. In Table 13.2, the **FWKEYx** will be read as 96h and should be written as A5h. These are the FCTLx passwords. The **FSSELx** bits are used to select the flash controller clock source. The constants for these bits are as follows:

Table 13.1

Flash memory control register 1 (FCTL1).

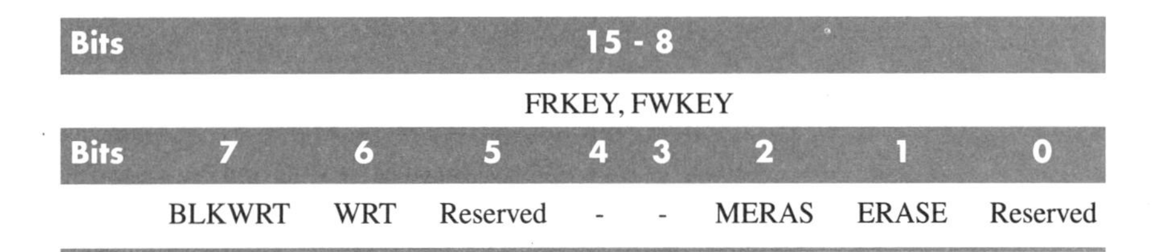

Bits	15 - 8							
	FRKEY, FWKEY							
Bits	**7**	**6**	**5**	**4**	**3**	**2**	**1**	**0**
	BLKWRT	WRT	Reserved	-	-	MERAS	ERASE	Reserved

Table 13.2

Flash memory control register 2 (FCTL2).

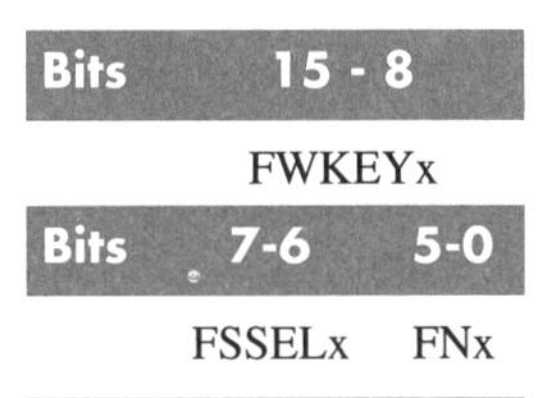

Bits	15 - 8	
	FWKEYx	
Bits	**7-6**	**5-0**
	FSSELx	FNx

Table 13.3 Flash memory control register 3 (FCTL3).

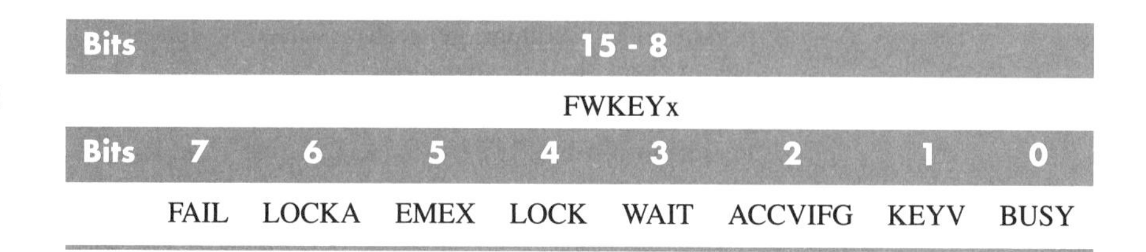

Bits	15 - 8							
	FWKEYx							
Bits	**7**	**6**	**5**	**4**	**3**	**2**	**1**	**0**
	FAIL	LOCKA	EMEX	LOCK	WAIT	ACCVIFG	KEYV	BUSY

FSSEL_0 (for selecting the ACLK), FSSEL_1 (for selecting the MCLK), FSSEL_2, and FSSEL_3 (for selecting the SMCLK). The **FNx** bits are used for the flash controller clock divider. The divider is 1 when FNx=00h. The divider is 64 (which is the maximum value) when FNx=3Fh.

The FCTL3 register, given in Table 13.3, is for the operation modes and failure handling. In Table 13.3, the FWKEYx will be read as 96h and should be written as A5h. These are the FCTLx passwords. The **FAIL** bit is set when an operation failure occurs. The **LOCKA** bit is used to unlock segment A in the information memory. This bit should be reset to allow the programmer to adjust segment A. The **EMEX** bit is for emergency exit. The **LOCK** bit is used to unlock the flash memory for writing and erasing operations. When the **WAIT** bit is reset, it indicates that the flash memory is not ready for a byte- or word-level writing operation. The **ACCVIF** stands for the access violation interrupt flag. The **KEYV** bit indicates the flash security key violation. When this bit is reset, it indicates that the FCTLx password is entered correctly. When it is set, it indicates that the FCTLx password is entered incorrectly. The **BUSY** bit indicates the status of the flash timing generator. When this bit is set, it indicates that the flash timing generator is busy.

13.3 Coding Practices for Flash Memory

In this section, we provide C and assembly code samples for flash memory programming. These are modified from TI sample codes. Next, we explain them in detail.

13.3.1 *Flash Memory in C*

The C code for flash programming is given in Listing 13.1. Initially, we set the digitally controlled oscillator (DCO) to 1 MHz. This clock will be used in the timing generator. The `erase_SegCD()` function erases the segments C and D separately. Here, the memory address of each segment is given separately. The memory address is 1040h for segment C. The memory address is 1000h for segment D. The erasing is done by simply assigning zero to the segment. Since only segment-based erasing can be done, all segment elements are erased by this operation. To note here, the erased bit value for the segment elements will be one. In the erasing operation, the ERASE bit in the FCTL1 register is reset after each operation. Therefore, to erase segment D it should be set after erasing segment C. The `write_SegC()` function writes values 0 to 63 to segment C at the byte level. Finally, the `copy_SegCD()` function copies the entries of segment C to segment D in reverse order. This is done to show that we can reach the segment elements at a byte level.

Listing 13.1 Flash memory processing in C.

```
#include <msp430.h>

void erase_SegCD(void);
void write_SegC(void);
void copy_C2D(void);

void main(void)
{
 WDTCTL = WDTPW|WDTHOLD;

 BCSCTL1 = CALBC1_1MHZ; // Set DCO to 1MHz
 DCOCTL = CALDCO_1MHZ;

 FCTL2 = FWKEY+FSSEL0+FN1;
 // MCLK/3 for Flash Timing Generator

 erase_SegCD(); // Erase segments C and D
 write_SegC();  // Write to segment C
 copy_C2D();    // Copy segment C to D

 while(1);
}

void erase_SegCD(void){
 char *Flash_ptrC;
 char *Flash_ptrD;

 Flash_ptrC = (char *) 0x1040;
 // Initialize Flash segment C pointer
 Flash_ptrD = (char *) 0x1000;
 // Initialize Flash segment D pointer

 FCTL1 = FWKEY+ERASE; // Set Erase bit
 FCTL3 = FWKEY;  // Clear Lock bit
 *Flash_ptrC = 0;

 FCTL1 = FWKEY+ERASE; // Set Erase bit
 *Flash_ptrD = 0;    // Dummy write to erase segments
 FCTL3 = FWKEY+LOCK;  // Set LOCK bit
}

void write_SegC(void){
 char *Flash_ptrC;
 char i;

 Flash_ptrC = (char *) 0x1040;
 // Initialize Flash pointer
 FCTL3 = FWKEY;  // Clear Lock bit
 FCTL1 = FWKEY+WRT;
 // Set WRT bit for write operation

 for (i=0; i<64; i++)
 *Flash_ptrC++ = i; // Write value to flash

 FCTL1 = FWKEY;    // Clear WRT bit
 FCTL3 = FWKEY+LOCK;  // Set LOCK bit
}

void copy_C2D(void){
```

```
 char *Flash_ptrC;
 char *Flash_ptrD;
 int i;

 Flash_ptrD = (char *) 0x1000;
 // Initialize Flash segment D pointer

 FCTL1 = FWKEY+WRT;
 // Set WRT bit for write operation
 FCTL3 = FWKEY;  // Clear Lock bit

 Flash_ptrC = (char *) 0x107F;

 for (i=0; i<64; i++)
 *Flash_ptrD++ = *Flash_ptrC--;
 // copy value segment C to segment D

 FCTL1 = FWKEY;  // Clear WRT bit
 FCTL3 = FWKEY+LOCK; // Set LOCK bit
}
```

13.3.2 *Flash Memory in Assembly*

The operations performed in Listing 13.1 are redone in assembly language in Listing 13.2. The subroutines here have the same name as in the C code. Therefore, the explanations given above also apply here.

Listing 13.2 Flash memory processing in assembly language.

```
 .cdecls C,LIST,"msp430.h"

 .text
 .retain
 .retainrefs

RESET
 mov.w #WDTPW|WDTHOLD,WDTCTL
 mov.w #__STACK_END,SP

 mov.b CALBC1_1MHZ,BCSCTL1
 mov.b CALDCO_1MHZ,DCOCTL ;Set DCO to 1MHz

 mov.w #FWKEY+FSSEL0+FN1,FCTL2
;Timing generator = MCLK/3

 call #Erase_SegCD
;Erase segments C and D

 call #Write_SegC
;Copy value to segment C

 call #CopyC2D
 jmp $

Erase_SegCD
```

```
 mov.w #FWKEY,FCTL3 ;Lock = 0

 mov.w #FWKEY+ERASE,FCTL1
;Erase bit = 1, allow interrupts
 mov.w #0,&1040h
;Dummy write to SegC to erase

 mov.w #FWKEY+ERASE,FCTL1
;Erase bit = 1, allow interrupts
 mov.w #0,&1000h
;Dummy write to SegD to erase
 ret

Write_SegC
 mov.w #FWKEY+WRT,FCTL1
;Write bit = 1, block interrupts
 mov.w #FWKEY,FCTL3 ;Lock = 0

 mov.w #1040h,R5
 mov.b #00h,R6

Prog_L1
 mov.b R6,0(R5)
 inc.w R5
 inc.b R6
 cmp.w #1080h,R5
 jne Prog_L1

 mov.w #FWKEY+LOCK,FCTL3 ;Lock = 1;
 ret

CopyC2D
;Copy Seg C to Seg D

 mov.w #FWKEY+WRT,FCTL1
;Write bit = 1, block interrupts
 mov.w #FWKEY,FCTL3 ;Lock = 0

 mov.w #1040h,R5
 mov.w #103Fh,R6

Prog_L2
 mov.b @R5+,0(R6)
 dec.w R6
 cmp.w #1080h,R5
 jne Prog_L2

 mov.w #FWKEY+LOCK,FCTL3 ;Lock = 1
 ret

;------------------------
;Stack Pointer definition
;------------------------
 .global __STACK_END
 .sect .stack

;-------------------
```

```
;Interrupt Vectors
;-------------------
 .sect RESET_VECTOR
 .short RESET
 .end
```

13.4 Flash Memory in Grace

The flash memory controller can be configured by the **Flash** block in the Device Overview Window (given in Fig. 5.11). First, we should check the "Enable Flash controller in my configuration" box. Then the flash can be configured as follows.

13.4.1 *The Basic User Mode*

In the basic user mode, shown in Fig. 13.1, the clock (for the timing generator) can be configured by its source and frequency divider. The oscillator frequency is directly taken from the basic clock module+ (BCM+). Then the clock divider is automatically set by Grace to set the Flash Timing Generator within a 257 to 476 kHz range. As in other blocks, interrupts can be enabled by checking the Flash

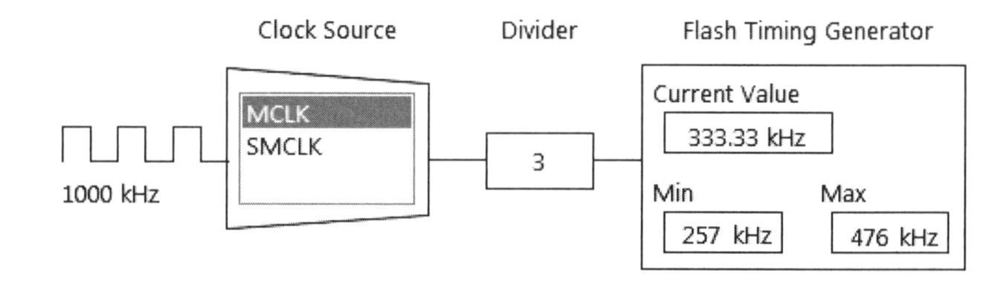

Figure 13.1 Basic user mode for the flash memory controller.

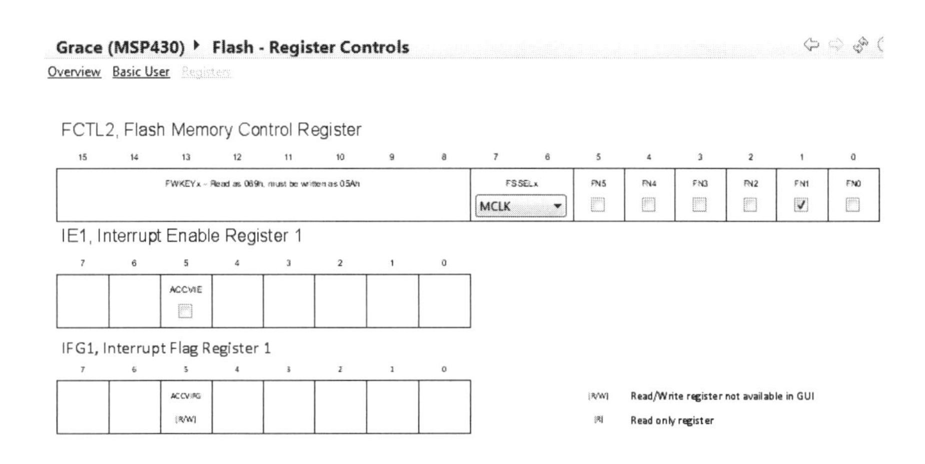

Figure 13.2

Register mode for the flash memory controller.

Ctrl Access Violation Int Enable box. The prototype interrupt service routine (ISR) for the interrupt can be generated by the Generate Interrupt Handler Code button.

13.4.2 *The Register Controls Mode*

In the register controls mode, given in Fig. 13.2, the FCTL2 register can be configured. Since the access violation handler is also present, the interrupt enable register (IE1) can be configured here. The interrupt flag register (IFG1) is also given here. However, it cannot be configured.

13.5 Summary

The MSP430G2553 has flash as the nonvolatile memory. This part of the memory holds the code and constants in our applications. In this chapter, we focused on the flash memory programming issues. There are several methods to program the flash. In this book, we only considered flash programming through the CPU. For more detail on other programming methods, we directed the reader to cited references. We provided sample C and assembly codes for the CPU-based flash programming. We also considered flash programming under Grace.

13.6 Problems

13.1 What is the starting address of segments A and B in the flash memory?

13.2 Write a program in C to erase the contents of segment B in the flash memory.

13.3 Repeat Prob. 13.2 in assembly language.

13.4 Write a program in C to erase the contents of segment B in the flash memory. Then, write the first 10 Fibonacci numbers to this area.

13.5 Repeat Prob. 13.4 in assembly language.

13.6 Research the FRAM technology.

14 Applications

Chapter Outline

In this chapter, we provide several applications for the MSP430. These are based on real-life problems. In these we aim to show the usefulness of the microcontroller in our daily lives. We provide the circuit layout and the equipment list for each application so that readers can implement them directly. We also suggest that you check the TI websites for other MSP430-based applications.

14.1 Car Door Alarm

The goal of this application is to learn how to use the digital input and output (I/O) pins of the MSP430G2553 microcontroller. As a real-world application, we examine a car door alarm system. In this section, we provide the equipment list, the layout of the circuit, and the procedure.

14.1.1 *Equipment List*

Following is a list of the equipment to be used in this application.

- Five LEDs
- Five 220-Ω resistors
- Five push buttons
- One 100-ηF capacitor

14.1.2 *Layout*

The layout of this application is shown in Fig. 14.1.

14.1.3 *System Design Specifications*

The design steps of the car door alarm system are as follows: In the first part of the application, we will assume that four push buttons are placed between the four car doors and the chassis. When a button is not pressed, it means that door is open and a warning should be given. This is done by an LED representing that door. When the same button is pressed, it means that door is closed now and the warning should be reset (or the LED should be turned off).

In the second part of the application, a lock button and lock warning LED will be added to the system. When the lock button is pressed, the car is locked and the lock warning LED is off. When the lock button is pressed again, it means that the car is unlocked now. Therefore, the lock warning LED should turn on. Initially, the car is not locked. Therefore, the lock warning LED must turn on. Also, the lock button cannot be used if all doors are not closed. As long as the car is locked, doors cannot be opened. Normally, mechanical systems are used for this purpose. In this application, this property is simulated by using LEDs. After the car is locked, even releasing the buttons (placed between the doors and the chassis) cannot turn on the LEDs. This way, we will assume that the doors are still closed. The user should press all buttons (placed between the doors and the chassis), then press the lock button in order to unlock the system again.

Hint: In the second part of the application, be careful about the time of pressing the lock button.

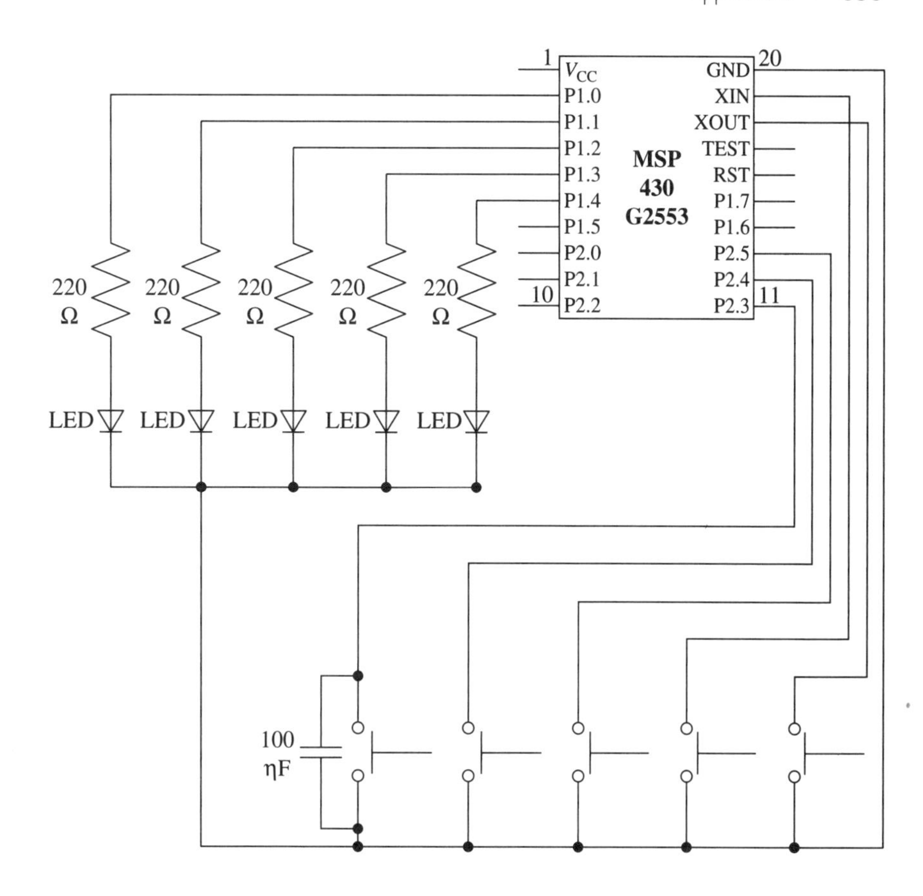

Figure 14.1

Layout of the car door alarm application.

14.2 Car Window Control

The goal of this application is to learn how to use the digital I/O pins of the MSP430 microcontroller. As a real-world application, we design a car window control system. In this section, we provide the equipment list, the layout of the circuit, and the procedure.

14.2.1 *Equipment List*

Following is a list of the equipment to be used in this application.

- One 12-V dc adaptor
- One LM7805 voltage regulator
- One 330-ηF capacitor
- One 10-μF electrolytic capacitor
- Two 100-ηF capacitors
- One stepper motor

- One ULN2003 motor driver
- Four push buttons

14.2.2 *Layout*

The layout of this application is shown in Fig. 14.2. For more information on the voltage supply block, please see Fig. 9.3.

14.2.3 *System Design Specifications*

In this application, we will design a car window control system by using a stepper motor and push buttons. The stepper motor is used to control a window. Initially, the window is assumed to be closed. The stepper motor will act in two states: either it will fully open the window or it will close it. The stepper motor will stop when one of these conditions is met. The number of stepper motor states to open the window will be decided by the user. Two of the push buttons will be used to

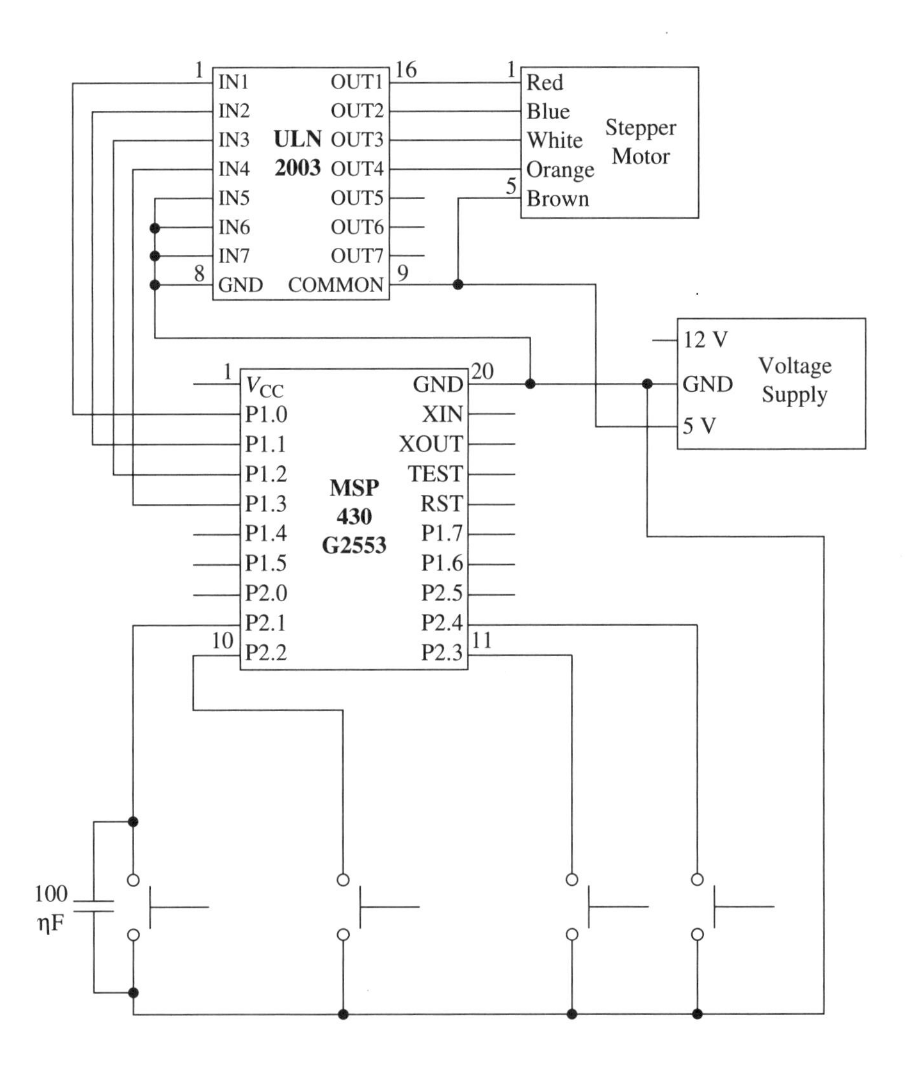

Figure 14.2 Layout of the car window control application.

control the direction of the stepper motor. When the first button is pressed, the stepper motor will start to rotate in one direction until the window is fully open. When the second button is pressed, the motor will rotate in the other direction until the window is closed. The user may press the other button when one of the buttons is pressed. The third button will be used to stop the rotation. Therefore, the window will be half open when it is pressed. The last button will be used for child protection. It will lock the system when it is pressed once. The system will be unlocked and can be used again when the last button is pressed again.

14.3 Car Park Tollgate

The goal of this application is to learn how to set and use port interrupts of the MSP430 microcontroller. As a real-world application, we design a car park tollgate system. In this section, we provide the equipment list, the layout of the circuit, and the procedure.

14.3.1 *Equipment List*

Following is a list of the equipment to be used in this application.

- One 12-V dc adaptor
- One LM7805 voltage regulator
- One 330-ηF capacitor
- One 10-μF electrolytic capacitor
- One 100-ηF capacitor
- One 16×2 character LCD (with a Samsung processor)
- One 10-kΩ potentiometer
- Two IR transmitter LEDs
- Two IR receiver LEDs
- Two 10-kΩ resistors
- Two 220-Ω resistors

The **IR transmitter LED** conducts current when V_{CC} is applied to its anode. Here, the cathode is connected to ground through a 220-Ω resistor. It emits IR light when this current is conducted. On the other hand, the **IR receiver LED** conducts current when a positive voltage is applied to its cathode through a 10-kΩ resistor and its anode is connected to ground. Also, it must absorb IR light to conduct current. This property is critical for the working logic of the IR sensor used in this application. In terms of logic level, when the light is absorbed by the receiver (with V_{CC} present, of course) it gives zero. Otherwise, it gives one.

14.3.2 *Layout*

The layout of this application is shown in Fig. 14.3. For more information on the voltage supply block, please see Fig. 9.3.

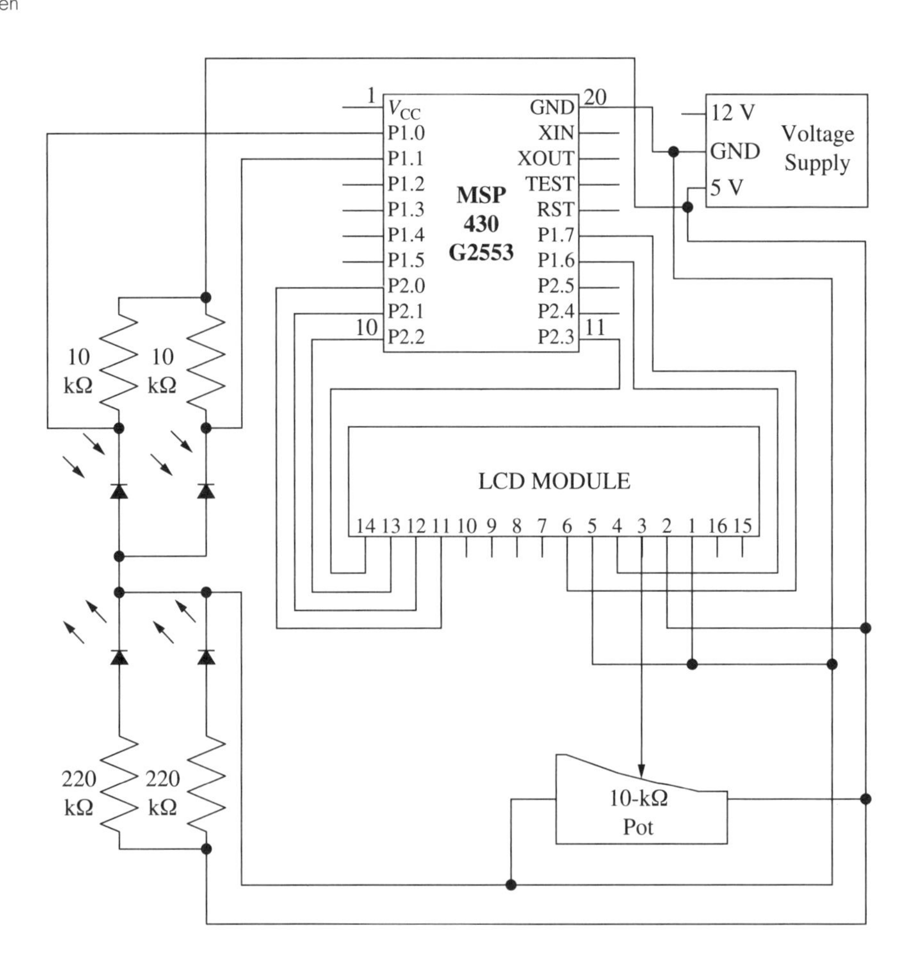

Figure 14.3 Layout of the car park tollgate application.

14.3.3 *System Design Specifications*

In this application, a car park tollgate system will be designed by using two infrared sensors and an LCD. One sensor can detect the movement, but cannot decide on its direction. Therefore, two sensors will be jointly used to detect the movement and its direction (entering to or exiting from the park). The LCD will be used to provide information on the status of the park. In the first line of the LCD, entering and exiting car numbers will be displayed. In the second line, the number of cars inside the park at that moment will be displayed. Be careful—when the number of cars inside the car park is zero, there will be no exiting process.

14.4 Digital Lock System

The goal of this application is to learn how to set and use the port interrupts of the MSP430 microcontroller. As a real-world application, we will design a digital lock system. In this section, we provide the equipment list, the layout of the circuit, and the procedure.

14.4.1 *Equipment List*

Following is a list of the equipment to be used in this application.

- One 12-V dc adaptor
- One LM7805 voltage regulator
- One 330-ηF capacitor
- Two 10-μF electrolytic capacitors
- One 1-μF electrolytic capacitor
- One 100-ηF capacitor
- One 16$\times$2 character LCD (with a Samsung processor)
- One 10-kΩ potentiometer
- Two LEDs (green and red)
- Two 220-Ω resistors
- One solenoid
- One ULN2003
- One MM74C922
- Two buzzers
- One 4$\times$3 keypad
- One push button

14.4.2 *Layout*

The layout of this application is shown in Fig. 14.4. For more information on the voltage supply block, please see Fig. 9.3.

14.4.3 *System Design Specifications*

In this application, we will design a digital lock system with a keypad, an LCD, and a solenoid. Initially, *Enter Your Password* is written on the first line of the LCD, and the system must wait in a suitable low-power mode. When there is an entry from the keypad, the system exits from the low-power mode and writes * on the second line of the LCD. Each * sign represents an entered number. If the * button on the keypad is pressed, the system erases the last entry. If the entered password is wrong, an *Access Denied* string is written to the second line of the LCD and the red LED is turned on. Also, the buzzer connected to the same pin with the red LED starts to beep. After 2 s, the system returns to the initial condition by turning off the red LED and stopping the buzzer. If the entered password is correct, an *Access Granted* string is written to the second line of the LCD. The solenoid is opened, and the green LED is turned on to indicate that the door is opened. Also, the second buzzer beeps for 2 s. After the door is opened, the user has two choices. First, the door can be closed by using the push button. Then the system returns to the initial condition by turning off the green LED and closing the solenoid again. Second, the password can be changed after pressing the # button on the keypad. Then a *Change Password* string is written to the first line of the LCD and after this process is completed. Meanwhile, the *Change Password*

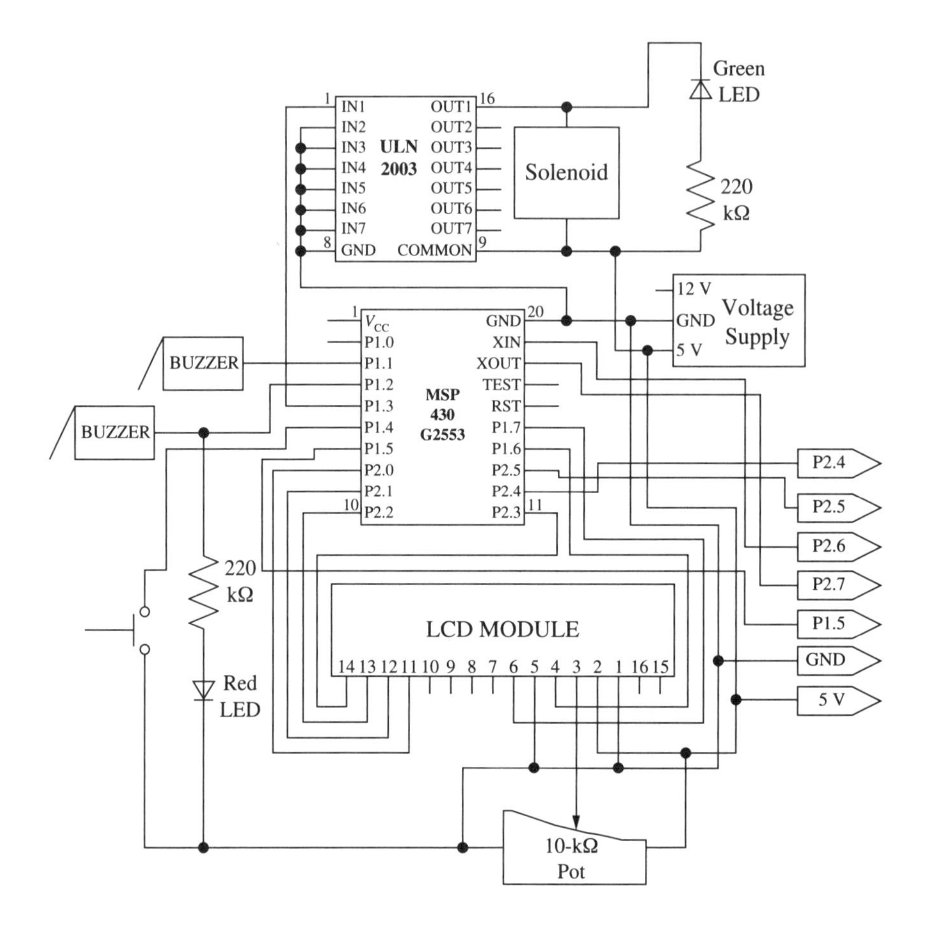

Figure 14.4 Layout of the digital lock application.

string will be changed to a *Password Changed* string. After this step, the system can return to the initial condition by pressing the push button.

14.5 Air Freshener Dispenser

The goal of this application is to learn how to set and use timers and low-power modes of the MSP430 microcontroller. As a real-world application, we will design an air freshener dispenser system. In this section, we provide the equipment list, the layout of the circuit, and the procedure.

14.5.1 *Equipment List*

Following is a list of the equipment to be used in this application.

- Five LEDs
- Five 220-Ω resistors
- Three push buttons
- Two 100-ηF capacitors

Figure 14.4 (*continued*)

Layout of the digital lock application.

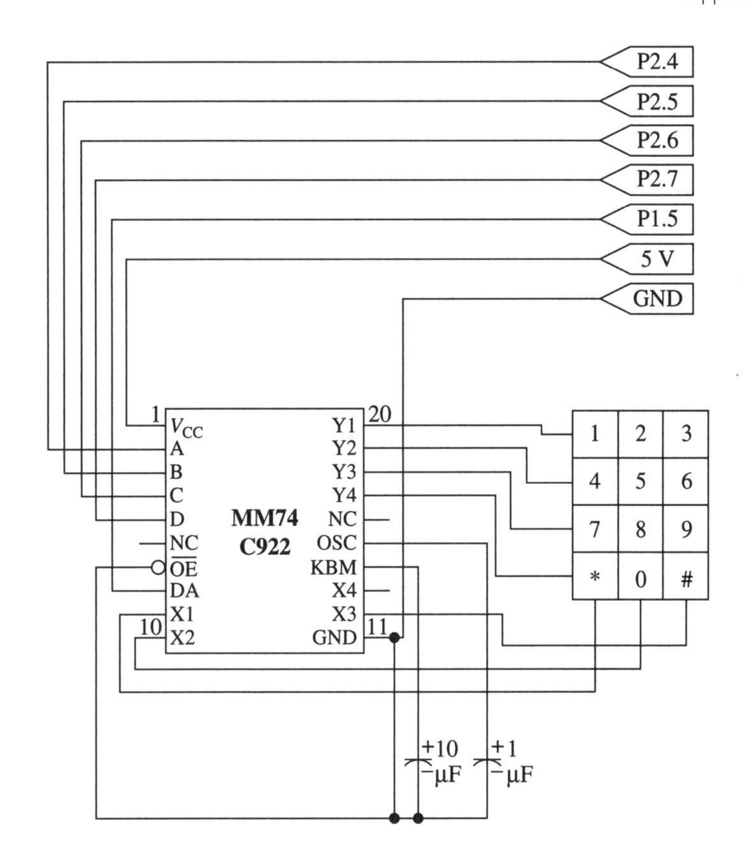

14.5.2 *Layout*

The layout of this application is shown in Fig. 14.5.

14.5.3 *System Design Specifications*

The design steps of the air freshener dispenser system are as follows: In the first part of the application, an air freshener dispenser with three different programs is implemented. These programs are called *short*, *medium*, and *long*. They correspond to the spraying of fresh odor in 5-, 10-, and 15-s periods. In an actual system, these should be in minutes. Also, the spraying operation should be done by a mechanism in an actual system. We simulate this operation by turning on an LED for one second. In our system, one push button will be used to switch between programs. Three LEDs will be associated with the programs. Therefore, selecting each program will turn on the associated LED. There is also an *instant spray* button. When it is pressed, the system will spray the odor and reset the counting process. Also, in this part, one of the three programs must be selected as the initial starting program.

In the second part, an *on/off* button and a warning LED will be added to the system. When the system is in the off state, all LEDs are turned off and the system goes into an appropriate low-power mode. All buttons except the on/off button will be unavailable in this state. When the system turns on by this button, the warning

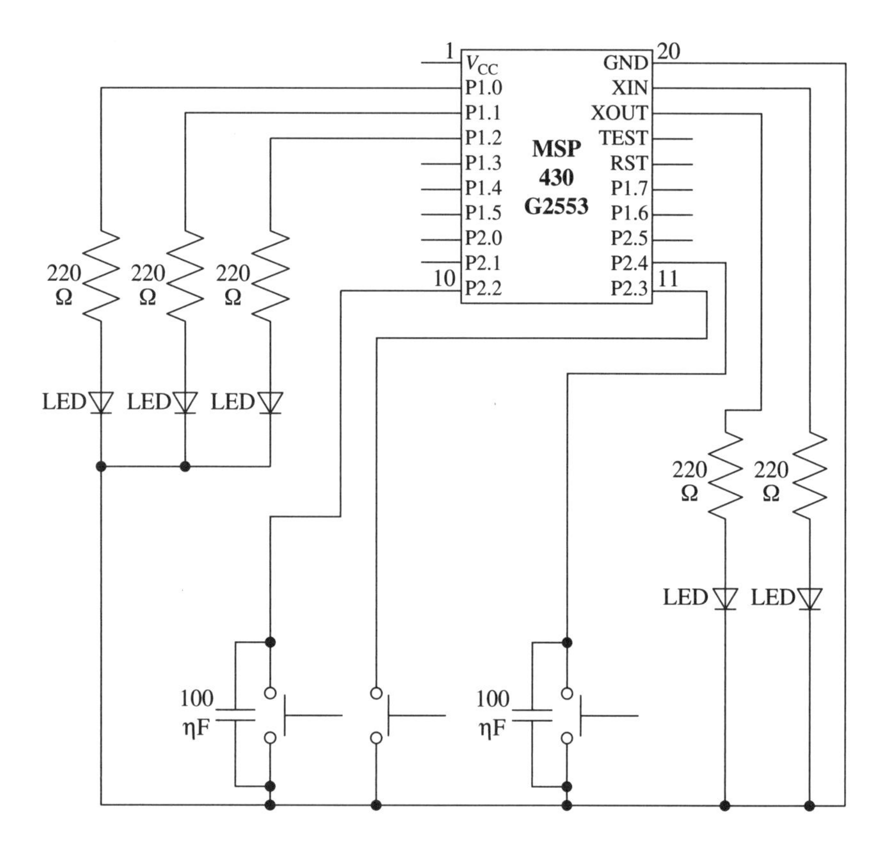

Figure 14.5 Layout of the air freshener dispenser application.

LED will turn on. This LED should blink for a 1-s period during operation. Initially, the system must be in the off state.

Hint: The watchdog timer can be used for the restarting process in the second part of the application.

14.6 Traffic Lights

The goal of this application is to learn how to set and use timers and low-power modes of the MSP430 microcontroller. As a real-world application, we design a traffic lights system. In this section, we provide the equipment list, the layout of the circuit, and the procedure.

14.6.1 *Equipment List*

Following is a list of the equipment to be used in this application.

- Five LEDs (two green, two red, one yellow)
- Five 220-Ω resistors
- One push button

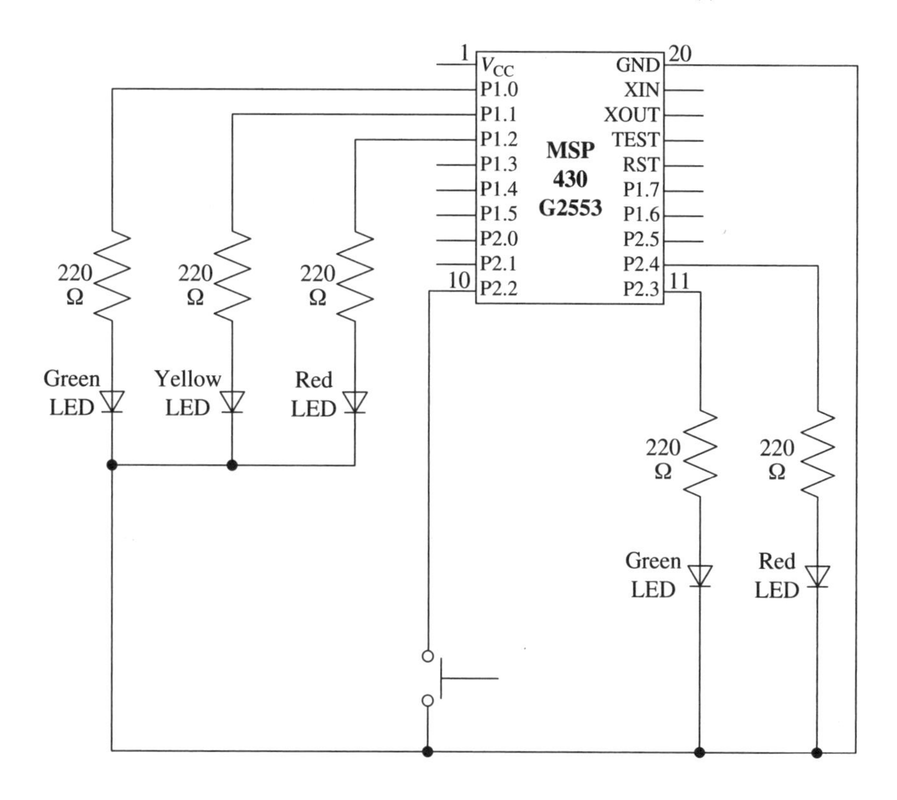

Figure 14.6

Layout of the traffic lights application.

14.6.2 *Layout*

The layout of this application is shown in Fig. 14.6.

14.6.3 *System Design Specifications*

In this application, we will design a traffic lights system for a street with a crosswalk. Three of the LEDs (green, yellow, and red) are for the cars. The other two LEDs (green and red) and the push button are for pedestrians. When the push button is not pressed, the system works in a loop as follows:

- **State 1:** The green LED for cars is turned on for 90 s. During this time, the red LED is turned on for pedestrians.
- **State 2:** The yellow LED for cars is turned on for 5 s. During this time, the red LED is turned on for pedestrians.
- **State 3:** The red LED for cars is turned on for 20 s. During this time, the green LED is turned on for pedestrians.
- **State 4:** The red and yellow LEDs for cars are turned on for 5 s. During this time, the red LED is turned on for pedestrians.

If a pedestrian pushes the button in State 1 after 60 s, the system will jump to State 2. If the button is pressed before 60 s, the system will wait until 60 s has passed.

Then it will jump to State 2. If the system is in State 3 or State 4, the push button will not be activated and cannot be used.

14.7 Sound Detector

The goal of this application is to learn how to use the ADC module of the MSP430 microcontroller. As a real-world application, we design a sound detector system. In this section, we provide the equipment list, the layout of the circuit, and the procedure.

14.7.1 *Equipment List*

Following is a list of the equipment to be used in this application.

- One 12-V dc adaptor
- One LM7805 voltage regulator
- One 330-ηF capacitor
- One 100-ηF capacitor
- One 47-ηF capacitor
- Four 10-μF electrolytic capacitors
- One 220-μF electrolytic capacitor
- One LM386 low-voltage audio power amplifier
- One electret microphone
- Two 7-segment displays
- Two 74HC595 shift registers
- Fourteen 220-Ω resistors
- One 10-Ω resistor
- One 10-kΩ resistor

14.7.2 *Layout*

The layout of this application is shown in Fig. 14.7. For more information on the voltage supply block, please see Fig. 9.3. In Fig. 14.7, all resistors connected to the 7-segment display are 220 Ω.

14.7.3 *System Design Specifications*

In this application, we will design a sound detector system. The sound will be converted to an electrical signal by the electret microphone. Then this signal will be amplified to the appropriate level with the audio amplifier. The audio amplifier will give approximately 2.5 V to its output when there is no sound in the environment. When the user snaps his or her finger, the system will detect it. The voltage level which detects the snap sound can be arranged by the user. After each detection, a short delay must be added to avoid detecting echoes. The number of snaps detected will be shown on two 7-segment displays as a two-digit number. When a new snap is detected, this number will be increased by one. When the snap count reaches 99, it will be reset (to 00). Initially, the 7-segment displays show 00. Shift registers

Figure 14.7

Layout of the sound detector application.

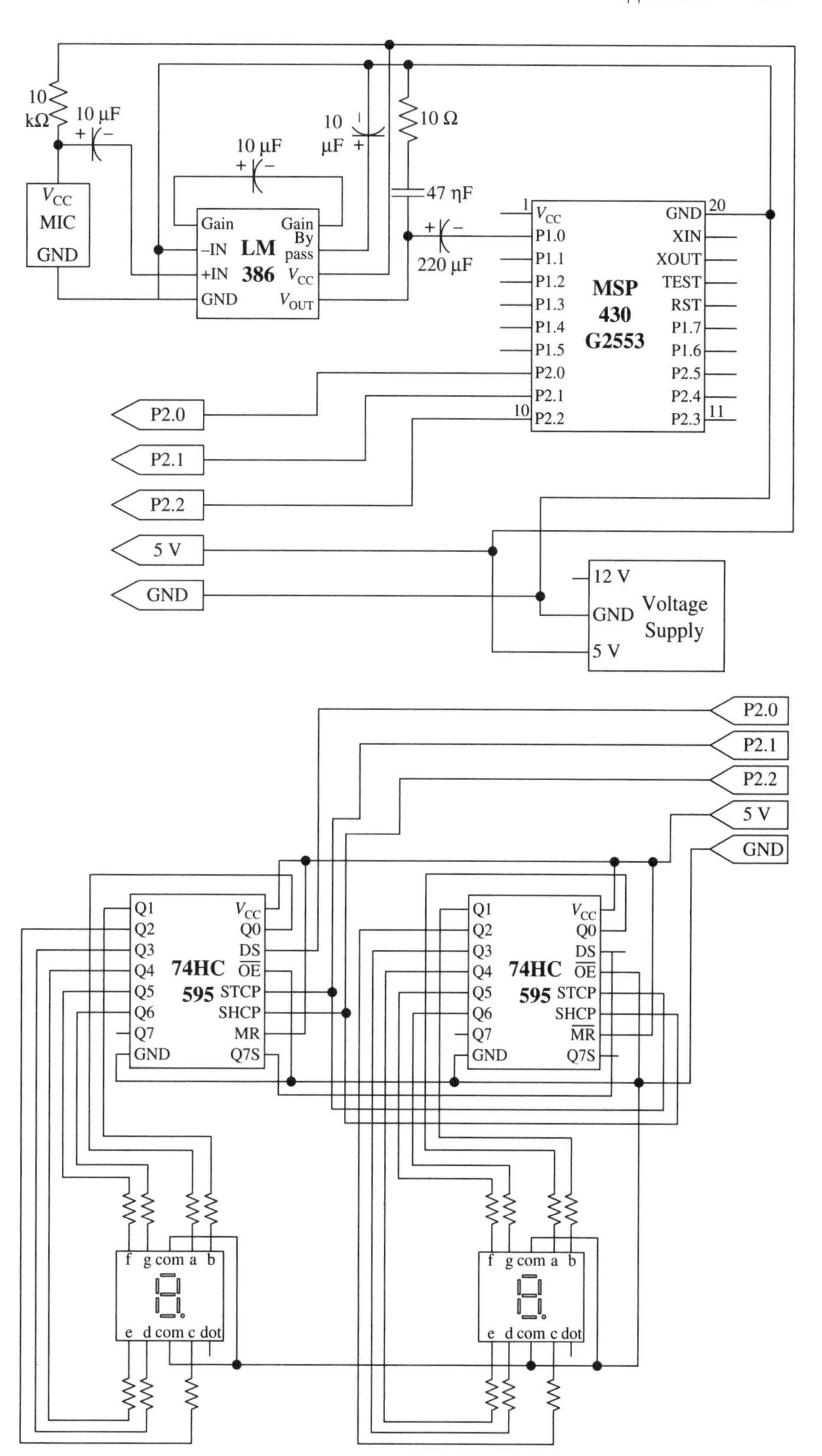

are connected as a cascade. Therefore, be careful about when the digit sequence is sent to them.

14.8 Obstacle-Avoiding Tank

The goal of this application is to learn how to use the ADC and PWM modules on the MSP430 microcontroller. As a real-world application, we will design an obstacle-avoiding tank. In this section, we provide the equipment list, the layout of the circuit, and the procedure.

14.8.1 *Equipment List*

Following is a list of the equipment to be used in this application.

- One 12-V dc adaptor
- One LM7805 voltage regulator
- One 330-ηF capacitor
- One 10-μF electrolytic capacitor
- Two 100-ηF capacitors
- One push button
- One red LED
- One 220-Ω resistor
- Two 12-V dc motors
- One L293D motor driver IC
- One GP2Y0A21YK proximity sensor

14.8.2 *Layout*

The layout of this application is shown in Fig. 14.8. For more information on the voltage supply block, please see Fig. 9.3.

14.8.3 *System Design Specifications*

In this application, we will design an obstacle-avoiding tank with a proximity sensor and two dc motors. GP2Y0A21YK is an analog proximity sensor which gives voltage values between 3.1 and 0.4 V for 10–80 cm distance values (please see device specific datasheet). The tank should check the obstacle distance every 0.2 s. The tank will move forward if there is no obstacle closer than 15 cm. This step is carried out by rotating the dc motors in the same direction with a suitable PWM signal. The tank should move backwards diagonally if there is an obstacle closer than 15 cm. This is achieved by stopping one of the dc motors and rotating the other in the reverse direction with a suitable PWM signal. During this phase, the tank should check the obstacle distance continuously. If there is no obstacle closer than 15 cm, the tank should continue to move in that direction. Then it should return to check the obstacle distance every 0.2 s. The system can also be turned off and on by a push button. This operation must be accomplished by a suitable low-power mode. The red LED should be turned on if the system is working.

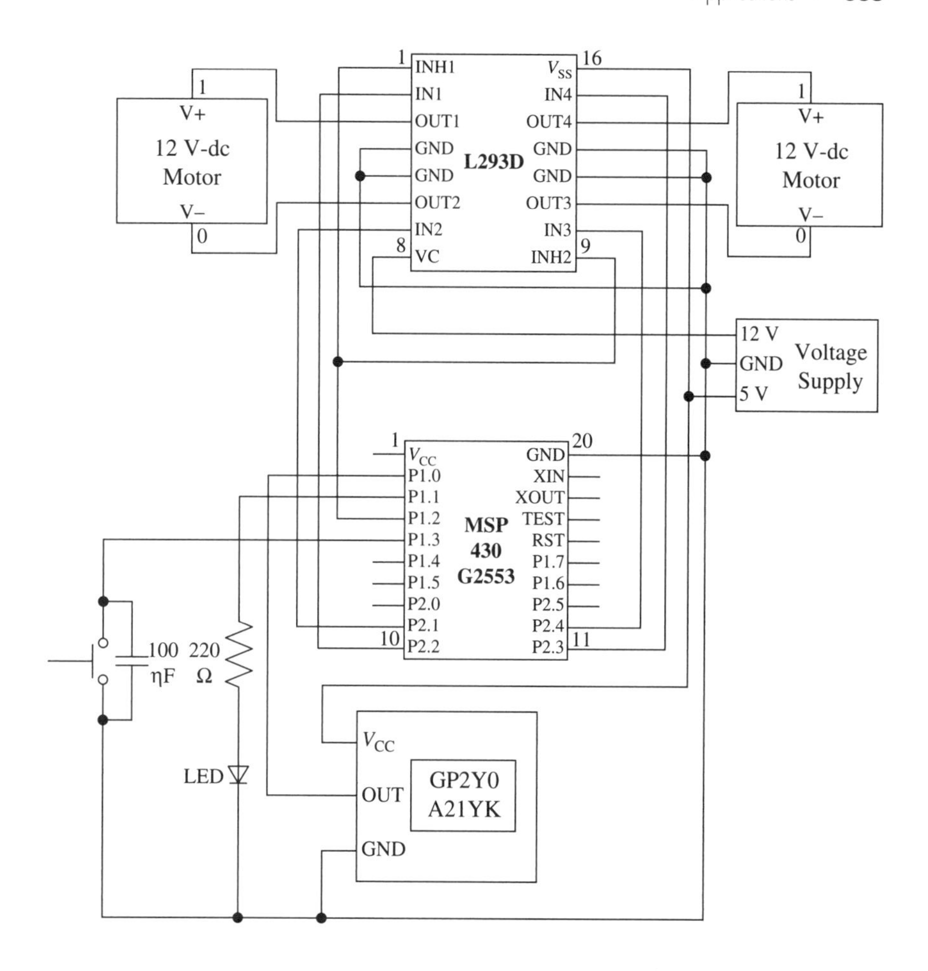

Figure 14.8 Layout of the obstacle avoiding tank application.

14.9 Car Parking Sensor System

The goal of this application is to learn how to use the ADC and PWM modules on the MSP430 microcontroller. As a real-world application, we design a car parking sensor system. In this section, we provide the equipment list, the layout of the circuit, and the procedure.

14.9.1 *Equipment List*

Following is a list of the equipment to be used in this application.

- One 12-V dc adaptor
- One LM7805 voltage regulator
- One 330-ηF capacitor
- One 10-μF electrolytic capacitor
- One 100-ηF capacitor
- One buzzer

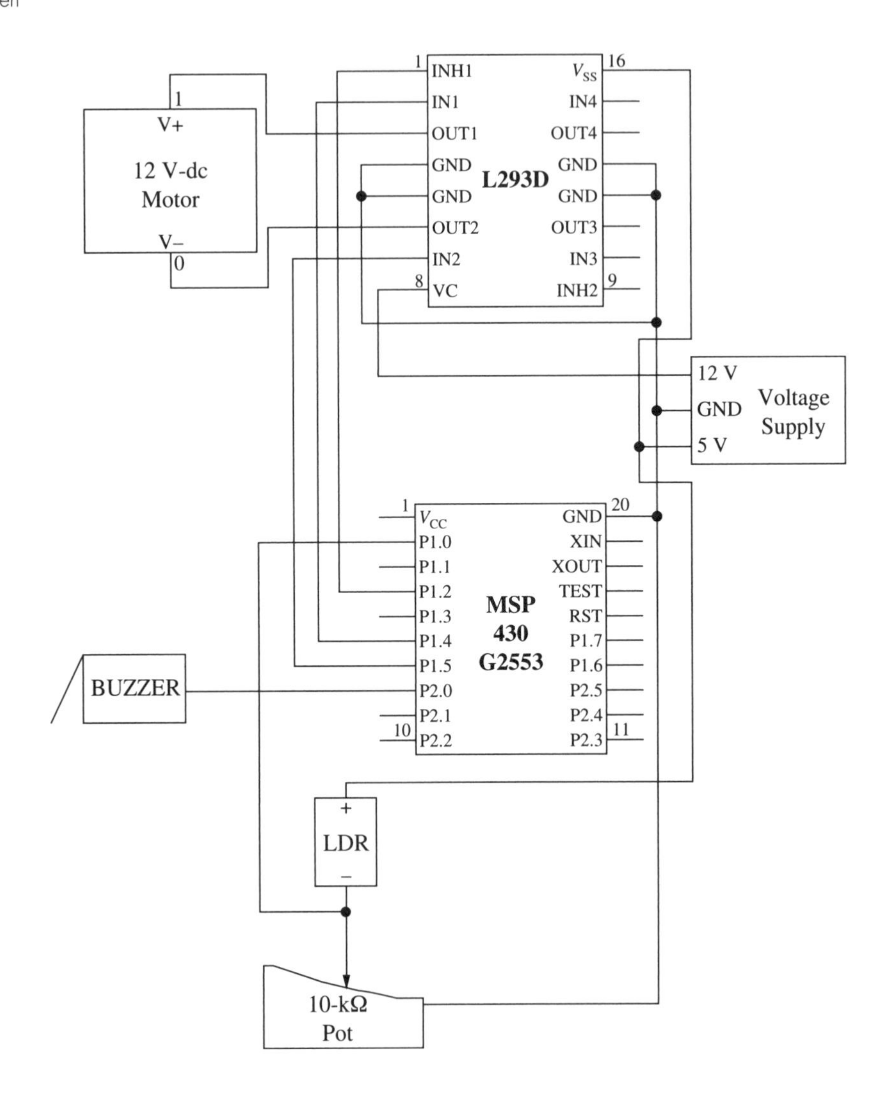

Figure 14.9 Layout of the car parking sensor application.

- One light-dependent resistor (LDR)
- One 12-V dc motor
- One L293D motor driver IC
- One 10-kΩ potentiometer

14.9.2 *Layout*

The layout of this application is shown in Fig. 14.9. For more information on the voltage supply block, please see Fig. 9.3.

14.9.3 *System Design Specifications*

In this application, we will design a car parking sensor system using an LDR (as a sensor), a buzzer (as warning), and a dc motor. First, we should calibrate the ADC input. Therefore, we should first construct the LDR potentiometer pair as shown in

Fig. 14.9. Then the calibration can be done by changing the potentiometer value. In calibration, the ADM10MEM value should be nearly 50h when there is no light on the LDR. The ADC10MEM value may change depending on the light conditions of the medium when there is full light on the LDR. Therefore, a maximum value must be chosen to eliminate this effect. For this application, the maximum value will be 200h. Finally, the ADC10MEM values must be mapped to a `variable` with the following constraints.

- If the ADC10MEM value is less than or equal to 50h, it is mapped to 0h in the `variable`.
- If the ADC10MEM value is greater than or equal to 200h, it is mapped to 200h in the `variable`.
- If the ADC10MEM value is between 50h and 200h, it is mapped to the interval [50h, 200h] in the `variable`.

Then this `variable` is used to generate a PWM by feeding it to the related timer register. The ADC register is 10 bits. The timer register (and the associated variable) is 16 bits. Therefore, the most significant 6 bits of `variable` and the timer register must be zero. Also, if `variable` equals 0h, which means that there is a danger of a crash, the motor should be stopped.

We should construct a lookup table for the buzzer warning part. This table will be a constant array with 32 elements which cover the interval [0h, 50h–200h]. The table will be used as follows:

- The buzzer will stop for the first element of the array (which covers the input between 0h and 50h).
- If the value is greater than 1C0h, it means there is no danger of a crash. Therefore, the buzzer will stop for the elements between 1C0h and 200h.
- The buzzer beeps with different gaps for the rest of the elements. These gaps can be obtained by using the `delay_ms()` function mentioned in Sec. 10.10. Gap values will be kept in another lookup table. A group of elements can share the same gap value. These gaps will be long when the obstacle is away, and they will start to decrease when approaching the obstacle.

14.10 Fire Alarm

The goal of this application is to learn how to use the ADC and timer modules on the MSP430 microcontroller. As a real-world application, we will design a fire alarm system. In this section, we provide the equipment list, the layout of the circuit, and the procedure.

14.10.1 *Equipment List*

Following is a list of the equipment to be used in this application.

- One 12-V dc adaptor
- One LM7805 voltage regulator

- One 330-ηF capacitor
- Two 100-ηF capacitors
- One 10-μF electrolytic capacitor
- One push button
- Two LEDs (green and red)
- Two 220-Ω resistors
- One relay
- One ULN2003
- One buzzer
- One MQ-2 gas sensor
- One 50-kΩ potentiometer

14.10.2 *Layout*

The layout of this application is shown in Fig. 14.10. For more information on the voltage supply block, please see Fig. 9.3.

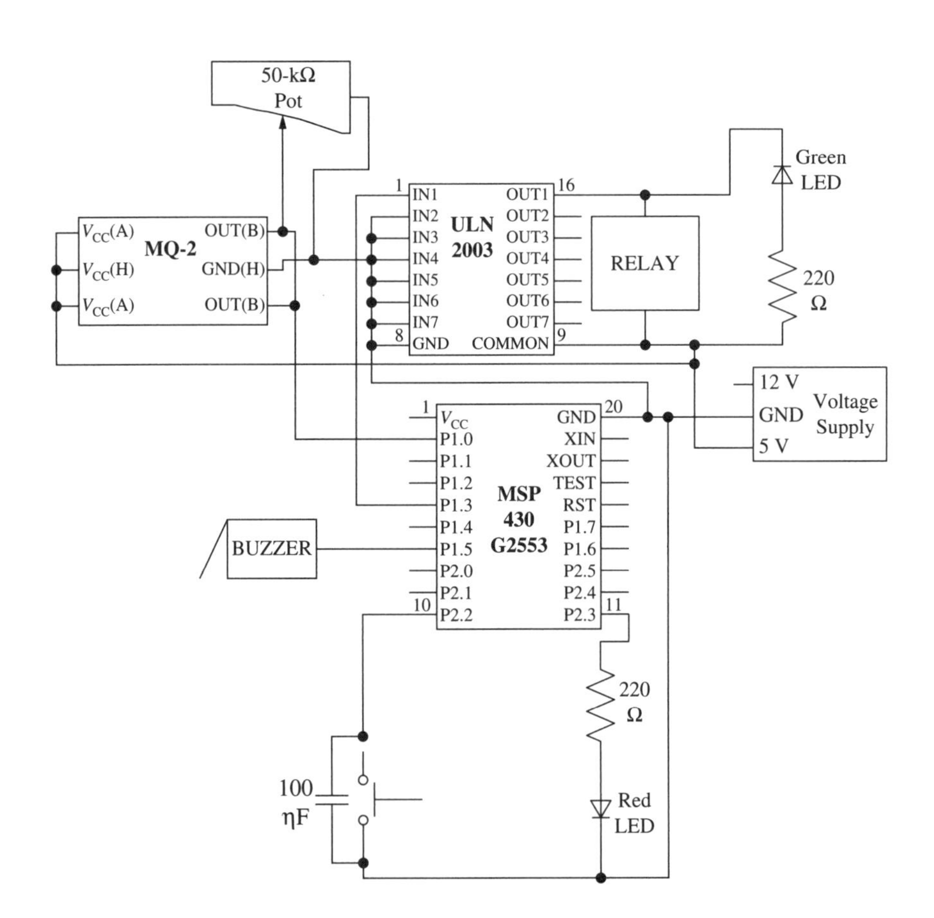

Figure 14.10 Layout of the fire alarm application.

14.10.3 *System Design Specifications*

In this application, we will design a fire alarm system with MQ-2, relay, and a buzzer. MQ-2 is an analog sensor which has high resistance to clean air. This resistance starts to drop when smoke exists in the environment. The sensitivity of the sensor can be arranged with a potentiometer connected between its output and ground pins. The system should check the smoke level every 5 s. In idle times, it should stay in a suitable low-power mode. If the smoke reaches a dangerous level (selected by the user), then the system should turn off the main electricity of the house by disconnecting the relay. The green LED will be turned off to show that the relay is disconnected. Also, the buzzer should start to beep with a 0.5-s interval. The system can be turned off and on by using a push button. This operation must be accomplished by a suitable low-power mode. If the system is working, the red LED will flash every 30 s.

14.11 Wave Generator

The goal of this application is to learn how to use an external DAC IC with the MSP430 microcontroller. As a real-world application, we will design a wave generator system. In this section, we provide the equipment list, the layout of the circuit, and the procedure.

14.11.1 *Equipment List*

Following is a list of the equipment to be used in this application.

- Two 12-V dc adaptors
- One LM7805 voltage regulator
- One 330-ηF capacitor
- Three 100-ηF capacitors
- One 10-μF electrolytic capacitor
- One UA741 OpAmp
- One 10-kΩ potentiometer
- One DAC0808 8-bit D/A converter
- One 220-Ω resistor
- One 5-kΩ resistor
- One 2.5-kΩ resistor
- Five push buttons
- One red LED

14.11.2 *Layout*

The layout of this application is shown in Fig. 14.11. For more information on the voltage supply block, please see Fig. 9.3.

Figure 14.11

Layout of the wave generator application.

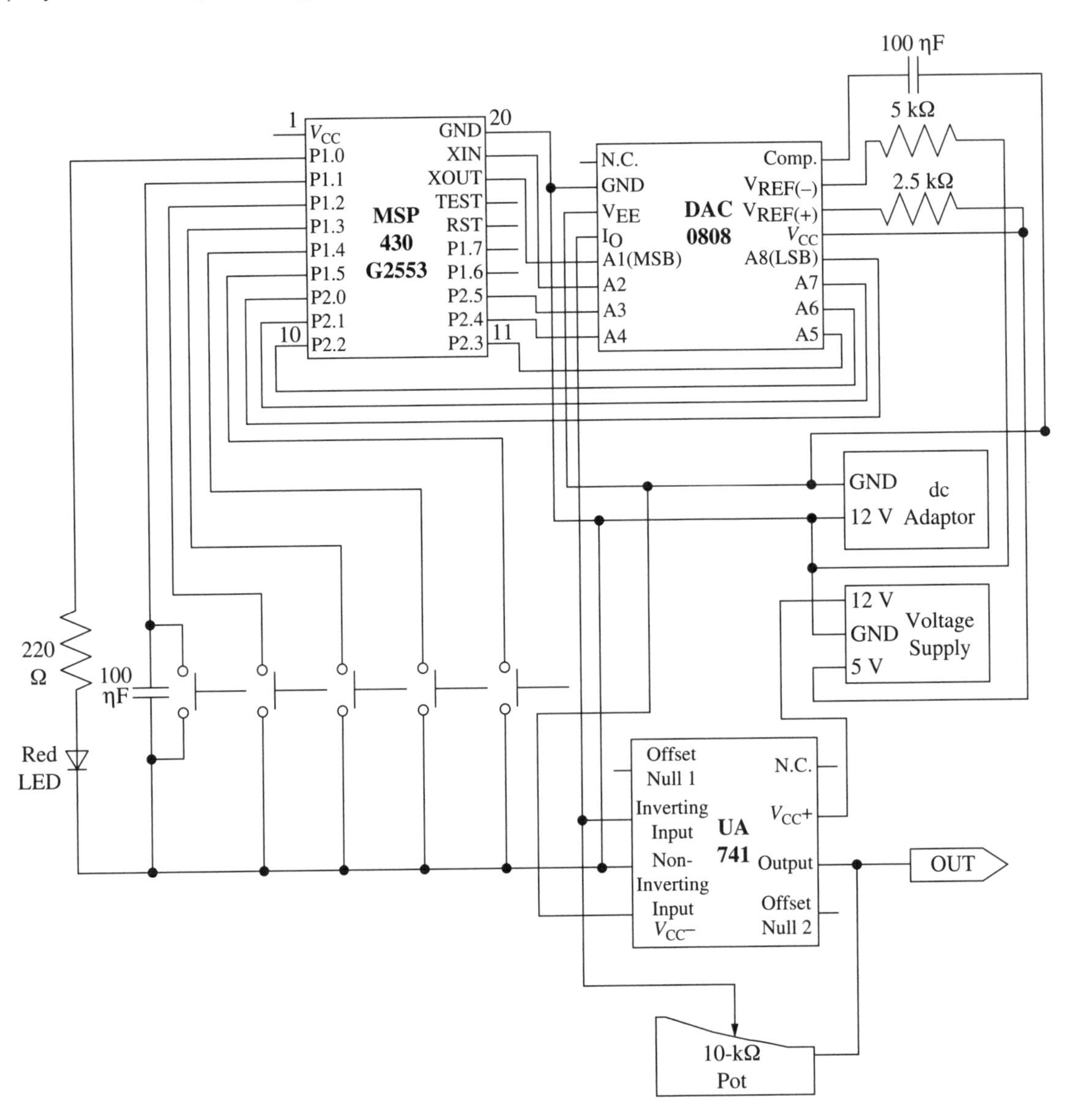

14.11.3 *System Design Specifications*

In this application, we will design a wave generator with an external DAC IC. The wave generator will have four signal options: sine, square, sawtooth, and triangle. These can be selected by four push buttons. Also, there will be another push button to turn off/on the system. This operation must be accomplished by a suitable low-power mode. The red LED will indicate the state of the system. Initially the system will be turned off. Hence, the red LED is turned off. When the system is turned on by pressing the turn on/off button, the red LED will turn on. The desired signal can be fed to output by pressing the related push button. Four lookup tables must be created within the code for four different signals. The user will decide on the properties of these signals (such as the period and the number of samples). DCO must be calibrated to 16 MHz to obtain higher frequencies. The user will have an option to change the amplitude of the generated signal by trimming the 10-kΩ potentiometer connected to the UA741 OpAmp.

14.12 Sports Watch

The goal of this application is to learn how to use the digital communication block on the MSP430 microcontroller. As a real-world application, we will design a sports watch. In this section, we provide the equipment list, the layout of the circuit, and the procedure.

14.12.1 *Equipment List*

Following is a list of the equipment to be used in this application.

- One 12-V dc adaptor
- One LM7805 voltage regulator
- One 330-ηF capacitor
- One 100-ηF capacitor
- One 10-μF electrolytic capacitor
- Two push buttons
- Two 10-kΩ resistors
- One 390-Ω resistor
- One 16×2 character LCD (with a Samsung processor)
- One 10-kΩ potentiometer
- One Hoperf HDPM01 sensor
- One 32-kHz crystal oscillator
- One 3.3-V Zener diode

14.12.2 *Layout*

The layout of this application is shown in Fig. 14.12. For more information on the voltage supply block, please see Fig. 9.3.

In Fig. 14.12, the MCLK pin of the HDPM01 sensor is connected to pin P1.0 of the MSP430 because this pin must be driven by a 32-kHz clock signal. The user can give the ACLK directly from pin P1.0. Also, be careful about the RS and E pins of the LCD. Until now, these pins were connected to pins P1.6 and P1.7 of the MSP430. However, in this application pins P1.6 and P1.7 are used by the I^2C mode. Therefore, pins P1.4 and P1.5 are connected to the RS and E pins of the LCD. The user should change the code sections related to these pins in the LCD header file given in Listing 10.21.

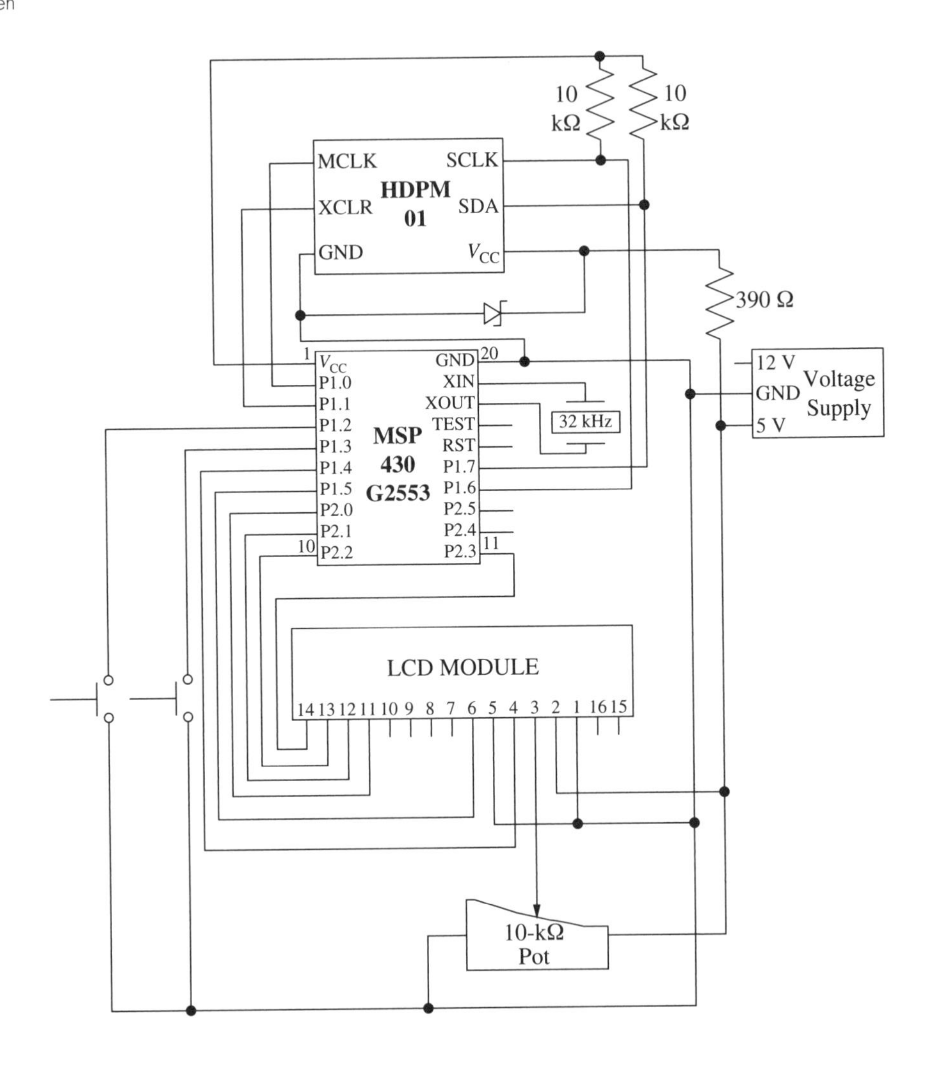

Figure 14.12 Layout of the sports watch application.

14.12.3 *System Design Specifications*

In this application, we will design a sports watch with an HDPM01 sensor and LCD. HDPM01 is a multifunctional sensor which detects temperature, air pressure, altitude, and location. Therefore, the designed sports watch will display these values. Two push buttons will be used to select the *temperature-pressure* and *altitude-compass* screens on the LCD. Initially, the watch should show the time. When one of the push buttons is pressed, the related data will be obtained from the sensor. It will be displayed on the LCD for 10 s. Then the watch will show the time again. An external crystal will be used in this application. Therefore, all timer-based operations should be using the ACLK supplied by the LFXT1 oscillator.

15 Appendix

Chapter Outline

15.1 MSP430 Intrinsic Functions

Listing 15.1 Header file containing MSP430 intrinsic functions.

```
/*---------------------------------------------------------*/
/* in430.h
/* Intrinsic function prototypes and convenience mapping */
/*  macros for migrating code from the IAR platform.     */
/*                                                        */
/*  Ver | dd mmm yyyy | Who  | Description of changes    */
/* =====|=============|======|===========================*/
/*  0.01| 06 Apr 2004 | A.D. | First Prototype            */
/*  0.02| 22 Jun 2004 | A.D. | File reformatted           */
/*                                                        */
/*---------------------------------------------------------*/

#ifndef __IN430_H
#define __IN430_H

/*---------------------------------------------------------*/
/* COMPILER INTRINSIC FUNCTIONS                            */
/*---------------------------------------------------------*/

void _enable_interrupts(void);
void _disable_interrupts(void);
unsigned short _bic_SR_register(unsigned short mask);
unsigned short _bic_SR_register_on_exit(unsigned short mask);
unsigned short _bis_SR_register(unsigned short mask);
unsigned short _bis_SR_register_on_exit(unsigned short mask);
unsigned short _get_SR_register(void);
unsigned short _get_SR_register_on_exit(void);
unsigned short _swap_bytes(unsigned short src);
void _nop(void);
void _never_executed(void);

/*---------------------------------------------------------*/
/* INTRINSIC MAPPING FOR IAR V1.XX                         */
/*---------------------------------------------------------*/

#define _EINT()           _enable_interrupts()
#define _DINT()           _disable_interrupts()
#define _BIC_SR(x)        _bic_SR_register(x)
#define _BIC_SR_IRQ(x)    _bic_SR_register_on_exit(x)
#define _BIS_SR(x)        _bis_SR_register(x)
#define _BIS_SR_IRQ(x)    _bis_SR_register_on_exit(x)
#define _SWAP_BYTES(x)    _swap_bytes(x)
#define _NOP()            _nop()

/*---------------------------------------------------------*/
/* INTRINSIC MAPPING FOR IAR V2.XX/V3.XX                   */
/*---------------------------------------------------------*/

#define __enable_interrupt()              _enable_interrupts()
#define __disable_interrupt()             _disable_interrupts()
#define __bic_SR_register(x)              _bic_SR_register(x)
#define __bic_SR_register_on_exit(x)      _bic_SR_register_on_exit(x)
#define __bis_SR_register(x)              _bis_SR_register(x)
```

```
#define __bis_SR_register_on_exit(x) _bis_SR_register_on_exit(x)
#define __get_SR_register()          _get_SR_register()
#define __get_SR_register_on_exit()  _get_SR_register_on_exit()
#define __swap_bytes(x)              _swap_bytes(x)
#define __no_operation()             _nop()

#endif /* __IN430_H */
```

15.2 MSP430G2553 Header File

Listing 15.2 MSP430G2553 header file.

```
/* ******************************************************
*
* Standard register and bit definitions for the
*Texas Instruments MSP430 microcontroller.
*
* This file supports assembler and C development for
* MSP430G2553 devices.
*
* Texas Instruments, Version 1.0
*
* Rev. 1.0, Setup
*
********************************************************

#ifndef __MSP430G2553
#define __MSP430G2553

#ifdef __cplusplus
extern "C" {
#endif

/*----------------------------------------------------*/
/* PERIPHERAL FILE MAP                                */
/*----------------------------------------------------*/

/* External references resolved by a device-specific linker
command file */
#define SFR_8BIT(address) extern volatile \
  unsigned char address
#define SFR_16BIT(address) extern volatile \
  unsigned int address

/* *************
*  STANDARD BITS
************** */

#define BIT0 (0x0001)
#define BIT1 (0x0002)
#define BIT2 (0x0004)
#define BIT3 (0x0008)
#define BIT4 (0x0010)
#define BIT5 (0x0020)
```

```
#define BIT6 (0x0040)
#define BIT7 (0x0080)
#define BIT8 (0x0100)
#define BIT9 (0x0200)
#define BITA (0x0400)
#define BITB (0x0800)
#define BITC (0x1000)
#define BITD (0x2000)
#define BITE (0x4000)
#define BITF (0x8000)

/* *********************
*  STATUS REGISTER BITS
********************* */

#define C      (0x0001)
#define Z      (0x0002)
#define N      (0x0004)
#define V      (0x0100)
#define GIE    (0x0008)
#define CPUOFF (0x0010)
#define OSCOFF (0x0020)
#define SCG0   (0x0040)
#define SCG1   (0x0080)

/* Low Power Modes coded with Bits 4-7 in SR */

/* Begin #defines for assembler */
#ifdef __ASM_HEADER__
#define LPM0 (CPUOFF)
#define LPM1 (SCG0+CPUOFF)
#define LPM2 (SCG1+CPUOFF)
#define LPM3 (SCG1+SCG0+CPUOFF)
#define LPM4 (SCG1+SCG0+OSCOFF+CPUOFF)
/* End #defines for assembler */

#else /* Begin #defines for C */
#define LPM0_bits (CPUOFF)
#define LPM1_bits (SCG0+CPUOFF)
#define LPM2_bits (SCG1+CPUOFF)
#define LPM3_bits (SCG1+SCG0+CPUOFF)
#define LPM4_bits (SCG1+SCG0+OSCOFF+CPUOFF)

#include "in430.h"

/* Enter Low Power Mode 0 */
#define LPM0      _bis_SR_register(LPM0_bits)
/* Exit Low Power Mode 0 */
#define LPM0_EXIT _bic_SR_register_on_exit(LPM0_bits)
/* Enter Low Power Mode 1 */
#define LPM1      _bis_SR_register(LPM1_bits)
/* Exit Low Power Mode 1 */
#define LPM1_EXIT _bic_SR_register_on_exit(LPM1_bits)
/* Enter Low Power Mode 2 */
#define LPM2      _bis_SR_register(LPM2_bits)
/* Exit Low Power Mode 2 */
```

```
#define LPM2_EXIT _bic_SR_register_on_exit(LPM2_bits)
/* Enter Low Power Mode 3 */
#define LPM3      _bis_SR_register(LPM3_bits)
 /* Exit Low Power Mode 3 */
#define LPM3_EXIT _bic_SR_register_on_exit(LPM3_bits)
/* Enter Low Power Mode 4 */
#define LPM4      _bis_SR_register(LPM4_bits)
/* Exit Low Power Mode 4 */
#define LPM4_EXIT _bic_SR_register_on_exit(LPM4_bits)
#endif /* End #defines for C */

/* *******************
*  PERIPHERAL FILE MAP
******************* */

/* **********************************************
*  SPECIAL FUNCTION REGISTER ADDRESSES + CONTROL BITS
********************************************** */

SFR_8BIT(IE1); /* Interrupt Enable 1 */
#define WDTIE  (0x01) /* Watchdog Interrupt Enable */
#define OFIE   (0x02) /* Osc. Fault  Interrupt Enable */
#define NMIIE  (0x10) /* NMI Interrupt Enable */
#define ACCVIE (0x20)
/* Flash Access Violation Interrupt Enable */

SFR_8BIT(IFG1); /* Interrupt Flag 1 */
#define WDTIFG  (0x01) /* Watchdog Interrupt Flag */
#define OFIFG   (0x02) /* Osc. Fault Interrupt Flag */
#define PORIFG  (0x04) /* Power On Interrupt Flag */
#define RSTIFG  (0x08) /* Reset Interrupt Flag */
#define NMIIFG  (0x10) /* NMI Interrupt Flag */

SFR_8BIT(IE2);   /* Interrupt Enable 2 */
#define UC0IE    IE2
#define UCA0RXIE (0x01)
#define UCA0TXIE (0x02)
#define UCB0RXIE (0x04)
#define UCB0TXIE (0x08)

SFR_8BIT(IFG2);   /* Interrupt Flag 2 */
#define UC0IFG    IFG2
#define UCA0RXIFG (0x01)
#define UCA0TXIFG (0x02)
#define UCB0RXIFG (0x04)
#define UCB0TXIFG (0x08)

/* ******
*  ADC10
****** */

/* Definition to show that Module is available */
#define __MSP430_HAS_ADC10__

SFR_8BIT(ADC10DTC0);
/* ADC10 Data Transfer Control 0 */
SFR_8BIT(ADC10DTC1);
```

```
/* ADC10 Data Transfer Control 1 */
SFR_8BIT(ADC10AE0);
/* ADC10 Analog Enable 0 */

SFR_16BIT(ADC10CTL0); /* ADC10 Control 0 */
SFR_16BIT(ADC10CTL1); /* ADC10 Control 1 */
SFR_16BIT(ADC10MEM);  /* ADC10 Memory */
SFR_16BIT(ADC10SA);
/* ADC10 Data Transfer Start Address */

/* ADC10CTL0 */

#define ADC10SC    (0x001)
/* ADC10 Start Conversion */
#define ENC        (0x002)
/* ADC10 Enable Conversion */
#define ADC10IFG   (0x004)
/* ADC10 Interrupt Flag */
#define ADC10IE    (0x008)
/* ADC10 Interrupt Enable */
#define ADC10ON    (0x010)
/* ADC10 On/Enable */
#define REFON      (0x020)
/* ADC10 Reference on */
#define REF2_5V    (0x040)
/* ADC10 Ref 0:1.5V / 1:2.5V */
#define MSC        (0x080)
/* ADC10 Multiple SampleConversion */
#define REFBURST   (0x100)
/* ADC10 Reference Burst Mode */
#define REFOUT     (0x200)
/* ADC10 Enable output of Ref. */
#define ADC10SR    (0x400)
/* ADC10 Sampling Rate 0:200ksps / 1:50ksps */
#define ADC10SHT0  (0x800)
/* ADC10 Sample Hold Select Bit: 0 */
#define ADC10SHT1  (0x1000)
/* ADC10 Sample Hold Select Bit: 1 */
#define SREF0      (0x2000)
/* ADC10 Reference Select Bit: 0 */
#define SREF1      (0x4000)
/* ADC10 Reference Select Bit: 1 */
#define SREF2      (0x8000)
/* ADC10 Reference Select Bit: 2 */
#define ADC10SHT_0 (0*0x800u)
/* 4 x ADC10CLKs */
#define ADC10SHT_1 (1*0x800u)
/* 8 x ADC10CLKs */
#define ADC10SHT_2 (2*0x800u)
/* 16 x ADC10CLKs */
#define ADC10SHT_3 (3*0x800u)
/* 64 x ADC10CLKs */

#define SREF_0 (0*0x2000u)
/* VR+ = AVCC and VR- = AVSS */
```

```
#define SREF_1 (1*0x2000u)
/* VR+ = VREF+ and VR- = AVSS */
#define SREF_2 (2*0x2000u)
/* VR+ = VEREF+ and VR- = AVSS */
#define SREF_3 (3*0x2000u)
/* VR+ = VEREF+ and VR- = AVSS */
#define SREF_4 (4*0x2000u)
/* VR+ = AVCC and VR- = VREF-/VEREF- */
#define SREF_5 (5*0x2000u)
/* VR+ = VREF+ and VR- = VREF-/VEREF- */
#define SREF_6 (6*0x2000u)
/* VR+ = VEREF+ and VR- = VREF-/VEREF- */
#define SREF_7 (7*0x2000u)
/* VR+ = VEREF+ and VR- = VREF-/VEREF- */

/* ADC10CTL1 */

#define ADC10BUSY  (0x0001)
/* ADC10 BUSY */
#define CONSEQ0    (0x0002)
/* ADC10 Conversion Sequence Select 0 */
#define CONSEQ1    (0x0004)
/* ADC10 Conversion Sequence Select 1 */
#define ADC10SSEL0 (0x0008)
/* ADC10 Clock Source Select Bit: 0 */
#define ADC10SSEL1 (0x0010)
/* ADC10 Clock Source Select Bit: 1 */
#define ADC10DIV0  (0x0020)
/* ADC10 Clock Divider Select Bit: 0 */
#define ADC10DIV1  (0x0040)
/* ADC10 Clock Divider Select Bit: 1 */
#define ADC10DIV2  (0x0080)
/* ADC10 Clock Divider Select Bit: 2 */
#define ISSH       (0x0100)
/* ADC10 Invert Sample Hold Signal */
#define ADC10DF    (0x0200)
/* ADC10 Data Format 0:binary 1:2's complement */
#define SHS0       (0x0400)
/* ADC10 Sample/Hold Source Bit: 0 */
#define SHS1       (0x0800)
/* ADC10 Sample/Hold Source Bit: 1 */
#define INCH0      (0x1000)
/* ADC10 Input Channel Select Bit: 0 */
#define INCH1      (0x2000)
/* ADC10 Input Channel Select Bit: 1 */
#define INCH2      (0x4000)
/* ADC10 Input Channel Select Bit: 2 */
#define INCH3      (0x8000)
/* ADC10 Input Channel Select Bit: 3 */

#define CONSEQ_0 (0*2u)
/* Single channel single conversion */
#define CONSEQ_1 (1*2u)
/* Sequence of channels */
```

```
#define CONSEQ_2 (2*2u)
/* Repeat single channel */
#define CONSEQ_3 (3*2u)
/* Repeat sequence of channels */

#define ADC10SSEL_0 (0*8u) /* ADC10OSC */
#define ADC10SSEL_1 (1*8u) /* ACLK */
#define ADC10SSEL_2 (2*8u) /* MCLK */
#define ADC10SSEL_3 (3*8u) /* SMCLK */

#define ADC10DIV_0 (0*0x20u)
/* ADC10 Clock Divider Select 0 */
#define ADC10DIV_1 (1*0x20u)
/* ADC10 Clock Divider Select 1 */
#define ADC10DIV_2 (2*0x20u)
/* ADC10 Clock Divider Select 2 */
#define ADC10DIV_3 (3*0x20u)
/* ADC10 Clock Divider Select 3 */
#define ADC10DIV_4 (4*0x20u)
/* ADC10 Clock Divider Select 4 */
#define ADC10DIV_5 (5*0x20u)
/* ADC10 Clock Divider Select 5 */
#define ADC10DIV_6 (6*0x20u)
/* ADC10 Clock Divider Select 6 */
#define ADC10DIV_7 (7*0x20u)
/* ADC10 Clock Divider Select 7 */

#define SHS_0 (0*0x400u) /* ADC10SC */
#define SHS_1 (1*0x400u) /* TA3 OUT1 */
#define SHS_2 (2*0x400u) /* TA3 OUT0 */
#define SHS_3 (3*0x400u) /* TA3 OUT2 */

#define INCH_0  (0*0x1000u)  /* Selects Channel 0 */
#define INCH_1  (1*0x1000u)  /* Selects Channel 1 */
#define INCH_2  (2*0x1000u)  /* Selects Channel 2 */
#define INCH_3  (3*0x1000u)  /* Selects Channel 3 */
#define INCH_4  (4*0x1000u)  /* Selects Channel 4 */
#define INCH_5  (5*0x1000u)  /* Selects Channel 5 */
#define INCH_6  (6*0x1000u)  /* Selects Channel 6 */
#define INCH_7  (7*0x1000u)  /* Selects Channel 7 */
#define INCH_8  (8*0x1000u)  /* Selects Channel 8 */
#define INCH_9  (9*0x1000u)  /* Selects Channel 9 */
#define INCH_10 (10*0x1000u) /* Selects Channel 10 */
#define INCH_11 (11*0x1000u) /* Selects Channel 11 */
#define INCH_12 (12*0x1000u) /* Selects Channel 12 */
#define INCH_13 (13*0x1000u) /* Selects Channel 13 */
#define INCH_14 (14*0x1000u) /* Selects Channel 14 */
#define INCH_15 (15*0x1000u) /* Selects Channel 15 */

/* ADC10DTC0 */
#define ADC10FETCH   (0x001)
/* This bit should normally be reset */
#define ADC10B1      (0x002) /* ADC10 block one */
#define ADC10CT      (0x004) /* ADC10 continuous transfer */
```

```
#define ADC10TB      (0x008) /* ADC10 two-block mode */
#define ADC10DISABLE (0x000) /* ADC10DTC1 */

/* ******************
* Basic Clock Module
******************* */

/* Definition to show that Module is available */
#define __MSP430_HAS_BC2__

SFR_8BIT(DCOCTL);  /* DCO Clock Frequency Control */
SFR_8BIT(BCSCTL1); /* Basic Clock System Control 1 */
SFR_8BIT(BCSCTL2); /* Basic Clock System Control 2 */
SFR_8BIT(BCSCTL3); /* Basic Clock System Control 3 */

#define MOD0 (0x01) /* Modulation Bit 0 */
#define MOD1 (0x02) /* Modulation Bit 1 */
#define MOD2 (0x04) /* Modulation Bit 2 */
#define MOD3 (0x08) /* Modulation Bit 3 */
#define MOD4 (0x10) /* Modulation Bit 4 */
#define DCO0 (0x20) /* DCO Select Bit 0 */
#define DCO1 (0x40) /* DCO Select Bit 1 */
#define DCO2 (0x80) /* DCO Select Bit 2 */

#define RSEL0  (0x01) /* Range Select Bit 0 */
#define RSEL1  (0x02) /* Range Select Bit 1 */
#define RSEL2  (0x04) /* Range Select Bit 2 */
#define RSEL3  (0x08) /* Range Select Bit 3 */
#define DIVA0  (0x10) /* ACLK Divider 0 */
#define DIVA1  (0x20) /* ACLK Divider 1 */
#define XTS    (0x40)
/* LFXTCLK 0:Low Freq. / 1: High Freq. */
#define XT2OFF (0x80) /* Enable XT2CLK */

#define DIVA_0 (0x00) /* ACLK Divider 0: /1 */
#define DIVA_1 (0x10) /* ACLK Divider 1: /2 */
#define DIVA_2 (0x20) /* ACLK Divider 2: /4 */
#define DIVA_3 (0x30) /* ACLK Divider 3: /8 */

#define DIVS0 (0x02) /* SMCLK Divider 0 */
#define DIVS1 (0x04) /* SMCLK Divider 1 */
#define SELS  (0x08)
/* SMCLK Source Select 0:DCOCLK / 1:XT2CLK/LFXTCLK */
#define DIVM0 (0x10) /* MCLK Divider 0 */
#define DIVM1 (0x20) /* MCLK Divider 1 */
#define SELM0 (0x40) /* MCLK Source Select 0 */
#define SELM1 (0x80) /* MCLK Source Select 1 */

#define DIVS_0 (0x00) /* SMCLK Divider 0: /1 */
#define DIVS_1 (0x02) /* SMCLK Divider 1: /2 */
#define DIVS_2 (0x04) /* SMCLK Divider 2: /4 */
#define DIVS_3 (0x06) /* SMCLK Divider 3: /8 */

#define DIVM_0 (0x00) /* MCLK Divider 0: /1 */
#define DIVM_1 (0x10) /* MCLK Divider 1: /2 */
#define DIVM_2 (0x20) /* MCLK Divider 2: /4 */
#define DIVM_3 (0x30) /* MCLK Divider 3: /8 */
```

```
#define SELM_0 (0x00) /* MCLK Source Select 0: DCOCLK */
#define SELM_1 (0x40) /* MCLK Source Select 1: DCOCLK */
#define SELM_2 (0x80)
/* MCLK Source Select 2: XT2CLK/LFXTCLK */
#define SELM_3 (0xC0) /* MCLK Source Select 3: LFXTCLK */

#define LFXT1OF (0x01)
/* Low/high Frequency Oscillator Fault Flag */
#define XT2OF   (0x02)
/* High frequency oscillator 2 fault flag */
#define XCAP0   (0x04) /* XIN/XOUT Cap 0 */
#define XCAP1   (0x08) /* XIN/XOUT Cap 1 */
#define LFXT1S0 (0x10) /* Mode 0 for LFXT1 (XTS = 0) */
#define LFXT1S1 (0x20) /* Mode 1 for LFXT1 (XTS = 0) */
#define XT2S0   (0x40) /* Mode 0 for XT2 */
#define XT2S1   (0x80) /* Mode 1 for XT2 */

#define XCAP_0 (0x00) /* XIN/XOUT Cap : 0 pF */
#define XCAP_1 (0x04) /* XIN/XOUT Cap : 6 pF */
#define XCAP_2 (0x08) /* XIN/XOUT Cap : 10 pF */
#define XCAP_3 (0x0C) /* XIN/XOUT Cap : 12.5 pF */

#define LFXT1S_0 (0x00)
/* Mode 0 for LFXT1 : Normal operation */
#define LFXT1S_1 (0x10) /* Mode 1 for LFXT1 : Reserved */
#define LFXT1S_2 (0x20) /* Mode 2 for LFXT1 : VLO */
#define LFXT1S_3 (0x30)
/* Mode 3 for LFXT1 : Digital input signal */

#define XT2S_0 (0x00) /* Mode 0 for XT2 : 0.4 - 1 MHz */
#define XT2S_1 (0x40) /* Mode 1 for XT2 : 1 - 4 MHz */
#define XT2S_2 (0x80) /* Mode 2 for XT2 : 2 - 16 MHz */
#define XT2S_3 (0xC0)
/* Mode 3 for XT2 : Digital input signal */

/* *************
* Comparator A
************** */

/* Definition to show that Module is available */
#define __MSP430_HAS_CAPLUS__

SFR_8BIT(CACTL1); /* Comparator A Control 1 */
SFR_8BIT(CACTL2); /* Comparator A Control 2 */
SFR_8BIT(CAPD);   /* Comparator A Port Disable */

#define CAIFG  (0x01)
/* Comp. A Interrupt Flag */
#define CAIE   (0x02)
/* Comp. A Interrupt Enable */
#define CAIES  (0x04)
/* Comp. A Int. Edge Select: 0:rising / 1:falling */
#define CAON   (0x08)
/* Comp. A enable */
#define CAREF0 (0x10)
/* Comp. A Internal Reference Select 0 */
```

```
#define CAREF1 (0x20)
/* Comp. A Internal Reference Select 1 */
#define CARSEL (0x40)
/* Comp. A Internal Reference Enable */
#define CAEX   (0x80)
/* Comp. A Exchange Inputs */

#define CAREF_0 (0x00)
/* Comp. A Int. Ref. Select 0 : Off */
#define CAREF_1 (0x10)
/* Comp. A Int. Ref. Select 1 : 0.25*Vcc */
#define CAREF_2 (0x20)
/* Comp. A Int. Ref. Select 2 : 0.5*Vcc */
#define CAREF_3 (0x30)
/* Comp. A Int. Ref. Select 3 : Vt*/

#define CAOUT   (0x01) /* Comp. A Output */
#define CAF     (0x02) /* Comp. A Enable Output Filter */
#define P2CA0   (0x04) /* Comp. A +Terminal Multiplexer */
#define P2CA1   (0x08) /* Comp. A -Terminal Multiplexer */
#define P2CA2   (0x10) /* Comp. A -Terminal Multiplexer */
#define P2CA3   (0x20) /* Comp. A -Terminal Multiplexer */
#define P2CA4   (0x40) /* Comp. A +Terminal Multiplexer */
#define CASHORT (0x80) /* Comp. A Short + and - Terminals */

#define CAPD0 (0x01)
/* Comp. A Disable Input Buffer of Port Register .0 */
#define CAPD1 (0x02)
/* Comp. A Disable Input Buffer of Port Register .1 */
#define CAPD2 (0x04)
/* Comp. A Disable Input Buffer of Port Register .2 */
#define CAPD3 (0x08)
/* Comp. A Disable Input Buffer of Port Register .3 */
#define CAPD4 (0x10)
/* Comp. A Disable Input Buffer of Port Register .4 */
#define CAPD5 (0x20)
/* Comp. A Disable Input Buffer of Port Register .5 */
#define CAPD6 (0x40)
/* Comp. A Disable Input Buffer of Port Register .6 */
#define CAPD7 (0x80)
/* Comp. A Disable Input Buffer of Port Register .7 */

/* ************
* Flash Memory
************* */

/* Definition to show that Module is available */
#define __MSP430_HAS_FLASH2__

SFR_16BIT(FCTL1); /* FLASH Control 1 */
SFR_16BIT(FCTL2); /* FLASH Control 2 */
SFR_16BIT(FCTL3); /* FLASH Control 3 */

#define FRKEY (0x9600) /* Flash key returned by read */
#define FWKEY (0xA500) /* Flash key for write */
#define FXKEY (0x3300) /* for use with XOR instruction */
```

```
#define ERASE  (0x0002)
/* Enable bit for Flash segment erase */
#define MERAS  (0x0004)
/* Enable bit for Flash mass erase */
#define WRT    (0x0040)
/* Enable bit for Flash write */
#define BLKWRT (0x0080)
/* Enable bit for Flash segment write */
/* old definition */
#define SEGWRT (0x0080)
/* Enable bit for Flash segment write */

/* Divide Flash clock by 1 to 64 using FN0 to FN5
according to: */
#define FN0 (0x0001)
/*  32*FN5 + 16*FN4 + 8*FN3 + 4*FN2 + 2*FN1 + FN0 + 1 */
#define FN1    (0x0002)
#ifndef FN2
#define FN2    (0x0004)
#endif
#ifndef FN3
#define FN3    (0x0008)
#endif
#ifndef FN4
#define FN4    (0x0010)
#endif
#define FN5    (0x0020)

/* Flash clock select 0 */
/* to distinguish from USART SSELx */

#define FSSEL0 (0x0040)
#define FSSEL1 (0x0080) /* Flash clock select 1 */

#define FSSEL_0 (0x0000) /* Flash clock select: 0 - ACLK */
#define FSSEL_1 (0x0040) /* Flash clock select: 1 - MCLK */
#define FSSEL_2 (0x0080) /* Flash clock select: 2 - SMCLK */
#define FSSEL_3 (0x00C0) /* Flash clock select: 3 - SMCLK */

#define BUSY    (0x0001) /* Flash busy: 1 */
#define KEYV    (0x0002) /* Flash Key violation flag */
#define ACCVIFG (0x0004) /* Flash Access violation flag */
#define WAIT    (0x0008) /* Wait flag for segment write */
#define LOCK    (0x0010)
/* Lock bit: 1 - Flash is locked (read only) */
#define EMEX    (0x0020) /* Flash Emergency Exit */
/* Segment A Lock bit: read = 1 - Segment is
locked (read only) */
#define LOCKA   (0x0040)
#define FAIL    (0x0080) /* Last Program or Erase failed */

/* ************************************************
*  DIGITAL I/O Port1/2 Pull up / Pull down Resistors
************************************************ */
```

```
/* Definition to show that Module is available */
#define __MSP430_HAS_PORT1_R__
/* Definition to show that Module is available */
#define __MSP430_HAS_PORT2_R__

SFR_8BIT(P1IN);    /* Port 1 Input */
SFR_8BIT(P1OUT);   /* Port 1 Output */
SFR_8BIT(P1DIR);   /* Port 1 Direction */
SFR_8BIT(P1IFG);   /* Port 1 Interrupt Flag */
SFR_8BIT(P1IES);   /* Port 1 Interrupt Edge Select */
SFR_8BIT(P1IE);    /* Port 1 Interrupt Enable */
SFR_8BIT(P1SEL);   /* Port 1 Selection */
SFR_8BIT(P1SEL2);  /* Port 1 Selection 2 */
SFR_8BIT(P1REN);   /* Port 1 Resistor Enable */

SFR_8BIT(P2IN);    /* Port 2 Input */
SFR_8BIT(P2OUT);   /* Port 2 Output */
SFR_8BIT(P2DIR);   /* Port 2 Direction */
SFR_8BIT(P2IFG);   /* Port 2 Interrupt Flag */
SFR_8BIT(P2IES);   /* Port 2 Interrupt Edge Select */
SFR_8BIT(P2IE);    /* Port 2 Interrupt Enable */
SFR_8BIT(P2SEL);   /* Port 2 Selection */
SFR_8BIT(P2SEL2);  /* Port 2 Selection 2 */
SFR_8BIT(P2REN);   /* Port 2 Resistor Enable */

/* ************************************************
*  DIGITAL I/O Port3 Pull up / Pull down Resistors
************************************************* */

/* Definition to show that Module is available */
#define __MSP430_HAS_PORT3_R__

SFR_8BIT(P3IN);    /* Port 3 Input */
SFR_8BIT(P3OUT);   /* Port 3 Output */
SFR_8BIT(P3DIR);   /* Port 3 Direction */
SFR_8BIT(P3SEL);   /* Port 3 Selection */
SFR_8BIT(P3SEL2);  /* Port 3 Selection 2 */
SFR_8BIT(P3REN);   /* Port 3 Resistor Enable */

/* **********
*  Timer0_A3
*********** */

/* Definition to show that Module is available */
#define __MSP430_HAS_TA3__

SFR_16BIT(TA0IV);     /* Timer0_A3 Interrupt Vector Word */
SFR_16BIT(TA0CTL);    /* Timer0_A3 Control */
SFR_16BIT(TA0CCTL0); /* Timer0_A3 Capture/Compare Control 0 */
SFR_16BIT(TA0CCTL1); /* Timer0_A3 Capture/Compare Control 1 */
SFR_16BIT(TA0CCTL2); /* Timer0_A3 Capture/Compare Control 2 */
SFR_16BIT(TA0R);      /* Timer0_A3 */
SFR_16BIT(TA0CCR0);   /* Timer0_A3 Capture/Compare 0 */
SFR_16BIT(TA0CCR1);   /* Timer0_A3 Capture/Compare 1 */
SFR_16BIT(TA0CCR2);   /* Timer0_A3 Capture/Compare 2 */
```

```
/* Alternate register names */
#define TAIV      TA0IV
/* Timer A Interrupt Vector Word */
#define TACTL     TA0CTL
/* Timer A Control */
#define TACCTL0   TA0CCTL0
/* Timer A Capture/Compare Control 0 */
#define TACCTL1   TA0CCTL1
/* Timer A Capture/Compare Control 1 */
#define TACCTL2   TA0CCTL2
/* Timer A Capture/Compare Control 2 */
#define TAR       TA0R
/* Timer A */
#define TACCR0    TA0CCR0
/* Timer A Capture/Compare 0 */
#define TACCR1    TA0CCR1
/* Timer A Capture/Compare 1 */
#define TACCR2    TA0CCR2
/* Timer A Capture/Compare 2 */
#define TAIV_     TA0IV_
/* Timer A Interrupt Vector Word */
#define TACTL_    TA0CTL_
/* Timer A Control */
#define TACCTL0_  TA0CCTL0_
/* Timer A Capture/Compare Control 0 */
#define TACCTL1_  TA0CCTL1_
/* Timer A Capture/Compare Control 1 */
#define TACCTL2_  TA0CCTL2_
/* Timer A Capture/Compare Control 2 */
#define TAR_      TA0R_
/* Timer A */
#define TACCR0_   TA0CCR0_
/* Timer A Capture/Compare 0 */
#define TACCR1_   TA0CCR1_
/* Timer A Capture/Compare 1 */
#define TACCR2_   TA0CCR2_
/* Timer A Capture/Compare 2 */

/* Alternate register names 2 */

#define CCTL0   TACCTL0
/* Timer A Capture/Compare Control 0 */
#define CCTL1   TACCTL1
/* Timer A Capture/Compare Control 1 */
#define CCTL2   TACCTL2
/* Timer A Capture/Compare Control 2 */
#define CCR0    TACCR0
/* Timer A Capture/Compare 0 */
#define CCR1    TACCR1
/* Timer A Capture/Compare 1 */
#define CCR2    TACCR2
/* Timer A Capture/Compare 2 */
#define CCTL0_ TACCTL0_
/* Timer A Capture/Compare Control 0 */
```

```
#define CCTL1_ TACCTL1_
/* Timer A Capture/Compare Control 1 */
#define CCTL2_ TACCTL2_
/* Timer A Capture/Compare Control 2 */
#define CCR0_  TACCR0_
/* Timer A Capture/Compare 0 */
#define CCR1_  TACCR1_
/* Timer A Capture/Compare 1 */
#define CCR2_  TACCR2_
/* Timer A Capture/Compare 2 */

#define TASSEL1 (0x0200)
/* Timer A clock source select 0 */
#define TASSEL0 (0x0100)
/* Timer A clock source select 1 */
#define ID1     (0x0080)
/* Timer A clock input divider 1 */
#define ID0     (0x0040)
/* Timer A clock input divider 0 */
#define MC1     (0x0020)
/* Timer A mode control 1 */
#define MC0     (0x0010)
/* Timer A mode control 0 */
#define TACLR   (0x0004)
/* Timer A counter clear */
#define TAIE    (0x0002)
/* Timer A counter interrupt enable */
#define TAIFG   (0x0001)
/* Timer A counter interrupt flag */

#define MC_0     (0*0x10u)
/* Timer A mode control: 0 -  Stop */
#define MC_1     (1*0x10u)
/* Timer A mode control: 1 -  Up to CCR0 */
#define MC_2     (2*0x10u)
/* Timer A mode control: 2 -  Continous up */
#define MC_3     (3*0x10u)
/* Timer A mode control: 3 -  Up/Down */
#define ID_0     (0*0x40u)
/* Timer A input divider: 0 -  /1 */
#define ID_1     (1*0x40u)
/* Timer A input divider: 1 -  /2 */
#define ID_2     (2*0x40u)
/* Timer A input divider: 2 -  /4 */
#define ID_3     (3*0x40u)
/* Timer A input divider: 3 -  /8 */
#define TASSEL_0 (0*0x100u)
/* Timer A clock source select: 0 -  TACLK */
#define TASSEL_1 (1*0x100u)
/* Timer A clock source select: 1 -  ACLK  */
#define TASSEL_2 (2*0x100u)
/* Timer A clock source select: 2 -  SMCLK */
#define TASSEL_3 (3*0x100u)
/* Timer A clock source select: 3 -  INCLK */
```

```
#define CM1     (0x8000)
/* Capture mode 1 */
#define CM0     (0x4000)
/* Capture mode 0 */
#define CCIS1   (0x2000)
/* Capture input select 1 */
#define CCIS0   (0x1000)
/* Capture input select 0 */
#define SCS     (0x0800)
/* Capture sychronize */
#define SCCI    (0x0400)
/* Latched capture signal (read) */
#define CAP     (0x0100)
/* Capture mode: 1 /Compare mode : 0 */
#define OUTMOD2 (0x0080)
/* Output mode 2 */
#define OUTMOD1 (0x0040)
/* Output mode 1 */
#define OUTMOD0 (0x0020)
/* Output mode 0 */
#define CCIE    (0x0010)
/* Capture/compare interrupt enable */
#define CCI     (0x0008)
/* Capture input signal (read) */
#define OUT     (0x0004)
/* PWM Output signal if output mode 0 */
#define COV     (0x0002)
/* Capture/compare overflow flag */
#define CCIFG   (0x0001)
/* Capture/compare interrupt flag */

#define OUTMOD_0 (0*0x20u)
/* PWM output mode: 0 -  output only */
#define OUTMOD_1 (1*0x20u)
/* PWM output mode: 1 -  set */
#define OUTMOD_2 (2*0x20u)
/* PWM output mode: 2 -  PWM toggle/reset */
#define OUTMOD_3 (3*0x20u)
/* PWM output mode: 3 -  PWM set/reset */
#define OUTMOD_4 (4*0x20u)
/* PWM output mode: 4 -  toggle */
#define OUTMOD_5 (5*0x20u)
/* PWM output mode: 5 -  Reset */
#define OUTMOD_6 (6*0x20u)
/* PWM output mode: 6 -  PWM toggle/set */
#define OUTMOD_7 (7*0x20u)
/* PWM output mode: 7 -  PWM reset/set */
#define CCIS_0   (0*0x1000u)
/* Capture input select: 0 -  CCIxA */
#define CCIS_1   (1*0x1000u)
/* Capture input select: 1 -  CCIxB */
```

```
#define CCIS_2   (2*0x1000u)
/* Capture input select: 2 -  GND */
#define CCIS_3   (3*0x1000u)
/* Capture input select: 3 -  Vcc */
#define CM_0     (0*0x4000u)
/* Capture mode: 0 -  disabled */
#define CM_1     (1*0x4000u)
/* Capture mode: 1 -  pos. edge */
#define CM_2     (2*0x4000u)
/* Capture mode: 1 -  neg. edge */
#define CM_3     (3*0x4000u)
/* Capture mode: 1 -  both edges */

/* T0_A3IV Definitions */
#define TA0IV_NONE   (0x0000) /* No Interrupt pending */
#define TA0IV_TACCR1 (0x0002) /* TA0CCR1_CCIFG */
#define TA0IV_TACCR2 (0x0004) /* TA0CCR2_CCIFG */
#define TA0IV_6      (0x0006) /* Reserved */
#define TA0IV_8      (0x0008) /* Reserved */
#define TA0IV_TAIFG  (0x000A) /* TA0IFG */

/* **********
*  Timer1_A3
*********** */

/* Definition to show that Module is available */
#define __MSP430_HAS_T1A3__

SFR_16BIT(TA1IV);
/* Timer1_A3 Interrupt Vector Word */
SFR_16BIT(TA1CTL);
/* Timer1_A3 Control */
SFR_16BIT(TA1CCTL0);
/* Timer1_A3 Capture/Compare Control 0 */
SFR_16BIT(TA1CCTL1);
/* Timer1_A3 Capture/Compare Control 1 */
SFR_16BIT(TA1CCTL2);
/* Timer1_A3 Capture/Compare Control 2 */
SFR_16BIT(TA1R);
/* Timer1_A3 */
SFR_16BIT(TA1CCR0);
/* Timer1_A3 Capture/Compare 0 */
SFR_16BIT(TA1CCR1);
/* Timer1_A3 Capture/Compare 1 */
SFR_16BIT(TA1CCR2);
/* Timer1_A3 Capture/Compare 2 */

/* Bits are already defined within the Timer0_Ax */

/* T1_A3IV Definitions */
#define TA1IV_NONE   (0x0000) /* No Interrupt pending */
#define TA1IV_TACCR1 (0x0002) /* TA1CCR1_CCIFG */
#define TA1IV_TACCR2 (0x0004) /* TA1CCR2_CCIFG */
#define TA1IV_TAIFG  (0x000A) /* TA1IFG */
```

```
/* ****
*  USCI
***** */

/* Definition to show that Module is available */
#define __MSP430_HAS_USCI__

SFR_8BIT(UCA0CTL0);    /* USCI A0 Control Register 0 */
SFR_8BIT(UCA0CTL1);    /* USCI A0 Control Register 1 */
SFR_8BIT(UCA0BR0);     /* USCI A0 Baud Rate 0 */
SFR_8BIT(UCA0BR1);     /* USCI A0 Baud Rate 1 */
SFR_8BIT(UCA0MCTL);    /* USCI A0 Modulation Control */
SFR_8BIT(UCA0STAT);    /* USCI A0 Status Register */
SFR_8BIT(UCA0RXBUF);   /* USCI A0 Receive Buffer */
SFR_8BIT(UCA0TXBUF);   /* USCI A0 Transmit Buffer */
SFR_8BIT(UCA0ABCTL);   /* USCI A0 LIN Control */
SFR_8BIT(UCA0IRTCTL); /* USCI A0 IrDA Transmit Control */
SFR_8BIT(UCA0IRRCTL); /* USCI A0 IrDA Receive Control */

SFR_8BIT(UCB0CTL0);    /* USCI B0 Control Register 0 */
SFR_8BIT(UCB0CTL1);    /* USCI B0 Control Register 1 */
SFR_8BIT(UCB0BR0);     /* USCI B0 Baud Rate 0 */
SFR_8BIT(UCB0BR1);     /* USCI B0 Baud Rate 1 */
SFR_8BIT(UCB0I2CIE);
/* USCI B0 I2C Interrupt Enable Register */
SFR_8BIT(UCB0STAT);    /* USCI B0 Status Register */
SFR_8BIT(UCB0RXBUF);   /* USCI B0 Receive Buffer */
SFR_8BIT(UCB0TXBUF);   /* USCI B0 Transmit Buffer */
SFR_16BIT(UCB0I2COA); /* USCI B0 I2C Own Address */
SFR_16BIT(UCB0I2CSA); /* USCI B0 I2C Slave Address */

// UART-Mode Bits
#define UCPEN   (0x80)
/* Async. Mode: Parity enable */
#define UCPAR   (0x40)
/* Async. Mode: Parity     0:odd / 1:even */
#define UCMSB   (0x20)
/* Async. Mode: MSB first  0:LSB / 1:MSB */
#define UC7BIT  (0x10)
/* Async. Mode: Data Bits  0:8-bits / 1:7-bits */
#define UCSPB   (0x08)
/* Async. Mode: Stop Bits  0:one / 1: two */
#define UCMODE1 (0x04)
/* Async. Mode: USCI Mode 1 */
#define UCMODE0 (0x02)
/* Async. Mode: USCI Mode 0 */
#define UCSYNC  (0x01)
/* Sync-Mode  0:UART-Mode / 1:SPI-Mode */

// SPI-Mode Bits
#define UCCKPH (0x80) /* Sync. Mode: Clock Phase */
#define UCCKPL (0x40) /* Sync. Mode: Clock Polarity */
#define UCMST  (0x08) /* Sync. Mode: Master Select */

// I2C-Mode Bits
#define UCA10    (0x80) /* 10-bit Address Mode */
```

```
#define UCSLA10  (0x40) /* 10-bit Slave Address Mode */
#define UCMM     (0x20) /* Multi-Master Environment */
//#define res    (0x10) /* reserved */
#define UCMODE_0 (0x00) /* Sync. Mode: USCI Mode: 0 */
#define UCMODE_1 (0x02) /* Sync. Mode: USCI Mode: 1 */
#define UCMODE_2 (0x04) /* Sync. Mode: USCI Mode: 2 */
#define UCMODE_3 (0x06) /* Sync. Mode: USCI Mode: 3 */

// UART-Mode Bits
#define UCSSEL1  (0x80)
/* USCI 0 Clock Source Select 1 */
#define UCSSEL0  (0x40)
/* USCI 0 Clock Source Select 0 */
#define UCRXEIE  (0x20)
/* RX Error interrupt enable */
#define UCBRKIE  (0x10)
/* Break interrupt enable */
#define UCDORM   (0x08)
/* Dormant (Sleep) Mode */
#define UCTXADDR (0x04)
/* Send next Data as Address */
#define UCTXBRK  (0x02)
/* Send next Data as Break */
#define UCSWRST  (0x01)
/* USCI Software Reset */

// SPI-Mode Bits
//#define res (0x20) /* reserved */
//#define res (0x10) /* reserved */
//#define res (0x08) /* reserved */
//#define res (0x04) /* reserved */
//#define res (0x02) /* reserved */

// I2C-Mode Bits
//#define res    (0x20) /* reserved */
#define UCTR     (0x10)
/* Transmit/Receive Select/Flag */
#define UCTXNACK (0x08)
/* Transmit NACK */
#define UCTXSTP  (0x04)
/* Transmit STOP */
#define UCTXSTT  (0x02)
/* Transmit START */
#define UCSSEL_0 (0x00)
/* USCI 0 Clock Source: 0 */
#define UCSSEL_1 (0x40)
/* USCI 0 Clock Source: 1 */
#define UCSSEL_2 (0x80)
/* USCI 0 Clock Source: 2 */
#define UCSSEL_3 (0xC0)
/* USCI 0 Clock Source: 3 */

#define UCBRF3 (0x80)
/* USCI First Stage Modulation Select 3 */
```

```
#define UCBRF2 (0x40)
/* USCI First Stage Modulation Select 2 */
#define UCBRF1 (0x20)
/* USCI First Stage Modulation Select 1 */
#define UCBRF0 (0x10)
/* USCI First Stage Modulation Select 0 */
#define UCBRS2 (0x08)
/* USCI Second Stage Modulation Select 2 */
#define UCBRS1 (0x04)
/* USCI Second Stage Modulation Select 1 */
#define UCBRS0 (0x02)
/* USCI Second Stage Modulation Select 0 */
#define UCOS16 (0x01)
/* USCI 16-times Oversampling enable */

#define UCBRF_0  (0x00)
/* USCI First Stage Modulation: 0 */
#define UCBRF_1  (0x10)
/* USCI First Stage Modulation: 1 */
#define UCBRF_2  (0x20)
/* USCI First Stage Modulation: 2 */
#define UCBRF_3  (0x30)
/* USCI First Stage Modulation: 3 */
#define UCBRF_4  (0x40)
/* USCI First Stage Modulation: 4 */
#define UCBRF_5  (0x50)
/* USCI First Stage Modulation: 5 */
#define UCBRF_6  (0x60)
/* USCI First Stage Modulation: 6 */
#define UCBRF_7  (0x70)
/* USCI First Stage Modulation: 7 */
#define UCBRF_8  (0x80)
/* USCI First Stage Modulation: 8 */
#define UCBRF_9  (0x90)
/* USCI First Stage Modulation: 9 */
#define UCBRF_10 (0xA0)
/* USCI First Stage Modulation: A */
#define UCBRF_11 (0xB0)
/* USCI First Stage Modulation: B */
#define UCBRF_12 (0xC0)
/* USCI First Stage Modulation: C */
#define UCBRF_13 (0xD0)
/* USCI First Stage Modulation: D */
#define UCBRF_14 (0xE0)
/* USCI First Stage Modulation: E */
#define UCBRF_15 (0xF0)
/* USCI First Stage Modulation: F */

#define UCBRS_0 (0x00)
/* USCI Second Stage Modulation: 0 */
#define UCBRS_1 (0x02)
/* USCI Second Stage Modulation: 1 */
#define UCBRS_2 (0x04)
/* USCI Second Stage Modulation: 2 */
```

```
#define UCBRS_3 (0x06)
/* USCI Second Stage Modulation: 3 */
#define UCBRS_4 (0x08)
/* USCI Second Stage Modulation: 4 */
#define UCBRS_5 (0x0A)
/* USCI Second Stage Modulation: 5 */
#define UCBRS_6 (0x0C)
/* USCI Second Stage Modulation: 6 */
#define UCBRS_7 (0x0E)
/* USCI Second Stage Modulation: 7 */

#define UCLISTEN (0x80) /* USCI Listen mode */
#define UCFE     (0x40) /* USCI Frame Error Flag */
#define UCOE     (0x20) /* USCI Overrun Error Flag */
#define UCPE     (0x10) /* USCI Parity Error Flag */
#define UCBRK    (0x08) /* USCI Break received */
#define UCRXERR  (0x04) /* USCI RX Error Flag */
#define UCADDR   (0x02) /* USCI Address received Flag */
#define UCBUSY   (0x01) /* USCI Busy Flag */
#define UCIDLE   (0x02) /* USCI Idle line detected Flag */

//#define res    (0x80) /* reserved */
//#define res    (0x40) /* reserved */
//#define res    (0x20) /* reserved */
//#define res    (0x10) /* reserved */
#define UCNACKIE (0x08)
/* NACK Condition interrupt enable */
#define UCSTPIE  (0x04)
/* STOP Condition interrupt enable */
#define UCSTTIE  (0x02)
/* START Condition interrupt enable */
#define UCALIE   (0x01)
/* Arbitration Lost interrupt enable */

#define UCSCLLOW  (0x40)
/* SCL low */
#define UCGC      (0x20)
/* General Call address received Flag */
#define UCBBUSY   (0x10)
/* Bus Busy Flag */
#define UCNACKIFG (0x08)
/* NAK Condition interrupt Flag */
#define UCSTPIFG  (0x04)
/* STOP Condition interrupt Flag */
#define UCSTTIFG  (0x02)
/* START Condition interrupt Flag */
#define UCALIFG   (0x01)
/* Arbitration Lost interrupt Flag */

#define UCIRTXPL5 (0x80)
/* IRDA Transmit Pulse Length 5 */
#define UCIRTXPL4 (0x40)
/* IRDA Transmit Pulse Length 4 */
#define UCIRTXPL3 (0x20)
/* IRDA Transmit Pulse Length 3 */
```

```
#define UCIRTXPL2 (0x10)
/* IRDA Transmit Pulse Length 2 */
#define UCIRTXPL1 (0x08)
/* IRDA Transmit Pulse Length 1 */
#define UCIRTXPL0 (0x04)
/* IRDA Transmit Pulse Length 0 */
#define UCIRTXCLK (0x02)
/* IRDA Transmit Pulse Clock Select */
#define UCIREN    (0x01)
/* IRDA Encoder/Decoder enable */

#define UCIRRXFL5 (0x80)
/* IRDA Receive Filter Length 5 */
#define UCIRRXFL4 (0x40)
/* IRDA Receive Filter Length 4 */
#define UCIRRXFL3 (0x20)
/* IRDA Receive Filter Length 3 */
#define UCIRRXFL2 (0x10)
/* IRDA Receive Filter Length 2 */
#define UCIRRXFL1 (0x08)
/* IRDA Receive Filter Length 1 */
#define UCIRRXFL0 (0x04)
/* IRDA Receive Filter Length 0 */
#define UCIRRXPL  (0x02)
/* IRDA Receive Input Polarity */
#define UCIRRXFE  (0x01)
 /* IRDA Receive Filter enable */

//#define res    (0x80) /* reserved */
//#define res    (0x40) /* reserved */
#define UCDELIM1 (0x20)
/* Break Sync Delimiter 1 */
#define UCDELIM0 (0x10)
/* Break Sync Delimiter 0 */
#define UCSTOE   (0x08)
/* Sync-Field Timeout error */
#define UCBTOE   (0x04)
/* Break Timeout error */
//#define res    (0x02) /* reserved */
#define UCABDEN  (0x01)
/* Auto Baud Rate detect enable */

#define UCGCEN (0x8000) /* I2C General Call enable */
#define UCOA9  (0x0200) /* I2C Own Address 9 */
#define UCOA8  (0x0100) /* I2C Own Address 8 */
#define UCOA7  (0x0080) /* I2C Own Address 7 */
#define UCOA6  (0x0040) /* I2C Own Address 6 */
#define UCOA5  (0x0020) /* I2C Own Address 5 */
#define UCOA4  (0x0010) /* I2C Own Address 4 */
#define UCOA3  (0x0008) /* I2C Own Address 3 */
#define UCOA2  (0x0004) /* I2C Own Address 2 */
#define UCOA1  (0x0002) /* I2C Own Address 1 */
#define UCOA0  (0x0001) /* I2C Own Address 0 */
```

```
#define UCSA9  (0x0200) /* I2C Slave Address 9 */
#define UCSA8  (0x0100) /* I2C Slave Address 8 */
#define UCSA7  (0x0080) /* I2C Slave Address 7 */
#define UCSA6  (0x0040) /* I2C Slave Address 6 */
#define UCSA5  (0x0020) /* I2C Slave Address 5 */
#define UCSA4  (0x0010) /* I2C Slave Address 4 */
#define UCSA3  (0x0008) /* I2C Slave Address 3 */
#define UCSA2  (0x0004) /* I2C Slave Address 2 */
#define UCSA1  (0x0002) /* I2C Slave Address 1 */
#define UCSA0  (0x0001) /* I2C Slave Address 0 */

/* **************
*  WATCHDOG TIMER
*************** */

/* Definition to show that Module is available */
#define __MSP430_HAS_WDT__

SFR_16BIT(WDTCTL); /* Watchdog Timer Control */
/* The bit names have been prefixed with "WDT" */
#define WDTIS0   (0x0001)
#define WDTIS1   (0x0002)
#define WDTSSEL  (0x0004)
#define WDTCNTCL (0x0008)
#define WDTTMSEL (0x0010)
#define WDTNMI   (0x0020)
#define WDTNMIES (0x0040)
#define WDTHOLD  (0x0080)

#define WDTPW    (0x5A00)

/* WDT-interval times [1ms] coded with Bits 0-2 */
/* WDT is clocked by fSMCLK (assumed 1MHz) */
/* 32ms interval (default) */
#define WDT_MDLY_32 (WDTPW+WDTTMSEL+WDTCNTCL)
/* 8ms     " */
#define WDT_MDLY_8 (WDTPW+WDTTMSEL+WDTCNTCL+WDTIS0)
/* 0.5ms   " */
#define WDT_MDLY_0_5 (WDTPW+WDTTMSEL+WDTCNTCL+WDTIS1)
/* 0.064ms " */
#define WDT_MDLY_0_064 (WDTPW+WDTTMSEL+WDTCNTCL \
  +WDTIS1+WDTIS0)

/* WDT is clocked by fACLK (assumed 32KHz) */
/* 1000ms   " */
#define WDT_ADLY_1000 (WDTPW+WDTTMSEL+WDTCNTCL+WDTSSEL)
/* 250ms    " */
#define WDT_ADLY_250 (WDTPW+WDTTMSEL+WDTCNTCL+WDTSSEL \
+WDTIS0)
/* 16ms     " */
#define WDT_ADLY_16 (WDTPW+WDTTMSEL+WDTCNTCL+WDTSSEL \
+WDTIS1)
/* 1.9ms    " */
#define WDT_ADLY_1_9 (WDTPW+WDTTMSEL+WDTCNTCL+WDTSSEL \
  +WDTIS1+WDTIS0)
```

```
/* Watchdog mode -> reset after expired time */
/* WDT is clocked by fSMCLK (assumed 1MHz) */
/* 32ms interval (default) */
#define WDT_MRST_32    (WDTPW+WDTCNTCL)
/* 8ms    " */
#define WDT_MRST_8     (WDTPW+WDTCNTCL+WDTIS0)
/* 0.5ms   " */
#define WDT_MRST_0_5   (WDTPW+WDTCNTCL+WDTIS1)
/* 0.064ms " */
#define WDT_MRST_0_064 (WDTPW+WDTCNTCL+WDTIS1+WDTIS0)

/* WDT is clocked by fACLK (assumed 32KHz) */
/* 1000ms  " */
#define WDT_ARST_1000 (WDTPW+WDTCNTCL+WDTSSEL)
/* 250ms   " */
#define WDT_ARST_250  (WDTPW+WDTCNTCL+WDTSSEL+WDTIS0)
/* 16ms    " */
#define WDT_ARST_16   (WDTPW+WDTCNTCL+WDTSSEL+WDTIS1)
/* 1.9ms   " */
#define WDT_ARST_1_9  (WDTPW+WDTCNTCL+WDTSSEL+WDTIS1 \
+WDTIS0)

/* INTERRUPT CONTROL */
/* These two bits are defined in the
Special Function Registers */

/* #define WDTIE  0x01 */
/* #define WDTIFG 0x01 */

/* ****************************
*  Calibration Data in Info Mem
***************************** */

#ifndef __DisableCalData

SFR_8BIT(CALDCO_16MHZ);
/* DCOCTL  Calibration Data for 16MHz */
SFR_8BIT(CALBC1_16MHZ);
/* BCSCTL1 Calibration Data for 16MHz */
SFR_8BIT(CALDCO_12MHZ);
/* DCOCTL  Calibration Data for 12MHz */
SFR_8BIT(CALBC1_12MHZ);
/* BCSCTL1 Calibration Data for 12MHz */
SFR_8BIT(CALDCO_8MHZ);
/* DCOCTL  Calibration Data for 8MHz */
SFR_8BIT(CALBC1_8MHZ);
/* BCSCTL1 Calibration Data for 8MHz */
SFR_8BIT(CALDCO_1MHZ);
/* DCOCTL  Calibration Data for 1MHz */
SFR_8BIT(CALBC1_1MHZ);
/* BCSCTL1 Calibration Data for 1MHz */

#endif /* #ifndef __DisableCalData */
```

```
/* ************************************
*  Interrupt Vectors (offset from 0xFFE0)
************************************ */

#define VECTOR_NAME(name)    name##_ptr
#define EMIT_PRAGMA(x)       _Pragma(#x)
#define CREATE_VECTOR(name)  \
  void (* const VECTOR_NAME(name))(void) = &name
#define PLACE_VECTOR(vector,section)  \
  EMIT_PRAGMA(DATA_SECTION(vector,section))
#define ISR_VECTOR(func,offset) CREATE_VECTOR(func);  \
 PLACE_VECTOR(VECTOR_NAME(func), offset)

#ifdef __ASM_HEADER__
#define PORT1_VECTOR       ".int02"
/* 0xFFE4 Port 1 */
#else
#define PORT1_VECTOR       (2 * 1u)
/* 0xFFE4 Port 1 */
#endif
#ifdef __ASM_HEADER__
#define PORT2_VECTOR       ".int03"
/* 0xFFE6 Port 2 */
#else
#define PORT2_VECTOR       (3 * 1u)
/* 0xFFE6 Port 2 */
#endif
#ifdef __ASM_HEADER__
#define ADC10_VECTOR       ".int05"
/* 0xFFEA ADC10 */
#else
#define ADC10_VECTOR       (5 * 1u)
/* 0xFFEA ADC10 */
#endif
#ifdef __ASM_HEADER__
#define USCIAB0TX_VECTOR   ".int06"
/* 0xFFEC USCI A0/B0 Transmit */
#else
#define USCIAB0TX_VECTOR   (6 * 1u)
/* 0xFFEC USCI A0/B0 Transmit */
#endif
#ifdef __ASM_HEADER__
#define USCIAB0RX_VECTOR   ".int07"
/* 0xFFEE USCI A0/B0 Receive */
#else
#define USCIAB0RX_VECTOR   (7 * 1u)
/* 0xFFEE USCI A0/B0 Receive */
#endif
#ifdef __ASM_HEADER__
#define TIMER0_A1_VECTOR   ".int08"
/* 0xFFF0 Timer0)A CC1, TA0 */
#else
```

```
#define TIMER0_A1_VECTOR   (8 * 1u)
/* 0xFFF0 Timer0)A CC1, TA0 */
#endif
#ifdef __ASM_HEADER__
#define TIMER0_A0_VECTOR   ".int09"
/* 0xFFF2 Timer0_A CC0 */
#else
#define TIMER0_A0_VECTOR   (9 * 1u)
/* 0xFFF2 Timer0_A CC0 */
#endif
#ifdef __ASM_HEADER__
#define WDT_VECTOR          ".int10"
/* 0xFFF4 Watchdog Timer */
#else
#define WDT_VECTOR          (10 * 1u)
/* 0xFFF4 Watchdog Timer */
#endif
#ifdef __ASM_HEADER__
#define COMPARATORA_VECTOR ".int11"
/* 0xFFF6 Comparator A */
#else
#define COMPARATORA_VECTOR (11 * 1u)
/* 0xFFF6 Comparator A */
#endif
#ifdef __ASM_HEADER__
#define TIMER1_A1_VECTOR   ".int12"
/* 0xFFF8 Timer1_A CC1-4, TA1 */
#else
#define TIMER1_A1_VECTOR   (12 * 1u)
/* 0xFFF8 Timer1_A CC1-4, TA1 */
#endif
#ifdef __ASM_HEADER__
#define TIMER1_A0_VECTOR   ".int13"
/* 0xFFFA Timer1_A CC0 */
#else
#define TIMER1_A0_VECTOR   (13 * 1u)
/* 0xFFFA Timer1_A CC0 */
#endif
#ifdef __ASM_HEADER__
#define NMI_VECTOR          ".int14"
/* 0xFFFC Non-maskable */
#else
#define NMI_VECTOR          (14 * 1u)
/* 0xFFFC Non-maskable */
#endif
#ifdef __ASM_HEADER__
#define RESET_VECTOR        ".reset"
/* 0xFFFE Reset [Highest Priority] */
#else
#define RESET_VECTOR        (15 * 1u)
/* 0xFFFE Reset [Highest Priority] */
#endif
```

```
/* ***************
*  End of Modules
*************** */

#ifdef __cplusplus
}
#endif /* extern "C" */

#endif /* #ifndef __MSP430G2553 */
```

References

1. Deitel, P., Deitel, H.: *C: How to Program*, 7th ed. Prentice Hall (2012).
2. Gaspar, P. D., Santo, A. E., Riberio, B.: *MSP430 Teaching ROM*. Texas Instruments (2009).
3. Kleitz, W.: *Digital Electronics: A Practical Approach with VHDL*, 9th ed. Prentice Hall (2012).
4. Lathi, B. P., Ding, Z.: *Modern Digital and Analog Communication Systems*, 4th ed. Oxford University Press (2009).
5. Mano, M. M., Ciletti, M. D.: *Digital Design*, 4th ed. Prentice Hall (2006).
6. Mano, M. M., Kimei, C.: *Logic and Computer Design Fundamentals*, 4th ed. Prentice Hall (2007).
7. NXP: *UM10204 I^2C-Bus Specification and User Manual*, rev. 5th ed. (2012).
8. Oppenheim, A. V., Schafer, R. W.: *Discrete-Time Signal Processing*, 3rd ed. Prentice Hall (2009).
9. Texas Instruments: *MSP430 Flash Memory Characteristics*, slaa334a ed. (2008).
10. Texas Instruments: *Understanding MSP430 Flash Data Retention*, slaa392 ed. (2008).
11. Texas Instruments: *Code Composer Studio v5.3 User's Guide for MSP430*, slau157u ed. (2012).
12. Texas Instruments: *MSP-EXP430G2 LaunchPad Experimenter Board User's Guide*, slau318c ed. (2012).
13. Texas Instruments: *MSP430 Assembly Language Tools v4.1 User's Guide*, slau131g ed. (2012).
14. Texas Instruments: *MSP430 Programming via the Bootstrap Loader User's Guide*, slau319 ed. (2012).
15. Texas Instruments: *MSP430 Programming via the JTAG Interface User's Guide*, slau320 ed. (2012).
16. Texas Instruments: *MSP430G2x53, MSP430G2x13 Mixed Signal Microcontroller*, slas735g ed. (2012).
17. Texas Instruments: *MSP430x2xx Family User's Guide*, slau144i ed. (2012).

Index

Note: Page numbers followed by *f* denote figures; page numbers followed by *t* denote tables.

V

W

X